# 243
## Topics in Current Chemistry

**Editorial Board:**
A. de Meijere · K.N. Houk · H. Kessler · J.-M. Lehn · S.V. Ley
S.L. Schreiber · J. Thiem · B.M. Trost · F. Vögtle · H. Yamamoto

# Topics in Current Chemistry
Recently Published and Forthcoming Volumes

**Anion Sensing**
Volume Editor: Stibor, I.
Vol. 255, 2005

**Organic Solid State Reactions**
Volume Editor: Toda, F.
Vol. 254, 2005

**DNA Binders and Related Subjects**
Volume Editors: Waring, J., Chaires, J.B.
Vol. 253, 2005

**Contrast Agents II**
Volume Editor: Krause, W.
Vol. 252, 2005

**Chalcogenocarboxylic Acid Derivatives**
Volume Editor: Kato, S.
Vol. 251, 2005

**New Aspects in Phosphorus Chemistry V**
Volume Editor: Majoral, J.-P.
Vol. 250, 2005

**Templates in Chemistry II**
Volume Editors: Schalley, C.A., Vögtle, F.,
Dötz, K.H.
Vol. 249, 2005

**Templates in Chemistry I**
Volume Editors: Schalley, C.A., Vögtle, F.,
Dötz, K.H.
Vol. 248, 2005

**Collagen**
Volume Editors: Brinckmann, J.,
Notbohm, H., Müller, P.K.
Vol. 247, 2005

**New Techniques in Solid-State NMR**
Volume Editor: Klinowski, J.
Vol. 246, 2005

**Functional Molecular Nanostructures**
Volume Editor: Schlüter, A.D.
Vol. 245, 2005

**Natural Product Synthesis II**
Volume Editor: Mulzer, J.
Vol. 244, 2005

**Natural Product Synthesis I**
Volume Editor: Mulzer, J.
Vol. 243, 2005

**Immobilized Catalysts**
Volume Editor: Kirschning, A.
Vol. 242, 2004

**Transition Metal and Rare Earth
Compounds III**
Volume Editor: Yersin, H.
Vol. 241, 2004

**The Chemistry of Pheromones
and Other Semiochemicals II**
Volume Editor: Schulz, S.
Vol. 240, 2005

**The Chemistry of Pheromones
and Other Semiochemicals I**
Volume Editor: Schulz, S.
Vol. 239, 2004

**Orotidine Monophosphate Decarboxylase**
Volume Editors: Lee, J.K., Tantillo, D.J.
Vol. 238, 2004

**Long-Range Charge Transfer in DNA II**
Volume Editor: Schuster, G.B.
Vol. 237, 2004

**Long-Range Charge Transfer in DNA I**
Volume Editor: Schuster, G.B.
Vol. 236, 2004

**Spin Crossover in Transition Metal
Compounds III**
Volume Editors: Gütlich, P., Goodwin, H.A.
Vol. 235, 2004

**Spin Crossover in Transition Metal
Compounds II**
Volume Editors: Gütlich, P., Goodwin, H.A.
Vol. 234, 2004

**Spin Crossover in Transition
Metal Compounds I**
Volume Editors: Gütlich, P., Goodwin, H.A.
Vol. 233, 2004

# Natural Product Synthesis I
Targets, Methods, Concepts

Volume Editor: Johann Mulzer

With contributions by
T. Bach · R. Bandichhor · B. Basler · S. Brandes · M. Hiersemann
H. Helmboldt · B. Nosse · O. Reiser · E. Ruijter · M. Sefkow
A. Spiegel · L.A. Wessjohann

The series *Topics in Current Chemistry* presents critical reviews of the present and future trends in modern chemical research. The scope of coverage includes all areas of chemical science including the interfaces with related disciplines such as biology, medicine and materials science. The goal of each thematic volume is to give the non specialist reader, whether at the university or in industry, a comprehensive overview of an area where new insights are emerging that are of interest to a larger scientific audience.

As a rule, contributions are specially commissioned. The editors and publishers will, however, always be pleased to receive suggestions and supplementary information. Papers are accepted for *Topics in Current Chemistry* in English.

In references *Topics in Current Chemistry* is abbreviated Top Curr Chem and is cited as a journal.

Visit the TCC content at springerlink.com

Library of Congress Control Number: 2004112161

ISSN 0340-1022
ISBN 3-540-21125-X **Springer Berlin Heidelberg New York**
DOI 10.1007/b10390

This work is subject to copyright. All rights are reserved, whether the whole or part of the material is concerned, specifically the rights of translation, reprinting, reuse of illustrations, recitation, broadcasting, reproduction on microfilms or in any other ways, and storage in data banks. Duplication of this publication or parts thereof is only permitted under the provisions of the German Copyright Law of September 9, 1965, in its current version, and permission for use must always be obtained from Springer-Verlag. Violations are liable to prosecution under the German Copyright Law.

**Springer is a part of Springer Science+Business Media**
springeronline.com
© Springer-Verlag Berlin Heidelberg 2005
Printed in Germany

The use of general descriptive names, registered names, trademarks, etc. in this publication does not imply, even in the absence of a specific statement, that such names are exempt from the relevant protective laws and regulations and therefore free for general use.

Cover design: KünkelLopka, Heidelberg/design & production GmbH, Heidelberg
Typesetting: Stürtz GmbH, Würzburg

Printed on acid-free paper    02/3020 ra – 5 4 3 2 1 0

## Volume Editor

Professor Dr. Johann Mulzer
Institut für Organische Chemie
Universität Wien
Währingerstr. 38
1090 Wien, Austria
*johann.mulzer@univie.ac.at*

## Editorial Board

Prof. Dr. Armin de Meijere
Institut für Organische Chemie
der Georg-August-Universität
Tammannstraße 2
37077 Göttingen, Germany
*E-mail: ameijer1@uni-goettingen.de*

Prof. Dr. Horst Kessler
Institut für Organische Chemie
TU München
Lichtenbergstraße 4
85747 Garching, Germany
*E-mail: kessler@ch.tum.de*

Prof. Steven V. Ley
University Chemical Laboratory
Lensfield Road
Cambridge CB2 1EW, Great Britain
*E-mail: svl1000@cus.cam.ac.uk*

Prof. Dr. Joachim Thiem
Institut für Organische Chemie
Universität Hamburg
Martin-Luther-King-Platz 6
20146 Hamburg, Germany
*E-mail: thiem@chemie.uni-hamburg.de*

Prof. Dr. Fritz Vögtle
Kekulé-Institut für Organische Chemie
und Biochemie der Universität Bonn
Gerhard-Domagk-Straße 1
53121 Bonn, Germany
*E-mail: voegtle@uni-bonn.de*

Prof. K.N. Houk
Department of Chemistry
and Biochemistry
University of California
405 Hilgard Avenue
Los Angeles, CA 90024-1589, USA
*E-mail: houk@chem.ucla.edu*

Prof. Jean-Marie Lehn
Institut de Chimie
Université de Strasbourg
1 rue Blaise Pascal, B.P.Z 296/R8
67008 Strasbourg Cedex, France
*E-mail: lehn@chimie.u-strasbg.fr*

Prof. Stuart L. Schreiber
Chemical Laboratories
Harvard University
12 Oxford Street
Cambridge, MA 02138-2902, USA
*E-mail: sls@slsiris.harvard.edu*

Prof. Barry M. Trost
Department of Chemistry
Stanford University
Stanford, CA 94305-5080, USA
*E-mail: bmtrost@leland.stanford.edu*

Prof. Hisashi Yamamoto
Arthur Holly Compton Distinguished
Professor
Department of Chemisty
The University of Chicago
5735 South Ellis Avenue
Chicago, IL 60637
733-702-5059, USA
*E-mail: yamamoto@uchicago.edu*

# Topics in Current Chemistry
Also Available Electronically

For all customers who have a standing order to Topics in Current Chemistry, we offer the electronic version via SpringerLink free of charge. Please contact your librarian who can receive a password for free access to the full articles by registering at:

springerlink.com

If you do not have a subscription, you can still view the tables of contents of the volumes and the abstract of each article by going to the SpringerLink Homepage, clicking on "Browse by Online Libraries", then "Chemical Sciences", and finally choose Topics in Current Chemistry.

You will find information about the

– Editorial Board
– Aims and Scope
– Instructions for Authors
– Sample Contribution

at springeronline.com using the search function.

# Preface

From its early days, the total synthesis of complex molecules, especially those that are natural products, has been the king's discipline in organic chemistry. The reasons for this are manifold: the challenge lying in a novel and intricate molecular architecture or the difficulty encountered when isolating the substance from its natural sources, or the possibility of finding a wide test ground for established methodology or the incentive to invent new methodology when the old one has failed, or simply the art and elegance which is so typical of a truly efficient synthetic sequence. In any case, everybody will agree that total synthesis is the best way to train young chemists and to prepare them for any kind of later employment.

In these two volumes, the contributions of a number of organic synthetic chemists from the German speaking area have been collected. It is the hope of the authors and the editor that these articles, which highlight all the various aspects of organic synthesis, will provide not only an insight into the basic strategy and tactics but also the purpose of organic syntheses.

Vienna, September 2004                                          Johann Mulzer

# Contents

Total Syntheses of Kelsoene and Preussin
B. Basler · S. Brandes · A. Spiegel · T. Bach . . . . . . . . . . . . . . . . . . . . . . .   1

Paraconic Acids – The Natural Products from *Lichen* Symbiont
R. Bandichhor · B. Nosse · O. Reiser . . . . . . . . . . . . . . . . . . . . . . . . . . . .  43

Recent Progress in the Total Synthesis of Dolabellane
and Dolastane Diterpenes
M. Hiersemann · H. Helmboldt . . . . . . . . . . . . . . . . . . . . . . . . . . . . . . .  73

Strategies for Total and Diversity-Oriented Synthesis
of Natural Product(-Like) Macrocycles
L.A. Wessjohann · E. Ruijter . . . . . . . . . . . . . . . . . . . . . . . . . . . . . . . . . 137

Enantioselective Synthesis of C(8)-Hydroxylated Lignans:
Early Approaches and Recent Advances
M. Sefkow . . . . . . . . . . . . . . . . . . . . . . . . . . . . . . . . . . . . . . . . . . . . . . . 185

Author Index Volumes 201–243 . . . . . . . . . . . . . . . . . . . . . . . . . . . . . . 225

Subject Index . . . . . . . . . . . . . . . . . . . . . . . . . . . . . . . . . . . . . . . . . . . . 237

# Contents of Volume 244
# Natural Product Synthesis II
## Targets, Methods, Concepts

Volume Editor: J. Mulzer
ISBN 3-540-21124-1


Marine Natural Products from *Pseudopterogorgia Elisabethae*:
Structures, Biosynthesis, Pharmacology and Total Synthesis
T.J. Heckrodt · J. Mulzer

Recent Advances in Vinylogous Aldol Reactions
and Their Applications in the Syntheses of Natural Products
M. Kalesse

Methanophenazine and Other Natural Biologically Active Phenazines
U. Beifuss · M. Tietze

Occurrence, Biological Activity,
and Convergent Organometallic Synthesis of Carbazole Alkaloids
H.-J. Knoelker

Recent Advances in Charge – Accelerated Aza-Claisen Rearrangements
U. Nubbemeyer

Synthetic Studies on the Pamamycin Macrodiolides
P. Metz

# Total Syntheses of Kelsoene and Preussin

Birte Basler · Sebastian Brandes · Anja Spiegel · Thorsten Bach (✉)

Lehrstuhl für Organische Chemie I, Technische Universität München,
Lichtenbergstrasse 4, 85747 Garching, Germany
*thorsten.bach@ch.tum.de*

| | | |
|---|---|---|
| 1 | Introduction. | 2 |
| 2 | Kelsoene. | 3 |
| 2.1 | Syntheses by Other Groups. | 3 |
| 2.1.1 | Intermolecular [2+2]-Photocycloaddition Approach. | 4 |
| 2.1.1.1 | Intermolecular [2+2]-Photocycloaddition as the Key Step. | 4 |
| 2.1.1.2 | Approaches to the Photocycloaddition Precursor | 4 |
| 2.1.1.3 | Completion of the Syntheses. | 7 |
| 2.1.2 | Homo-Favorskii Approach. | 10 |
| 2.1.2.1 | Homo-Favorskii Rearrangement as the Key Step. | 10 |
| 2.1.2.2 | Synthesis of the Precursor and Completion of the Synthesis | 11 |
| 2.2 | Intramolecular [2+2]-Photocycloaddition Approach. | 13 |
| 2.2.1 | Background. | 13 |
| 2.2.2 | Retrosynthesis. | 14 |
| 2.2.3 | Model Studies and First Attempt. | 15 |
| 2.2.4 | Successful Second Approach | 17 |
| 3 | Preussin. | 20 |
| 3.1 | Syntheses by Other Groups. | 21 |
| 3.1.1 | Nucleophilic Attack on L-Phenylalanine-derived Electrophiles | 21 |
| 3.1.2 | Syntheses Using Other L-Amino Acids | 26 |
| 3.1.3 | Sugars and Meso-Compounds as Building Blocks | 27 |
| 3.1.4 | Alternative Approaches. | 31 |
| 3.2 | Paternò-Büchi Approach. | 34 |
| 3.2.1 | Background. | 34 |
| 3.2.2 | Retrosynthesis. | 36 |
| 3.2.3 | Synthesis. | 37 |
| 4 | Conclusion. | 39 |
| | References. | 40 |

**Abstract** Total syntheses of the natural products kelsoene and preussin are comprehensively reviewed. Kelsoene is a sesquiterpene with a unique tricyclo[6.2.0.0$^{2,6}$]decane skeleton. It contains six stereogenic centers the selective construction of which has been addressed differently in the five syntheses known to date. Three syntheses employ an intermolecular [2+2]-photocycloaddition reaction as key step. One synthesis is based on a homo-Favorskii rearrangement and one on an intramolecular [2+2]-photocycloaddition. Preussin is a pyrrolidine alkaloid with three stereogenic centers which are all located within the central heterocyclic core (C-2, C-3, C-5). So far, 18 total syntheses of preussin have been completed. Seven syntheses include the nucleophilic attack on an L-phenylalanine derived electrophile as key step, five use α-amino- or α-hydroxycarboxylic acids as chiral pool building blocks. Two syntheses are based on sugars as chiral starting materials and two are based on the desymmetrization of *meso*-compounds. In addition, there are two syntheses which use a chiral auxiliary to establish the first stereogenic element en route to preussin.

**Keywords** Alkaloids · Natural products · Photochemistry · Terpenes · Total synthesis

**List of Abbreviations**
| | |
|---|---|
| Am | Amyl |
| ds | Diastereoselectivity |
| im | Imidazole |
| MSH | O-(Mesitylenesulfonyl)hydroxyl-amine |
| NOE | Nuclear Overhauser effect/enhancement |
| NOESY | Nuclear Overhauser enhancement spectroscopy |
| PPL | Pig pancreatic lipase |
| SES | Trimethylsilyl ethyl sulfonyl |
| TIBAL | Triisobutylaluminum |
| TPAP | Tetrapropylammonium perruthenate |

# 1
# Introduction

The title compounds kelsoene (*rac*-1) and (+)-preussin (2) have recently been synthesized in our laboratories using a stereoselective photochemical reaction as key step (Fig. 1). It is the purpose of this review to give a more detailed account on this work. In addition, other successful synthetic strategies to kelsoene and preussin will be comprehensively discussed. This intention has defined the way the content of the review was arranged. In two individual sections the target molecules are presented. Each section provides a short introduction to the target, a review on the syntheses conducted by other groups and finally an account of our own contribution.

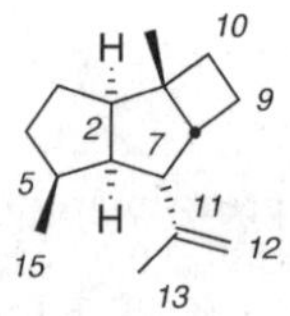

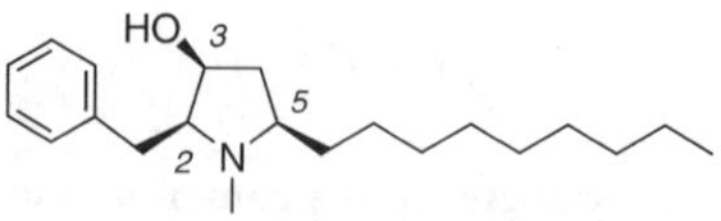

**Fig. 1** Chemical structures of (+)-kelsoene (1) and (+)-preussin (2)

## 2
## Kelsoene

(+)-Kelsoene (1) was first isolated from the sponge *Cymbastela hooperi* Van Soest, Desqueyroux-Faundez, Wright, König (Axinellidae, Halichondrida) collected from Kelso reef, Great Barrier Reef, Australia [1]. Its constitution and relative configuration was elucidated by König and Wright. It is a sesquiterpene with a tricyclo[6.2.0.0$^{2,6}$]decane skeleton rarely encountered in natural products. The compound was later also found in the liverworts *Ptychanthus striatus* [2], *Calypogeia muelleriana* [3], and *Tritomaria quinquedentata* [4]. Proof of the absolute configuration was obtained from total synthesis (see below) while initial NMR studies [5] had led to the conclusion that the natural product possessed the configuration of *ent*-1. Labeling studies (Scheme 1, ○=$^{13}$C label) with [2–$^{13}$C]-mevalonate indicated that the bio-

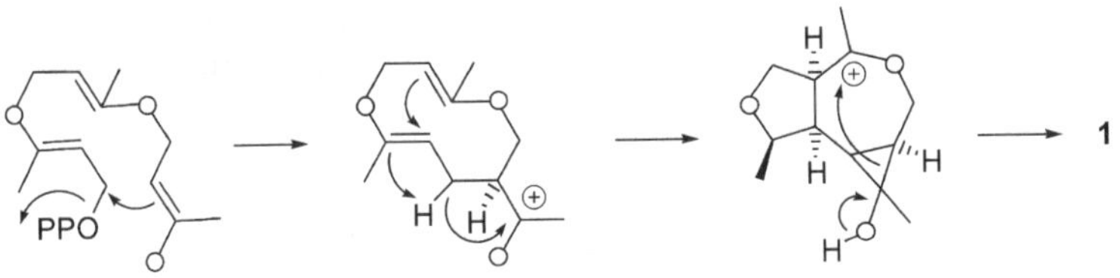

**Scheme 1**

synthesis proceeds from farnesyl diphosphate (3) via the germacradienyl cation (4) and the alloaromadendranyl cation (5) [2, 6].

From a synthetic point of view the stereoselective construction of the tricyclodecane skeleton and the installation of the methyl group at C-5 and of the 2-propenyl group at C-7 pose significant challenges which were addressed differently in the five syntheses of kelsoene known to date.

### 2.1
### Syntheses by Other Groups

The following sections give an overview of the total syntheses of kelsoene that have been reported by other groups. With respect to the key step by which the tricyclic framework is constituted, they can be divided into two groups, the [2+2]-photocycloaddition approach and the homo-Favorskii strategy.

### 2.1.1
### Intermolecular [2+2]-Photocycloaddition Approach

#### 2.1.1.1
#### Intermolecular [2+2]-Photocycloaddition as the Key Step

The first total synthesis of kelsoene was achieved by Mehta and Srinivas [7, 8]. The tricyclic scaffold was established by an intermolecular [2+2]-photocycloaddition of diquinane enone *rac*-6 and 1,2-dichloroethylene (7) as the key step (Scheme 2). As a consequence of the steric hindrance implemented

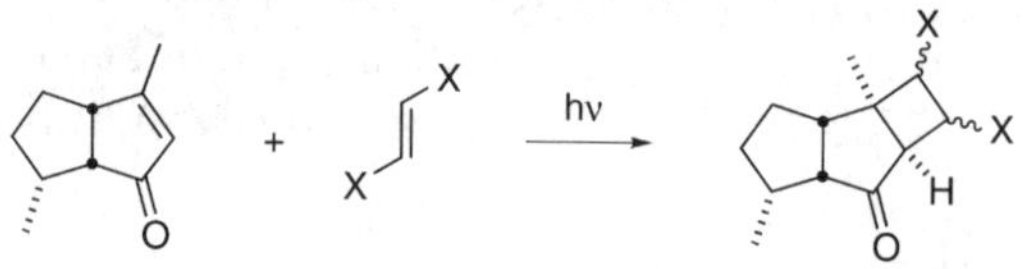

*rac*-6 / *ent*-6 / 6    7 (X = Cl)    *rac*-9 / *ent*-9 / 9 (X = Cl)  85%
*rac*-6                  8 (X = H)     *rac*-10 (X = H)               90%

**Scheme 2**

in the bicyclic ring system of 6, the alkene 7 is forced to attack exclusively from the *exo*-face of 6. The perfect facial diastereoselectivity results in the formation of the tricyclodecane 9 the framework of which is connected in the requisite *cis-anti-cis* fashion [7]. In subsequent syntheses by others [9–11] and in the enantioselective synthesis by Mehta and Srinivas [12] an identical or almost identical photochemical key step was employed. In all syntheses, diquinane enone 6, its enantiomer *ent*-6, or the racemate *rac*-6 was used as the photoactive compound, the reaction partner was 1,2-dichloroethylene (7) [7–10, 12] or ethylene (8) [11].

The fact that all approaches employing the intermolecular photocycloaddition key step used the same precursor for the construction of the four-membered ring renders enone 6 the key intermediate of the different synthetic strategies. It is therefore sensible to compare first the different strategies to synthesize precursor 6. Afterwards, the different ways to complete the syntheses of kelsoene will be discussed.

#### 2.1.1.2
#### Approaches to the Photocycloaddition Precursor

In the first total synthesis of kelsoene (*rac*-1) [7, 8], commercially available 1,5-cyclooctadiene (11) was chosen as the starting material (Scheme 3). Oxidative cyclization with a mixture of $PdCl_2$ and $Pb(OAc)_4$ in acetic acid led to the formation of a diquinane diacetate [13] that was saponified to give

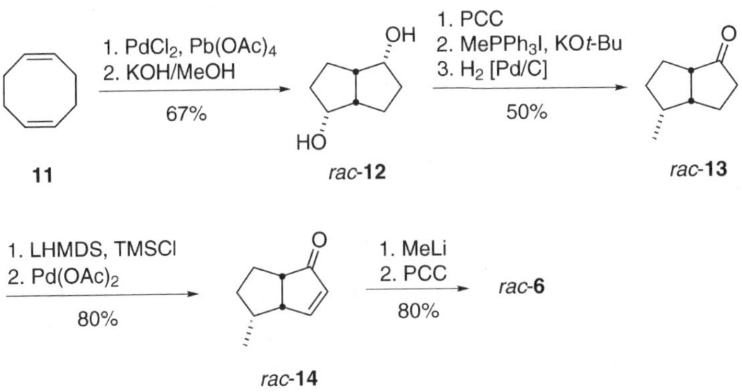

**Scheme 3**

2,6-dihydroxybicyclo[3.3.0]octane (*rac*-12). Oxidation of the two hydroxy groups followed by selective monomethylenation and catalytic hydrogenation furnished the *endo*-methyl diastereoisomer *rac*-13 as the major product (dr=80/20). The observed diastereoselectivity was explained by the fact that the catalytic hydrogenation proceeded preferentially from the more easily accessible convex face of the diquinane framework [7].

Ketone *rac*-13 was transformed into the corresponding silylenolether and by Pd(II)-mediated Saegusa oxidation [14] into α,β-unsaturated ketone *rac*-14. By alkylative enone transposition comprising methyl lithium addition and pyridinium chlorochromate (PCC) oxidation [15], *rac*-14 was finally converted into the racemic photocycloaddition precursor *rac*-6. In conclusion, the bicyclic irradiation precursor *rac*-6 was synthesized in a straightforward manner from simple 1,5-cyclooctadiene (11) in nine steps and with an overall yield of 21%.

In a succeeding publication, the same authors reported on an enantioselective approach to diquinane enones 6 and *ent*-6 by combining the above-described synthesis with an enzymatic kinetic resolution (Scheme 4) [12]. After lipase-catalyzed enantioselective transesterification of diol *rac*-12,

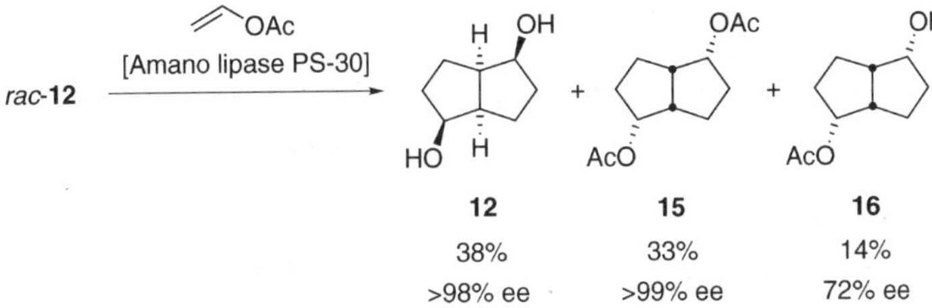

**Scheme 4**

enantiomerically pure diol **12** and diacetate **15** as well as monoacetate **16** were isolated in 38%, 33%, and 14% yield, respectively.

Continuing the synthesis of *rac*-**6** with enantiomerically pure diols **12** and *ent*-**12** (after saponification of **15** with KOH), both enantiomers **6** and *ent*-**6** were accessible. This allowed for an enantioselective synthesis of natural kelsoene (**1**) and its enantiomer (*ent*-**1**) (see below) with only one additional step as compared to the synthesis of the racemate (*rac*-**1**).

The approach of Schulz et al. to enantiomerically pure diquinane enone *ent*-**6** employed (*R*)-(+)-pulegone (**17**) as chiral pool starting material [9, 10] (Scheme 5). Bromination and Favorskii rearrangement of **17** generated a

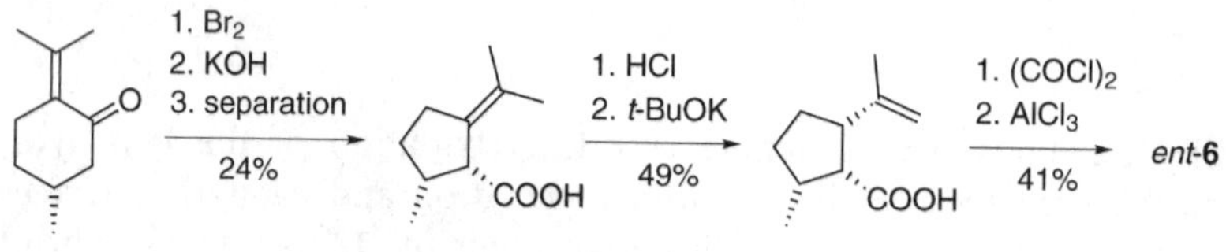

**Scheme 5**

mixture of *cis*- (**18**) and *trans*-pulegonic acids [16] that was separated by column chromatography. Acid-induced lactonization of the desired *cis*-substituted acid **18** followed by elimination with a bulky base led to the formation of the all-*cis*-substituted acid **19** with a terminal double bond [16, 17].

The photocycloaddition precursor *ent*-**6** was obtained from **19** by transformation into the corresponding acid chloride and AlCl$_3$-mediated intramolecular acylation of the double bond. While the conciseness of this strategy is appealing, drawbacks are the low yields achieved in the individual reaction steps giving *ent*-**6** in an overall yield of only 5%.

Racemic diquinane enone *rac*-**6** was prepared by Piers and Orellana starting from cyclopentenone (Scheme 6) [11]. After the preparation of the heterocuprate from stannane **20**, conjugate addition to cyclopentenone in the presence of BF$_3$·Et$_2$O provided carbonyl compound **21**. It was expected that conversion of **21** by intramolecular alkylation and subsequent hydrogenation should provide the desired *endo*-substituted diquinane *rac*-**13**. While other hydrogenation methods proved to be rather unselective, reduction in the presence of Wilkinson's catalyst finally resulted in the formation of *rac*-**13** with good facial diastereoselectivity [11].

The introduction of the double bond of *rac*-**14** was performed by conversion of *rac*-**13** into its α-phenylselenide, subsequent peroxide oxidation, and elimination. Following the synthesis reported by Mehta and Srinivas, an alkylative enone transposition was used as the last step towards irradiation

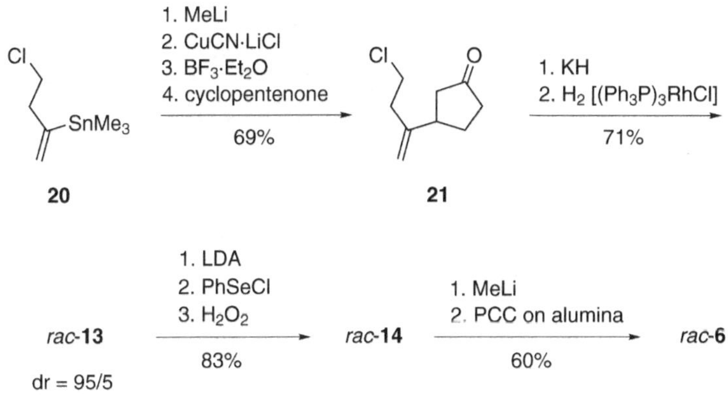

## Scheme 6

precursor *rac-6*. Not taking into account the preparation of 4-chloro-2-trimethylstannylbut-1-ene (**20**), this synthetic route offers another elegant access to *rac-6* with a reported overall yield of 24% exceeding those of the other syntheses.

In conclusion, the three groups which applied the intermolecular photocycloaddition as the key step in their approach to kelsoene (**1**) reported different strategies to synthesize the irradiation precursor **6** in racemic or enantiomerically pure form. After the photocycloaddition step the syntheses of kelsoene were completed in different ways. The next section describes the different strategies employed in the second half of the way to kelsoene.

### 2.1.1.3
### Completion of the Syntheses

The first report on the construction of the complete 5-5-4-fused tricyclic framework of kelsoene and later on the first total synthesis of racemic kelsoene *rac-1* was published by Srinivas and Mehta in 1999 [7, 8]. As discussed in the last section, the insertion of an enzymatic kinetic resolution step allowed for the analogous synthesis in an enantioselective manner [12]. The intermolecular [2+2]-photocycloaddition as the key step in their approach was performed by irradiating diquinane enone **6** with 1,2-dichloroethylene (**7**) (Scheme 2). This led to the formation of a mixture of *cis-* and *trans-*dichlorosubstituted cycloaddition products **9** in 85% yield. The perfect facial diastereoselectivity resulting from the attack of alkene **7** on the *exo*-face of **6** led to the exclusive formation of the desired *cis-anti-cis* connected tricyclodecane **9**. After protection of the carbonyl group as an acetal, the chlorine atoms were removed by reductive dehalogenation with sodium naphthalenide. Originally, the resulting cyclobutene was then hydrogenated and deprotected to yield cyclobutane **10** (Scheme 7).

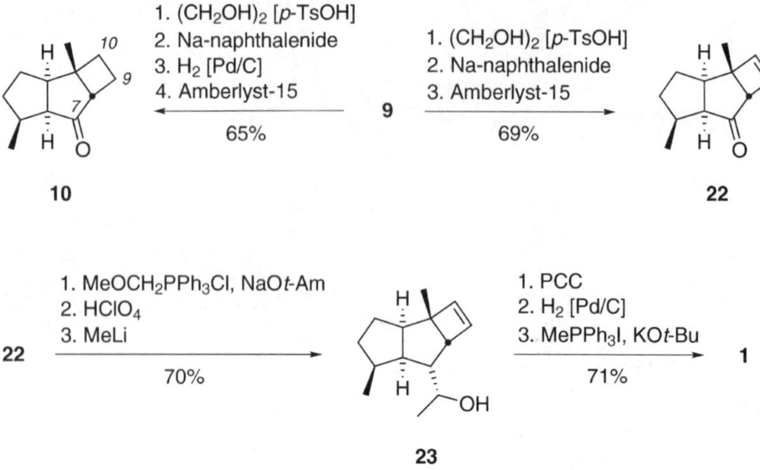

**Scheme 7**

Unfortunately, all attempts to elaborate this intermediate product to the target molecule by adding a side chain at C-7 failed. The carbonyl group in 10 did not react as expected with a variety of nucleophilic reagents and ylides [7]. This behavior was explained by the sterically encumbered environment around the carbonyl functionality blocking any attack. In order to decrease the steric hindrance, it seemed necessary to remove the *endo*-hydrogen atoms at C-9 and C-10 which shield the bottom face of the carbonyl group [8]. Cyclobutene 22 was expected to provide the requisite properties and was therefore tested in a homologation attempt. Indeed, the Wittig reaction of ketone 22 with methoxymethylene triphenylphosphane proceeded successfully. After cleavage of the resulting enolether with perchloric acid and addition of methyl lithium, alcohol 23 was obtained as a mixture of diastereoisomers. A three-step sequence including oxidation of the alcohol to the corresponding ketone, hydrogenation of the double bond and final methylenation of the carbonyl group, completed the first total synthesis of enantiomerically pure kelsoene (1). In addition, *ent*-1 was synthesized starting from *ent*-12. By comparison of the specific rotations the (1*R*,2*S*,5*R*,6*R*,7*R*,8*S*)-configuration was assigned to the natural product.

In a study which was conducted simultaneously to the work in the Mehta group and which also aimed to prove the absolute configuration of natural kelsoene (1), Schulz et al. used a stereoselective approach starting from the enantiomerically pure chiral pool material (*R*)-pulegone 17 [9, 10] (see above). The final steps of their synthesis of the unnatural enantiomer of kelsoene (*ent*-1) were similar to the above-described first total synthesis of natural kelsoene (1) (Scheme 8). Taking into account the steric limitations of the system as communicated by Srinivas and Mehta, diquinane enone *ent*-6

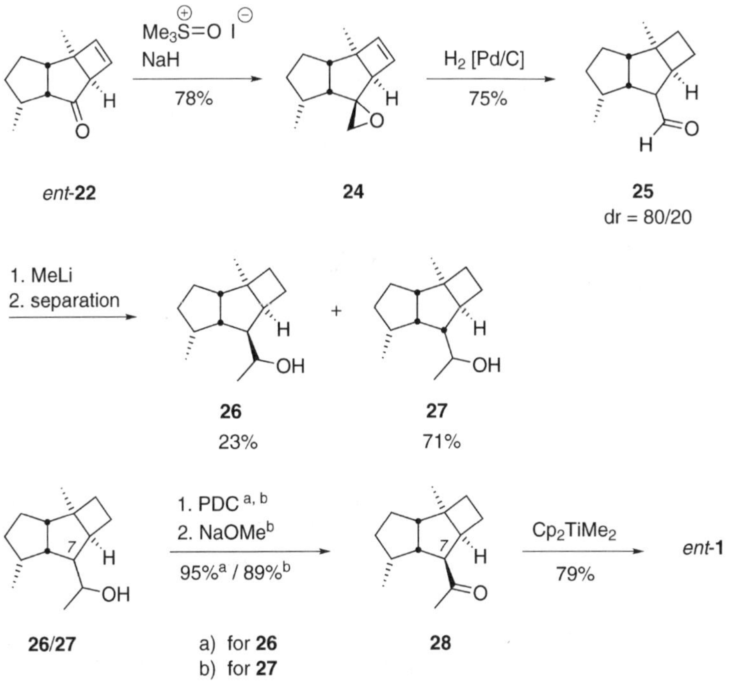

**Scheme 8**

was transformed into the cyclobutene derivative *ent*-22 [8]. In the following step, the homologation procedure was modified converting *ent*-22 with trimethylsulfoxonium iodide into epoxide 24.

Upon hydrogenation of 24 a 1,2-rearrangement of the epoxide occurred generating aldehyde 25 as a mixture of diastereoisomers. After reaction with methyl lithium, the diastereomeric alcohols 26 and 27 were separated and isolated in yields of 23% and 71%. While alcohol 26 as the minor diastereoisomer could be oxidized with pyridinium dichromate (PDC) and methylenated to give the enantiomer of kelsoene (*ent*-1), its diastereoisomer 27 with the inverse configuration at C-7 required a supplementary epimerization step with sodium methanolate. The enantiomerically pure *ent*-1 allowed for the determination of the absolute configuration of natural kelsoene (1) [9, 10]. The previously reported assignment based on NMR-correlation experiments [5] was corrected.

In contrast to the preceding syntheses of kelsoene, Piers and Orellana used ethylene 8 in the [2+2]-photocycloaddition key step to furnish tricyclic ketone *rac*-10 diastereoselectively (Scheme 2) [11]. In accordance with the report by Srinivas and Mehta, attempts to homologate the sterically encumbered ketone *rac*-10 using the Wittig, Magnus [18, 19] or Taguchi [20] pro-

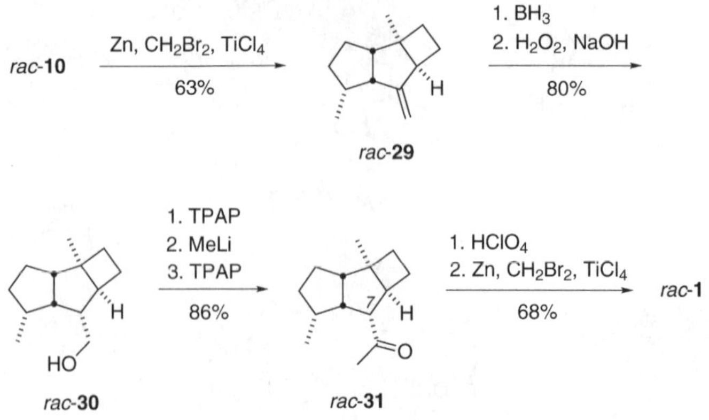

**Scheme 9**

cedure were unsuccessful [11]. However, alkene *rac*-29 was isolated in satisfying yields by treatment of ketone *rac*-10 with the Lombardo reagent [21, 22] (Scheme 9). The latter reagent is known to be effective in methylenations of hindered ketones.

Hydroboration occurred from the less hindered top face of *rac*-29 and resulted in the formation of alcohol *rac*-30. After a three-step sequence which included oxidation with tetrapropylammonium perruthenate (TPAP), methyl lithium addition and repeated oxidation with TPAP, ketone *rac*-31 was isolated. Finally, epimerization of the stereogenic center at C-7 to the correct configuration and methylenation with the Lombardo reagent led to the formation of racemic kelsoene (*rac*-1).

## 2.1.2
## Homo-Favorskii Approach

### 2.1.2.1
### Homo-Favorskii Rearrangement as the Key Step

In contrast to the [2+2]-photocycloaddition which is a widely used method to generate four-membered rings, Koreeda and Zhang used a thermal rearrangement key step to construct the tricyclic framework of kelsoene [23]. In earlier work, it was shown that upon treatment with base, $\gamma$-keto-*p*-toluenesulfonate *rac*-32 is converted to a 44/56 mixture of bicyclo[3.1.1]heptanone *rac*-33 and bicyclo[3.2.0]heptanone *rac*-34 (Scheme 10) [24–26].

While ketone *rac*-33 is the result of an intramolecular substitution of the tosylate leaving group by the intermediate enolate, *rac*-34 is the product of a stereospecific homoallyl rearrangement commonly referred to as the homo-Favorskii rearrangement [24–26]. Presumably due to their relatively high strain energies, bicyclo[3.1.1]heptanones rearrange acid-induced to bicy-

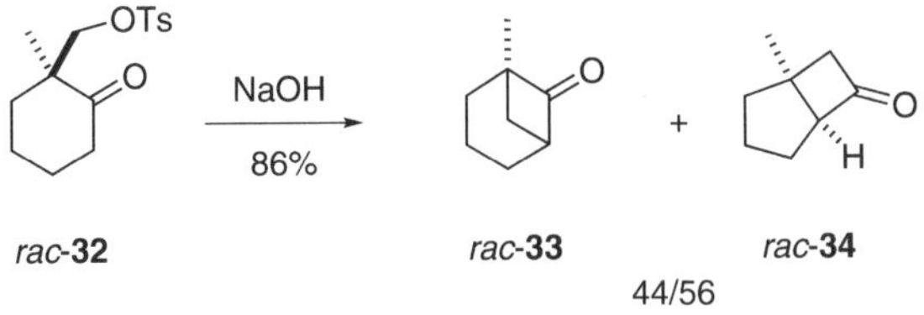

*rac*-32 → *rac*-33 + *rac*-34 (NaOH, 86%, 44/56)

**Scheme 10**

clo[3.2.0]heptanones [26, 27], so that both reaction pathways of the base-catalyzed rearrangement offer an access to bicyclo[3.2.0]heptanone derivatives. Thus, the homo-Favorskii rearrangement was expected to be well suited for the construction of the tricyclic carbon framework of racemic kelsoene (*rac*-1).

### 2.1.2.2
### Synthesis of the Precursor and Completion of the Synthesis

The preparation of the requisite γ-keto-*p*-toluenesulfonate *rac*-35 as homo-Favorskii precursor commenced with commercially available 2,5-dihydroanisole (36) that was protected and epoxidized to acetal *rac*-37 (Scheme 11). Regioselective opening of the epoxide with *p*-chlorophenylselenide followed by sequential oxidation to the selenoxide and thermal elimination generated an allylic alcohol that was protected to give pivaloate *rac*-38.

In order to append the butynyl group, *rac*-38 was reacted with the corresponding cyanocuprate in high yields. Pd(0)-mediated enyne-cyclization of *rac*-39 to the bicyclic diene *rac*-40 proceeded smoothly. Regio- and stereoselective reduction of the exocyclic double bond proved to be difficult, but was finally achieved with a mixture of cobaltdichloride and lithium borohydride. Acetal cleavage with subsequent enolate methylation yielded stereoselectively enone *rac*-41 which was subjected to a hydroxymethylation followed by tosylation of the generated alcohol. The completion of the synthesis of precursor *rac*-35 required a stereocontrolled conjugate addition of the 2-propenyl group to enone *rac*-42. While standard cuprate addition procedures gave unsatisfying results, the Kharasch conditions [28] proved to be well suited, and *rac*-35 was isolated in a yield of 68%.

With the appropriate precursor *rac*-35 in hand, the rearrangement key step was examined. As expected, basic treatment of γ-keto-tosylate *rac*-35 resulted in the formation of a 56/44 mixture of the formal substitution product *rac*-43 and *rac*-44 as the product of homoallylic rearrangement (Scheme 12), in a combined yield of 95%.

Subjection of the mixture of cyclobutanones *rac*-43 and *rac*-44 to mild acidic conditions resulted in the rearrangement of cyclobutanone *rac*-43 to its more stable isomer *rac*-45, whereas *rac*-44 remained unchanged. The ob-

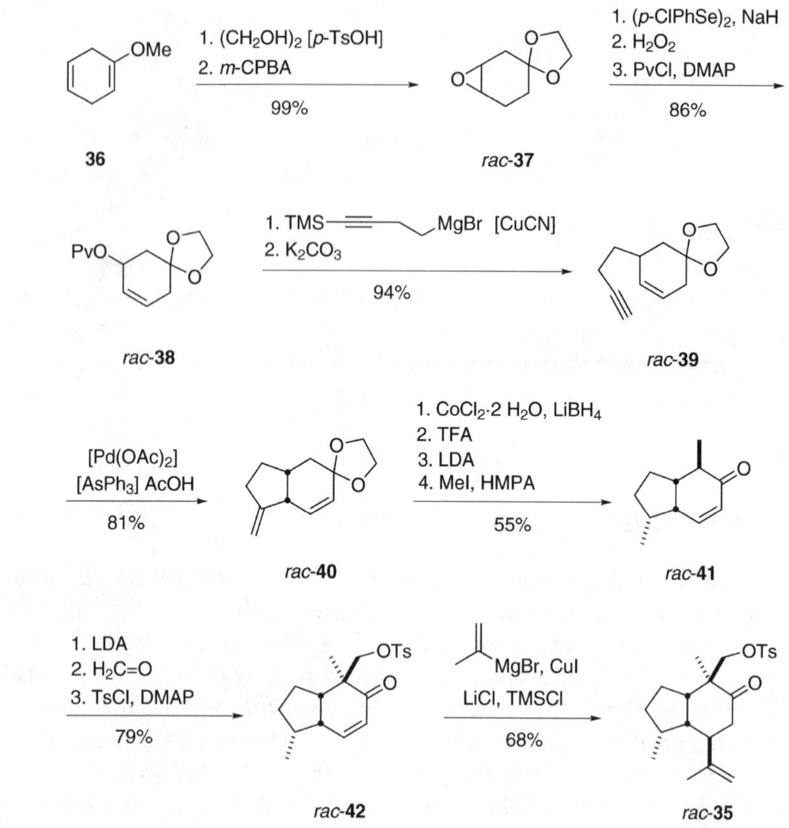

**Scheme 11**

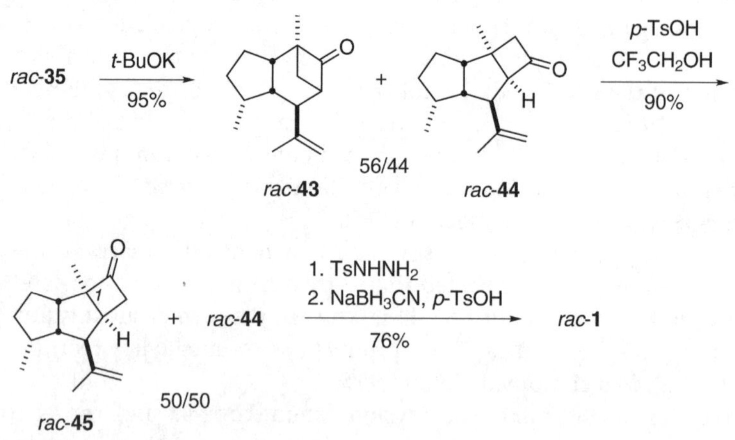

**Scheme 12**

served isomerization of *rac*-43 was suggested to occur via a cationic or non-classical cationic species and proceeded with retention of the stereochemistry at the quaternary stereogenic center C-1. Finally, by transformation of the cyclobutanones *rac*-44 and *rac*-45 into their tosylhydrazones and subsequent reduction, racemic kelsoene (*rac*-1) was isolated in 16 steps with an overall yield of 13%. In more recent work [29], Koreeda et al. have achieved the enantioselective preparation of pivaloate 38 by deracemization of a cyclic allyl carbonate [30]. Starting from (*S*)-configurated 38, (+)-kelsoene (1) was accessible in enantiomerically pure form.

## 2.2
### Intramolecular [2+2]-Photocycloaddition Approach

### 2.2.1
### Background

When the first publication on kelsoene by König and Wright appeared in 1997, there was a research project in our group directed towards the intramolecular [2+2]-photocycloaddition reaction of vinylglycine-derived *N*-allyloxazolidinones such as 46 [31, 32]. The goal was to prepare enantiomerically pure 3-azabicyclo[3.2.0]heptanes which were at that time considered interesting pharmacophores [33]. We could show that the [2+2]-photocycloaddition of the oxazolidinones 46 proceeds in high yields and with good facial diastereoselectivity. Cu(OTf)$_2$ or CuOTf were the preferred catalysts for the alkyl substituted (R=H, Me) substrates [34, 35] whereas the reaction with the phenyl substituted system (R=Ph) could also be promoted by a sensitizer, such as acetophenone [36]. In all cases the products 47 showed the *anti-cis* configuration depicted in Scheme 13. The facial diastereoselectivity was explained by the conformation 46' of the photoactive starting materials.

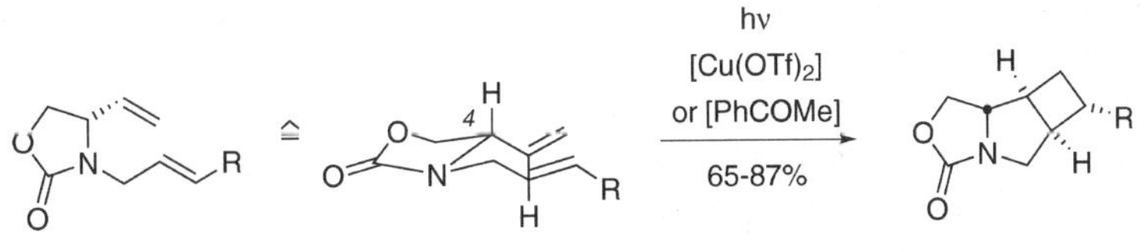

**Scheme 13**

The conformation is favored due to minimized 1,3-allylic strain between the hydrogen atom at C-4 and the terminal *cis*-hydrogen atom of the double bond.

## 2.2.2
### Retrosynthesis

The skeleton of **47** is a heterocyclic tricyclo[6.2.0.0$^{2,6}$]decane and the similarity to the tricyclic kelsoene is obvious. In the course of the above-mentioned studies we had become curious whether the high facial diastereocontrol in the photocycloaddition reaction could be extended to other bridged 1,6-hexadienes. Kelsoene was an ideal test case. The retrosynthetic strategy for kelsoene along an intramolecular [2+2]-photocycloaddition pathway appeared straightforward. To avoid chemoselectivity problems the precursor to kelsoene should not contain additional double bonds. Alcohol **48**, the hydroxy group of which was possibly to be protected, seemed to be a suitable substrate for the photocycloaddition (Scheme 14). Access to the 1,2,3-substi-

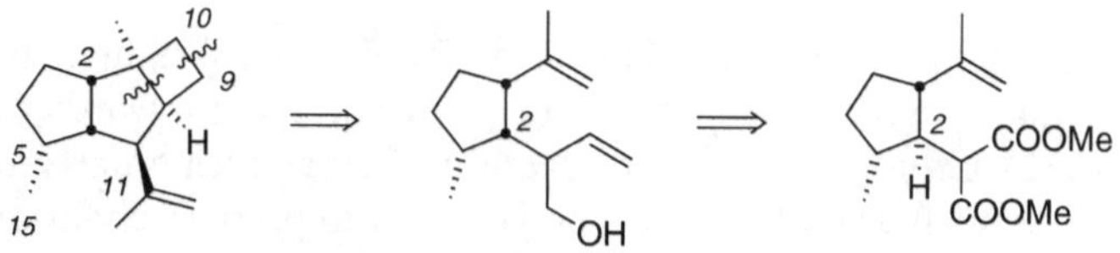

**Scheme 14**

tuted cyclopentane motif of **48** was envisaged from the known ene reaction product **49** [37]. Although the original report suggested that it was difficult to obtain malonate **49** in enantiomerically pure form we considered it a good starting material which contained most of the required carbon atoms with the right topology.

An advantage of the intramolecular [2+2]-photocycloaddition strategy as compared to its intermolecular counterpart is the fact that two rings are formed in a single step. A disadvantage is the conformational freedom of the system and the potentially more complex synthesis of the precursor. With regard to the stereogenic center at carbon atom C-7 in kelsoene we considered it likely that it can be epimerized at the ketone (*rac*-**31**→*rac*-**28**) stage. Carbonyl olefination would complete the synthesis. It later turned out that this notion was correct. Piers and Orellana were the first to use these transformations in their kelsoene synthesis (see above).

Figure 2 depicts the relevant cyclopentane conformations for either diastereomeric alcohol **48**. Conformations **A** and **C** would lead to the desired *cis-anti-cis* tricyclodecane skeleton of kelsoene whereas conformations **B** and **D** would be responsible for the formation of the undesired *cis-syn-cis* product. We were convinced that the influence of the stereogenic center to which the hydroxymethyl group is attached (*) was marginal as compared to

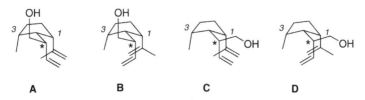

**Fig. 2** Possible conformations A,B and C,D of the two diastereomeric 1,6-dienes 148

the interaction between the 2-propenyl group at C-1 and the methyl group at C-3 or the hydrogen atom at C-4. Based on the analogy to **46** we hoped for the 1,3-allylic strain to be still dominant although we were aware of the fact that the additional methyl group at the vinyl substituent might change the situation dramatically.

## 2.2.3
## Model Studies and First Attempt

As our group was still fairly small in 1997 it took quite a while until we could get serious about the kelsoene synthesis. The synthetic work did not commence before the summer of 1999, a few weeks before the first synthesis of kelsoene was published by Mehta and Srinivas. We planned to move into two directions. On the one hand, model studies on simple cyclopentanes should give us a flavor for the facial diastereoselectivity to be expected in the photocycloaddition. On the other hand, preliminary studies as to the conversion **49**→**48** were to be conducted. As it turned out the latter experiments were more facile than expected whereas the former study was more complicated than originally thought [38]. According to Tietze's protocol [37], enantiomerically pure (−)-β-citronellene (**50**) was converted to the racemization prone aldehyde **51** (Scheme 15). Following the original procedure we obtained the racemic Knoevenagel product *rac*-**52** which was immediately taken into the ene reaction to yield the all-*trans* cyclopentane product *rac*-**49**. Attempts to suppress the racemization were postponed and work on the synthetic route was continued instead. To this end, the stereogenic center at cyclopentane carbon atom C-2 was to be inverted. Following a known strategy [39], an α-selenylation of compound *rac*-**49** and subsequent elimination led to the α,β-unsaturated malonate *rac*-**53** which was diastereoselectively reduced with NaBH$_4$ in EtOH. The relative configuration of the product was confirmed by NOE studies. Reduction of the diester to the diol *rac*-**54** was facile using lithium aluminum hydride in Et$_2$O. Attempts to conduct the two reduction steps in a single operation proved less efficient than the two-step protocol. Single acetylation of one alcohol group was possible by an enzyme catalyzed transacylation using pig pancreatic lipase (PPL). The product *rac*-**55** was obtained as a mixture of diastereoisomers. Based on the

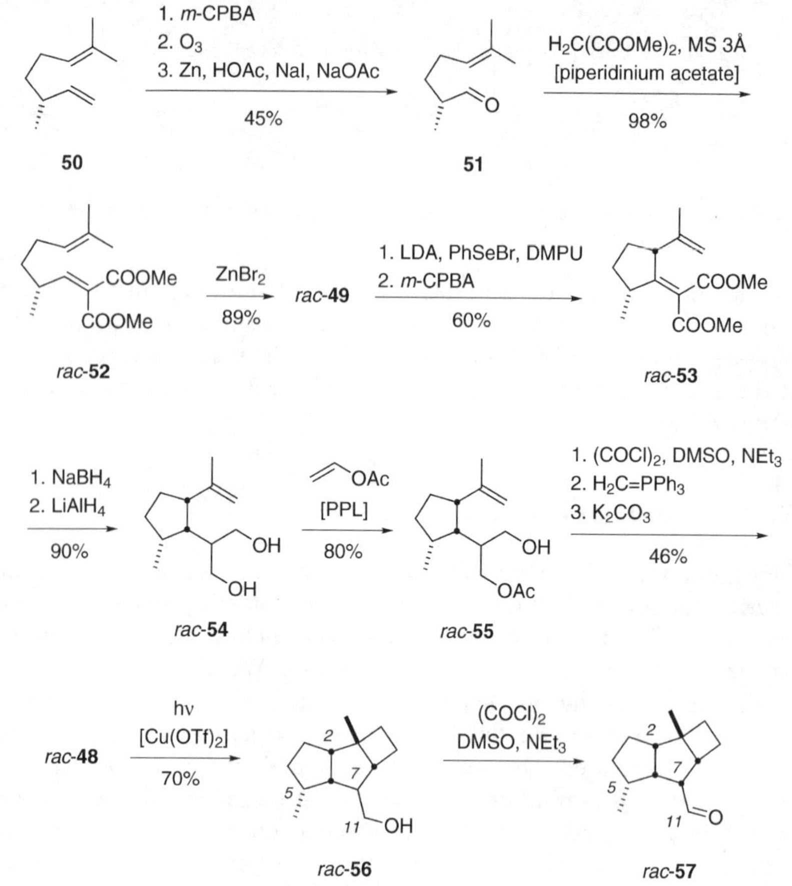

**Scheme 15**

considerations mentioned above we continued without separating the diastereoisomers. The free alcohol was converted into an olefin by Swern oxidation and carbonyl olefination. Removal of the acetyl group yielded the precursor *rac*-48 which was suited for the photocycloaddition.

As in many other cases we used Cu(OTf)$_2$ as the precatalyst which is more stable and easier to handle than CuOTf [40]. The photocycloaddition delivered two diastereoisomers which were suspected to differ in the relative configuration at C-7. Indeed, elimination of water resulted in a single, diastereomerically pure product with an exocyclic olefinic double bond between C-7 and C-11. Along the same line of evidence aldehyde *rac*-57 was obtained as a single diastereoisomer presumably because an epimerization at C-7 had occurred. The configuration assignment of compounds *rac*-56 and *rac*-57 proved difficult. NOESY data suggested a *cis-syn-cis* arrangement

Total Syntheses of Kelsoene and Preussin

in the central ring of tricyclo[6.2.0.0$^{2,6}$]decane *rac*-57. As we were not fully convinced, the aldehyde rac-57 was further converted to the presumed ketone *rac*-28 (Scheme 8) which had become known due to the work of Mehta et al. The ketone we obtained was not identical to ketone *rac*-28.

Simultaneously to the synthetic studies described above, our model studies had progressed. Although the synthetic challenge in this part of the project was marginal the structure elucidation of the final products was complicated [41]. Starting material for the syntheses was 1-acetyl-1-cyclopentene (58) which was converted into the *trans*-cyclopentane *rac*-59 by a Sakurai reaction and a carbonyl olefination (Scheme 16). The synthesis of *cis*-cyclo-

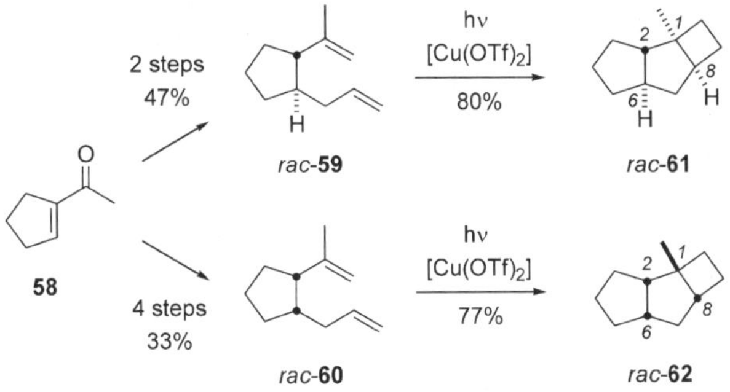

**Scheme 16**

pentane *rac*-60 was achieved by the sequence Sakurai reaction, silyl enol ether formation, diastereoselective protonation, and Lombardo olefination. The photocycloaddition proceeded smoothly for both substrates and yielded the products as single diastereoisomers. The proof of the *trans-anti-cis* configuration for product *rac*-61 was possible by NOESY measurements and comparison with related compounds. The structure proof for compound *rac*-62, however, was not possible by NOESY experiments. We finally relied on a tedious chemical correlation which gave unambiguous proof for the undesired *cis-syn cis* structure of product *rac*-62 [41].

## 2.2.4
### Successful Second Approach

From the results described above it was clear that our synthetic plan had to be modified significantly if the synthesis of kelsoene was to be successful. We were willing to take on the challenge and started to think about a modification. Key to our plan was the transformation *rac*-59→*rac*-61 in the model studies. The *trans-anti-cis* configuration of the product would be acceptable if we succeeded in inverting the stereogenic center at carbon atom C-6

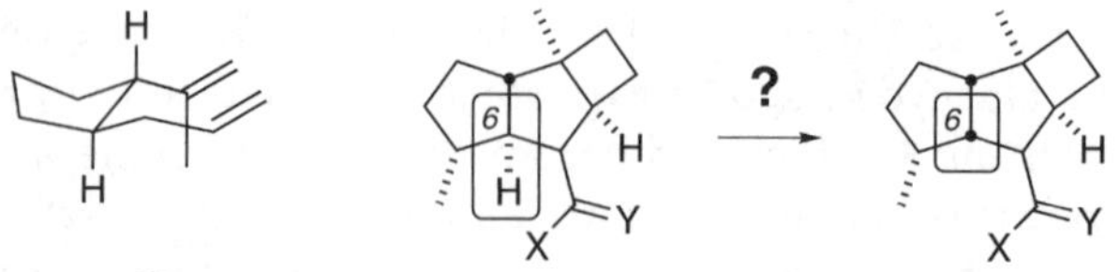

**Scheme 17**

at a later stage. Scheme 17 depicts the cyclopentane conformation which is responsible for the formation of product *rac*-61 and illustrates the decisive inversion of configuration in a potential kelsoene precursor. Literature precedence for an inversion in a congested situation like in Scheme 17 did not exist. It was clear, however, that an elimination/hydrogenation reaction would be a viable solution. Preliminary studies revealed that the formation of the thermodynamically favored enolate of a ketone (X=Me, Y=O) is difficult in the tricyclic system. Common strategies to form $\alpha,\beta$-unsaturated ketones from ketones were consequently not applicable. On the other hand there existed scattered literature precedence for the $\alpha$-bromination of an aldehyde [42, 43], for the transformation of an $\alpha$-bromoaldehyde into an $\alpha$-bromoketone [44], and for the hydro-bromo-elimination to an $\alpha,\beta$-unsaturated ketone [45, 46].

Following known pathways we started to construct the required photocycloaddition precursor from malonate *rac*-49 which was reduced to diol *rac*-63 (Scheme 18). Monoacetylation to alcohol *rac*-64, oxidation and olefination yielded the acetate *rac*-65 as a mixture of diastereoisomers. It turned out that it was advantageous to use the acetate as photocycloaddition substrate instead of the alcohol. The alcohol gave lower yields due to the formation of a heterocyclic by-product which is presumably formed in a Lewis-acid-catalyzed process. The acetate photocycloaddition *rac*-65→*rac*-66 was best conducted with CuOTf as the catalyst. Saponification and Swern oxidation gave an aldehyde in diastereomerically pure form which could be readily $\alpha$-brominated to the $\alpha$-bromoaldehyde *rac*-67. The subsequent steps proceeded nicely and yielded the required $\alpha,\beta$-unsaturated ketone *rac*-68. The final obstacle on our way to the known ketone *rac*-28 was the hydrogenation. It was a pleasant surprise to note that this transformation proceeded smoothly under standard conditions and with high facial diastereoselectivity (dr=95/5). The expected preference for an attack of hydrogen *anti* to the methyl group at C-5 and *syn* to the hydrogen atom at C-2 was borne out experimentally. The primary ketone epimerized already upon standing to the more stable ketone *rac*-28. To facilitate epimerization we followed the protocol by Piers and Orellana who had employed $HClO_4$ in $CDCl_3$ as an acidic catalyst [11]. The structure of ketone *rac*-28 was proven by comparison of its spectroscopic properties with those reported in the literature. Although

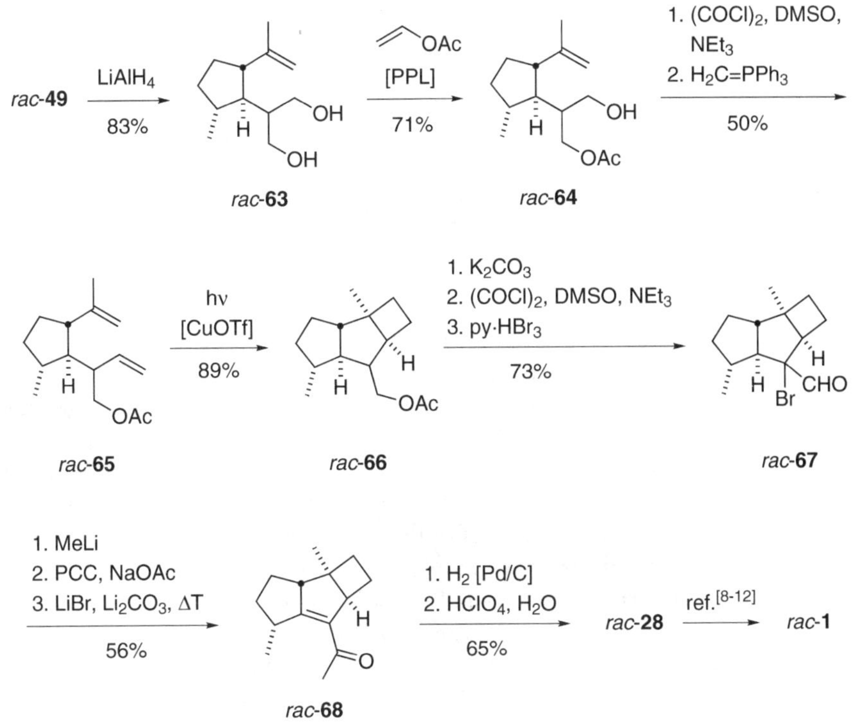

**Scheme 18**

the synthesis was formally complete intermediate *rac*-28 was converted to *rac*-(±)-kelsoene (*rac*-1) employing a Wittig reaction [8] as the final step of the synthesis. As noted in our preliminary publication [47] we observed a lipophilic by-product in the reactions which was not separable. Due to the limited amount of material we could not attempt other methylenation procedures, but the use of Ti-based methodology is strongly recommended.

A final issue which we addressed was the potential access to enantiomerically pure kelsoene. The first and obvious solution would be to start with enantiomerically pure malonate **49** or *ent*-**49**. Our attempts to conduct the Knoevenagel condensation under racemization free conditions were not successful, however. A second option resulted from the known ability of enzymes to distinguish between enantiotopic groups. In preliminary experiments on the monoacetylation of diols we had used the cyclopentane diol **69** as test substrate (Scheme 19). The isolated product **70** was enantiomerically enriched (93% ee). Based on the known propensity of PPL to acylate predominantly the *pro-R* hydroxymethyl group in 2-substituted 1,3-propandiols [48, 49] we assigned the (R)-configuration to this product. If the reagent-controlled preference exerted by the enzyme was extremely high one could

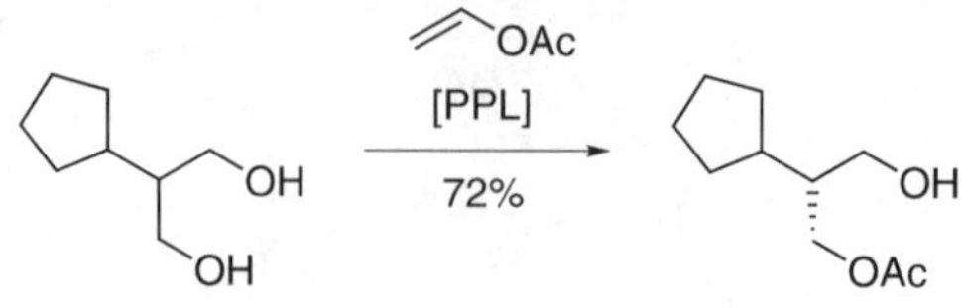

**Scheme 19**

expect in a diastereomeric product mixture such as **55** or **64** that each diastereoisomer is significantly enantiomerically enriched.

Although the separation of the diastereomeric alcohols **55** or **64** was not possible by flash chromatography we succeeded in a separation by preparative HPLC. The enantiomeric excess of the individual diastereoisomers was determined after saponification to the diol **54** or **63** by chiral GC. It turned out that the enantiomeric excess of both acetates was only 7% ee. The value was so low that we did not make an effort to continue with only marginally enantiomerically enriched material after HPLC separation.

In summary, our own synthesis of kelsoene comprised 18 steps starting from (–)-β-citronellene (**50**) and 14 steps starting from the previously described malonate *rac*-**49**. Despite its powerful photochemical key step *rac*-**65**→*rac*-**66** the subsequent epimerization at carbon atom C-6 is a major drawback. An alternative enantioselective access to key intermediate **65**/*ent*-**65** was not pursued.

## 3
## Preussin

Preussin (**2**) was first isolated from fermentation broths of *A. ochraceus* ATCC 22947 and was described as the antifungal agent L-657,398 by researchers at Merck, Sharp & Dohme [50]. The absolute and relative configuration of the compound was determined shortly after by Johnson, Phillipson, and Kahle using NMR techniques [51]. They had isolated the same compound from fermentations of *Preussia* sp. and had given it the more mellifluous name preussin. Natural (+)-preussin is (2*S*,3*S*,5*R*)-configurated (Fig. 1). Whereas the biosynthetic origin of preussin has not been investigated in detail its biological properties caused considerable attention. Beside the originally reported antifungal activity, preussin turned out to be a selective cell growth inhibitor [52]. More detailed studies by Müller et al. revealed that preussin induces apoptosis in several human tumor cells [53]. It was found in vitro to be a potent inhibitor of cyclin E kinase (CDK2-cyclin E) with a 50% inhibitory concentration of ca. 500 nmol/l. Preussin inhibits cell cycle progression into the S phase. Remarkably, the induction of apoptosis was not blocked by high levels of B cell lymphoma-2 (Bcl-2) which usually confers resistance to chemotherapeutic agents.

Up to now (June 2003) 18 total syntheses of preussin have been reported. As in the preceding section, the syntheses by other groups will be discussed first and we will subsequently provide a more extensive account on our own work. The considerably large number of known syntheses limits the amount of space we could give to the discussion of each individual route. We have tried to group certain strategies together and apologize to the researchers whose contributions are not discussed in detail.

## 3.1
## Syntheses by Other Groups

### 3.1.1
### Nucleophilic Attack on L-Phenylalanine-derived Electrophiles

L-Phenylalanine and derivatives thereof are starting materials to nearly half of the preussin syntheses reported so far. In six of these syntheses the pyrrolidine ring is formed by a ring closure between C-5 and the nitrogen atom. In addition, there is another scission in the ring between C-3 and C-4. The latter step is in five of these six syntheses a nucleophilic attack on L-phenylalaninal. Another synthesis using a one carbon elongated starting material derived from L-phenylalanine is also discussed in this section, although there is no nucleophilic attack on a carbonyl group.

The most recent approach to (+)-preussin conducted by Okue, Watanabe and Kitahara used the nucleophilic attack of an enolate on L-phenylalaninal (Scheme 20) and is one of the shortest syntheses of (+)-preussin [54].

Starting from methyl (S)-[1-(methoxy)methylcarbamoyl-2-phenylethyl]carbamate (71) aldehyde 72 was generated via lithium aluminum hydride reduction of the Weinreb amide. A chelation-controlled aldol reaction using the zinc enolate of 2-undecanone (73) provided the *syn*-alcohol 74 and its *anti*-diastereoisomer in a diasteromeric ratio of *syn*:*anti*=91/9. After protection of the alcohol as its TBDMS-ether 75, the compound was subjected to $Et_3GeH$ reduction in the presence of $BF_3 \cdot OEt_2$. In the course of the reductive cyclization the alcohol was also deprotected, yielding a single diastereoisomer 76 of the desired pyrrolidine. The methyl carbamate was then reduced to (+)-preussin (2) which was obtained in 16% overall yield.

The same group has used a closely related strategy to synthesize all eight stereoisomers of preussin by non-stereoselective reactions. They tested the bioactivities of the preussin isomers against cell growth of the fission yeast *cdc25* mutant and found, that all stereoisomers were almost equally bioactive [55].

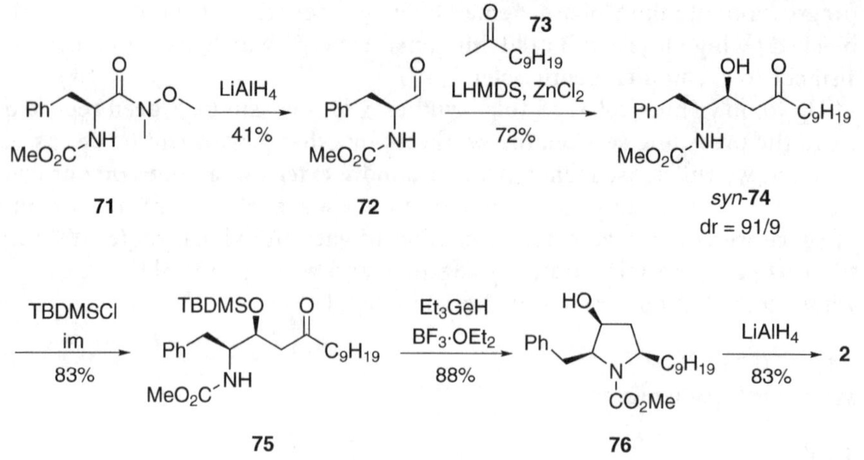

**Scheme 20**

Much earlier than Kitahara et al., an even shorter synthesis of preussin was reported by Overhand and Hecht [56]. Up to date the Hecht synthesis has remained one of the most straightforward and concise routes to the target (Scheme 21). The synthesis did not include a nucleophilic attack on

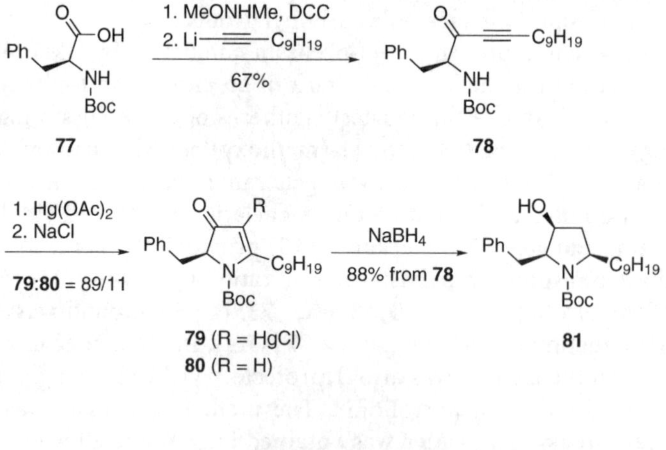

**Scheme 21**

phenylalaninal, but on the Weinreb amide of phenylalanine. The retrosynthetic approach shows the same disconnections as do the other syntheses in this section.

Starting from $N$-(Boc)-($S$)-phenylalanine (**77**) the Weinreb amide was formed by a DCC-mediated condensation with $N,O$-dimethylhydroxylamine. Treatment of the Weinreb amide with undecynyl lithium yielded the ynone

**78**. When subjected to mercury acetate the ynone **78** underwent a 5-*endo-dig* cyclization and after work-up with aqueous sodium chloride, furnished the pyrrolinones **79** and **80** in a ratio of 89/11. The mixture of pyrrolinones was reduced directly with sodium borohydride to the *N*-Boc-pyrrolidinol **81** which was obtained as a single diastereoisomer. Reduction of the carbamate with lithium aluminum hydride gave (+)-preussin (**2**) in 37% overall yield.

Using methyl *N*-(diphenylmethylene)-L-phenylalaninate (**82**) as starting material (Scheme 22), McGrane and Livinghouse were able to synthesize (+)-preussin (**2**) in eight steps with an overall yield of 15% [57].

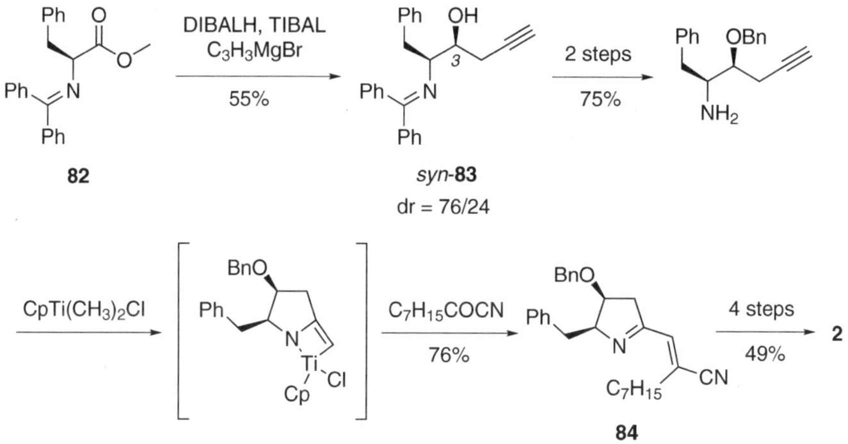

**Scheme 22**

To form the stereocenter at C-3 a direct reduction-alkynylation sequence was applied, that provided the diastereomeric homopropargylic alcohols **83** in a ratio of *syn:anti*=76/24. The major isomer *syn-*83 was isolated in 55% yield. The key step of the synthesis was an intramolecular imidotitanium-alkyne [2+2] cycloaddition/acyl cyanide condensation. With this sequence the pyrrolidine ring was formed and all the carbon atoms of the alkyl side chain were established in acrylonitrile **84**. The reduction of the imine double bond proceeded stereoselectively and the nitrile group was removed reductively en route to the target compound.

Veeresa and Datta [58] used a Swern oxidation to generate an aldehyde from *N*-Boc-protected phenylalaninol (**85**). The aldehyde was in situ attacked by allylmagnesium bromide (Scheme 23).

This chelation-controlled addition generated the homoallylic alcohols **86** with a diastereoselectivity of *syn:anti*=86/14. In contrast to the other syntheses described so far, the ring closing reaction was a $S_N2$ reaction but not a reductive cyclization. After cleavage of the acetonide in **87** the *N*-Boc-protected primary amine displaced the mesylate, thereby forming the pyrrolidine ring **81**. Altogether this synthesis consisted of ten steps, starting from L-phenylalanine, and yielded (+)-preussin (**2**) in 9%.

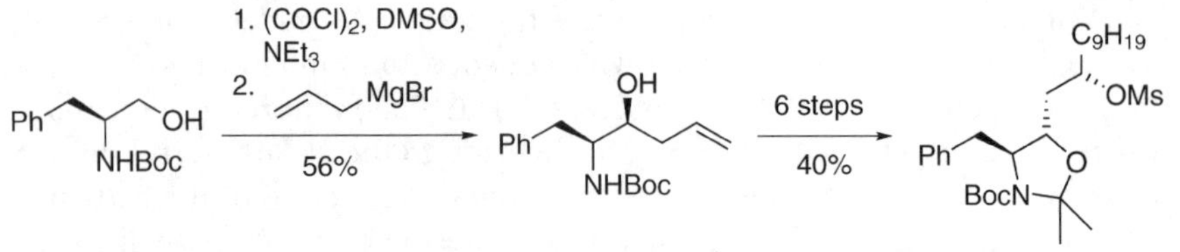

The key step in the approach of Jurczak et al. [59] was a chelation-controlled addition of allyltrimethylsilane to the *N*-Cbz-protected phenylalaninal **88** in the presence of SnCl$_4$ (Scheme 24). The preussin C-3/C-4 bond was

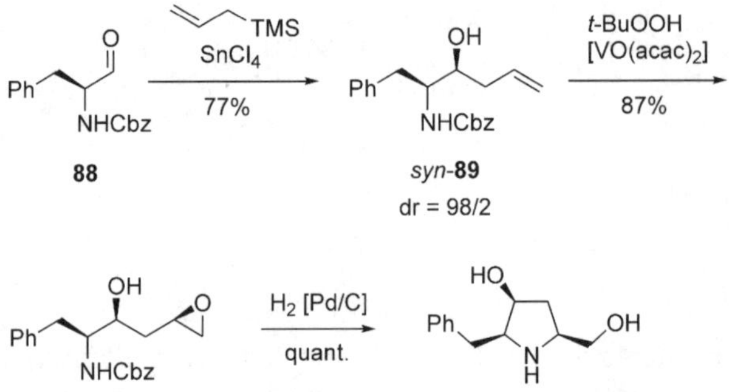

**Scheme 24**

formed with excellent facial diastereoselectivity (*syn:anti*=98/2), yielding the homoallyllic alcohol *syn*-**89**.

After epoxidation of the terminal olefin in *syn*-**89** the pyrrolidine **91** was formed by reductive cleavage of the Cbz-protection and concomitant S$_N$2 cyclization of the free amine to epoxide **90**. In five additional steps (+)-preussin (**2**) was synthesized with an overall yield of 19%. After *N*-methoxycarbonylation and oxidation of the alcohol to an aldehyde the alkyl side chain was introduced by a Wittig reaction.

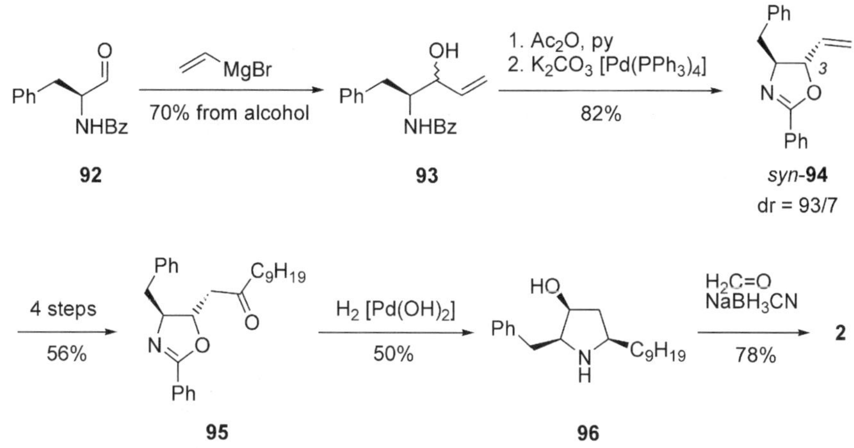

**Scheme 25**

Ham et al. [60] used an unselective attack of a vinyl Grignard reagent on the *N*-benzoyl-protected phenylalaninal **92** to generate alcohols **93** (Scheme 25). A *trans*-selective, Pd(0)-catalyzed oxazoline formation starting from the homoallylic amide **93** was subsequently employed to build up the (*S*)-configuration at carbon atom C-3 of oxazoline **94**.

Hydrogenolysis of the oxazoline **95** also led to cyclization and reduction of the resulting aminoketone, forming the pyrrolidine **96**. The pyrrolidine was transformed into (+)-preussin (**2**), which was obtained in ten steps from L-*N*-benzoylphenylalaninol, with an overall yield of 13%.

A slightly differing approach towards preussin, as compared to the other syntheses in this chapter, was used by Beier and Schaumann [61] (Scheme 26).

The synthesis commenced with (1*R*)-[(1′*S*)-(benzyloxycarbonylamino)-2-phenylethyl]oxirane (**97**), a starting material derived from L-phenylalanine by a five step sequence including an elongation by one carbon atom [62]. Two epoxide opening reactions were used as key steps to form preussin in twelve steps and with an overall yield of 28% (starting from **97**). Ring opening of the epoxide **97** with lithiated allyl phenyl sulfide gave the diastereomeric sulfides **98** as a 50/50 mixture (Scheme 26). The sulfides were transformed in two steps to the diastereomeric sulfoxides **99**. The reaction with trimethyl phosphite yielded the allylic alcohol **100** by a [2,3]-sigmatropic rearrangement. Epoxidation of the allylic alcohol according to the Sharpless protocol gave a mixture of the pyrrolidine **101** and the epoxide **102**, which could also be transformed to the pyrrolidine **101** via a two step sequence. (+)-Preussin (**2**) was obtained in five additional steps. After diol cleavage the alkyl side chain was attached to the aldehyde by a Wittig reaction.

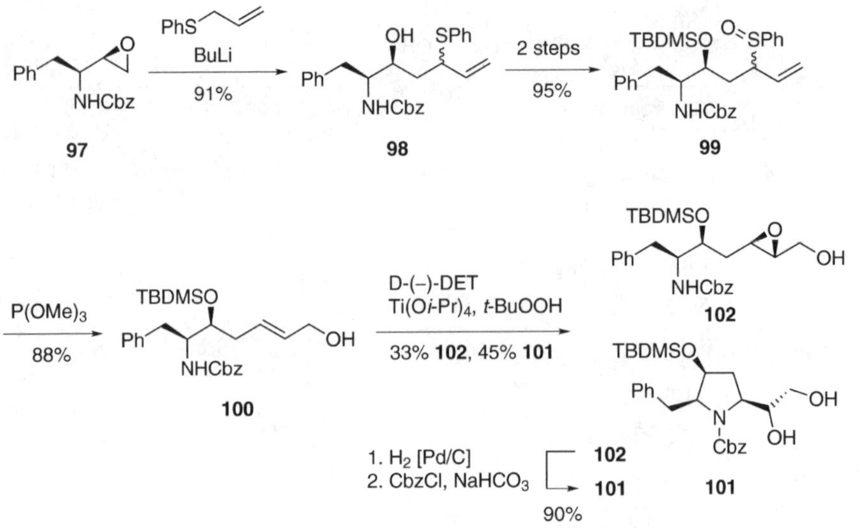

Scheme 26

## 3.1.2
## Syntheses Using Other L-Amino Acids

Common to all syntheses in this and the next section is the formation of the bond between C-2 and the nitrogen atom of preussin. A second retrosynthetic disconnection can be found at different positions of the pyrrolidine.

Besides eight syntheses, which started from L-phenylalanine or a derivative, two approaches towards (+)-preussin employed L-serine or L-aspartic acid. In one of the more recent approaches towards preussin Craig and coworkers [63] used l-serine (**103**) as starting material (Scheme 27).

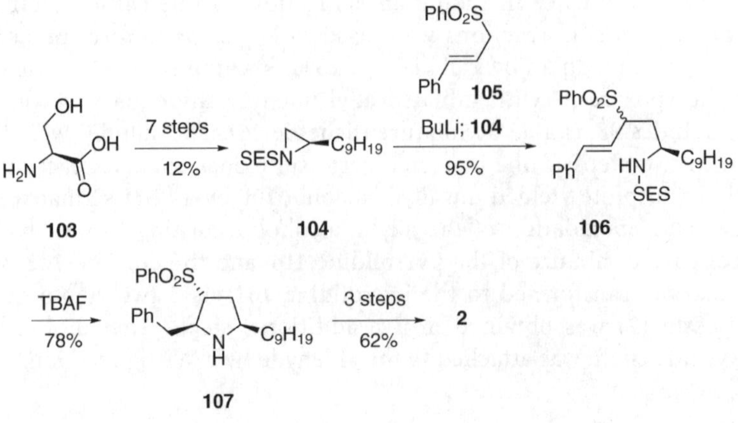

Scheme 27

They transformed L-serine in seven steps to the *N*-2-(trimethylsilyl)ethylsulfonyl(SES)-protected aziridine **104**. The latter reacted with the lithio-anion of (*E*)-3-phenyl-1-(phenylsulfonyl)-2-propene (**105**) to a diastereomeric mixture of the sulfones **106** (Scheme 27). This mixture was treated with TBAF, yielding the pyrrolidine **107** as a single diastereoisomer with 2,3-*trans*-2,5-*cis*-configuration previously observed for 5-*endo-trig* cyclizations of this type. In three additional steps the pyrrolidine **107** was transformed to (+)-preussin (**2**), with 5% yield overall and in twelve steps altogether.

Y. Yamamoto and co-workers used L-aspartic acid 4-methyl ester (**108**) as their starting material for the synthesis of preussin [64]. The ester was transformed in nine steps to the TBDPS-protected aminoalcohol **109** (Scheme 28). Allylation of the *N*-Boc-protected amine using allyl bromide

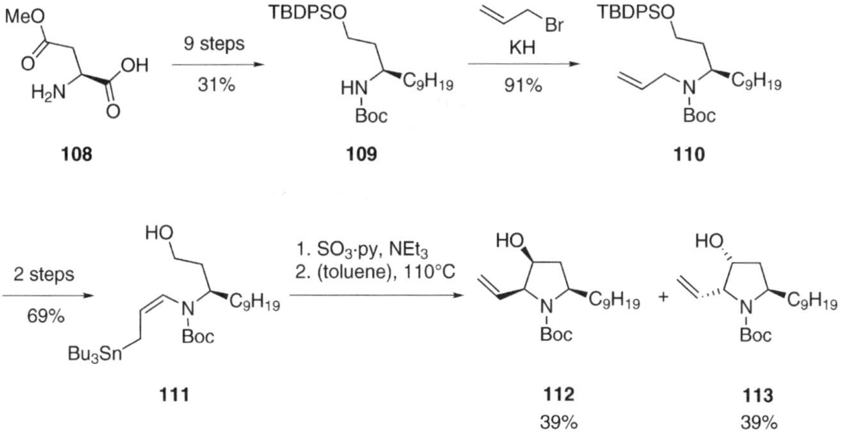

**Scheme 28**

gave carbamate **110**. The latter compound was transformed in two steps to the γ-aminoallylstannane **111**, which upon oxidation added intramolecularly to the aldehyde upon heating. The diastereomeric pyrrolidines **112** and **113** were obtained as a 50/50 mixture. (+)-Preussin was synthesized from **112** in seven additional steps. The total yield was 1% in 21 steps altogether.

## 3.1.3
### Sugars and Meso-Compounds as Building Blocks

Two syntheses and one formal synthesis of (+)-preussin have been reported using sugars as chiral pool building blocks. In these syntheses parts of the sugar backbone have been implemented into the preussin skeleton. All four pyrrolidine carbon atoms stem from the carbohydrate. As a consequence, the two carbon-nitrogen bonds (N/C-2 and N/C-5) of the pyrrolidine ring

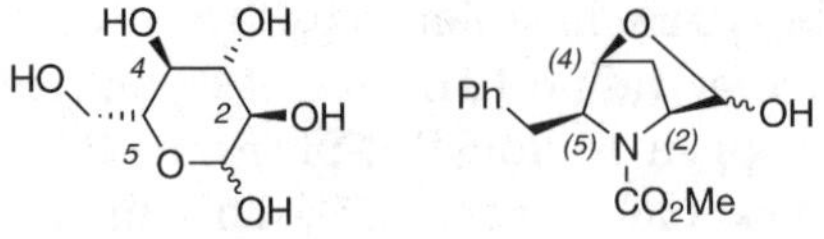

**Fig. 3** Comparison of the relevant stereogenic centers in d-glucose (**114**) and in preussin intermediate **115**

had to be formed in these syntheses. Two other syntheses which did not use sugars as chiral pool starting materials have the same retrosynthetic disconnections and will be discussed in this section as well. In both strategies the desymmetrization of a *meso*-compound is a key feature.

The first total synthesis of (+)-preussin was published in 1991 by Pak and Lee [65] using a derivative of D-glucose (**114**) as starting material. Figure 3 compares the stereogenic centers present in **114** and in an intermediate **115** to preussin.

Up to now, the Pak synthesis is the highest yielding synthesis of (+)-preussin with a total yield of 42% over eleven steps, starting from 5,6-anhydro-3-deoxy-1,2-*O*-isopropylidene-α-D-allofuranose (**116**). The stereogenic centers at C-2, C-4, and C-5 of D-glucose were elaborated into the three stereogenic centers of (+)-preussin (cf. Fig. 3).

Starting with a copper catalyzed epoxide ring opening, using phenyl magnesium chloride, a secondary alcohol was generated, which in turn was transformed into an electrophile by tosylation (Scheme 29). The tosylate **117**

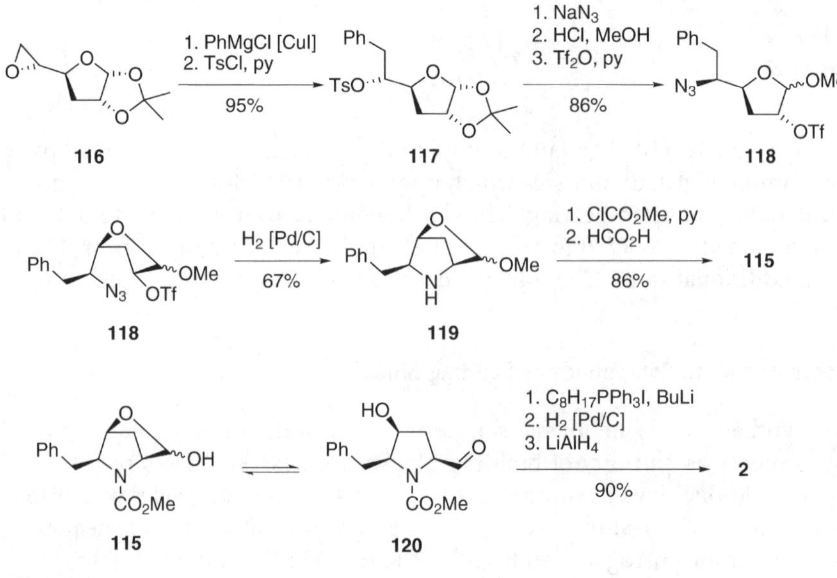

**Scheme 29**

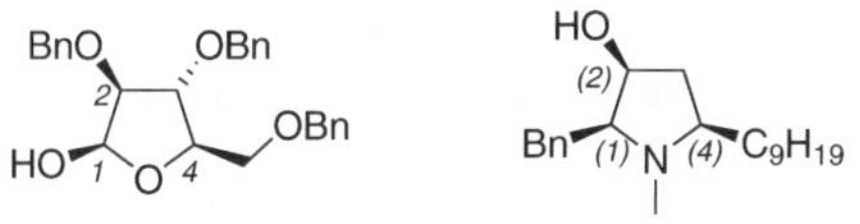

**Fig. 4** Comparison of the relevant stereogenic centers in the d-arabinose derivative **121** and in preussin (**2**)

was reacted with sodium azide and the acetonide group was removed with methanolic hydrogen chloride, yielding a mixture of anomers in a ratio of $\alpha{:}\beta=16/84$. The separated anomers were taken individually through the next four steps of the synthesis. To this end, each anomer was subjected to triflating conditions yielding the anomeric triflates **118**. Upon reduction of the azide to the primary amine a nucleophilic displacement of the triflate took place, yielding the bicyclic amines **119**. Methoxycarbonylation and subsequent demethylation with formic acid gave a mixture of the cyclic hemiacetal **115** and aldehyde **120**. The mixture was subjected to a Wittig reaction using the ylide derived from octyltriphenylphosphonium iodide and butyl lithium, furnishing a mixture of isomers ($Z{:}E=90/10$). The isomers were hydrogenated and afterwards the methoxycarbonyl group was reduced to yield (+)-preussin (**2**).

The second total synthesis using a sugar derivative as chiral pool starting material was published by Yoda et al. [66] and is based on 2,3,5-tri-O-benzyl-β-D-arabinofuranose (**121**) (Fig. 4). Via 13 steps they were able to synthesize (+)-preussin with an overall yield of 18%.

De Armas and co-workers [67] chose the epoxide **122**, a derivative of D-mannose, as their starting material for what they claim a formal synthesis of (+)-preussin (Scheme 30, PMB=*para*-methoxybenzyl). In 14 steps and 8%

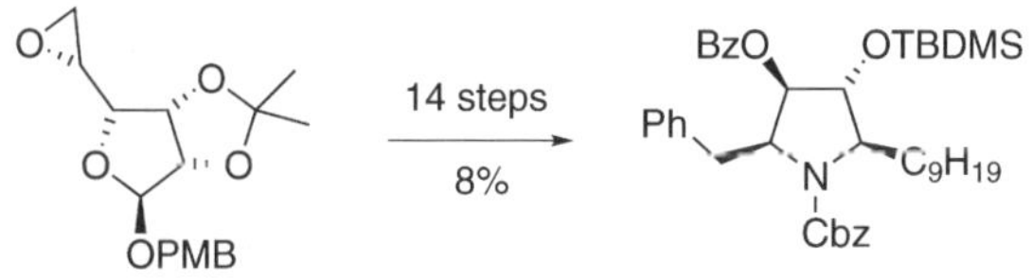

**Scheme 30**

yield they synthesized the previously unknown precursor **123** of (+)-preussin.

Dong and Lin [68, 69] did not start with a sugar as chiral building block, but a product from an asymmetric Sharpless epoxidation commonly used for the desymmetrization of *meso*-compounds [70, 71]. They transformed (2R,3S)-1,2-epoxy-3-benzyloxy-pent-4-ene (124) in 13 steps to (+)-preussin, with an overall yield of 18%.

The key step in this synthesis was an oxidative cyclization of the primary alcohol 125 upon PDC oxidation, yielding the lactam 126 (Scheme 31). The

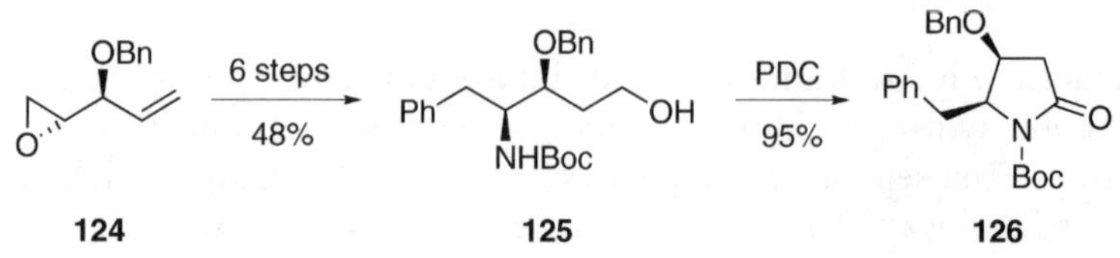

**Scheme 31**

lactam was transformed in six steps to (+)-preussin. The alkyl side chain was introduced by Grignard addition to lactam 126. The stereogenic center at C-5 of preussin was established by a diastereoselective ketone reduction followed by an intramolecular nucleophilic mesylate displacement.

Verma and Ghosh [72] desymmetrized a σ-symmetric 3-dimethyl(phenyl)silyl substituted glutaric anhydride (127) with Evans' oxazolidinone 128 (Scheme 32) as one of the key steps in their synthesis of (+)-preussin.

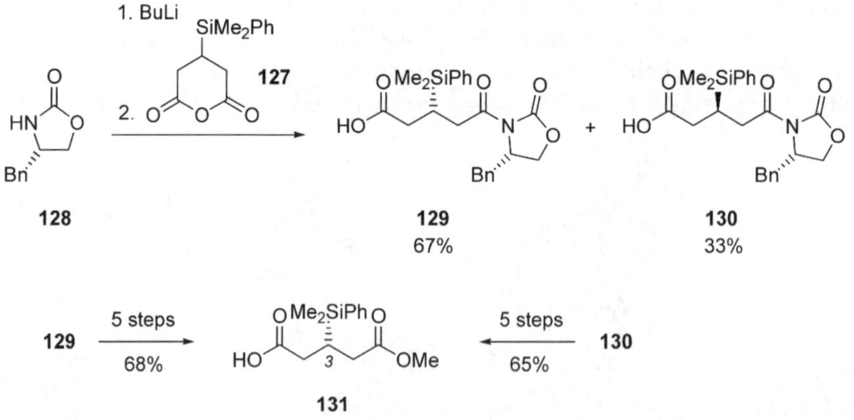

**Scheme 32**

They were able to synthesize the first homochiral intermediate 131 from both diastereomeric acids 129 and 130 formed in the desymmetrization step [73]. The stereogenic center, to which the silyl group is attached, was carried through the synthesis and ended up as carbon atom C-3 in preussin. The alkyl chain was introduced by Grignard attack on a mixed anhydride derived from acid **131**. Following this strategy preussin was obtained in 21 consecutive steps with an overall yield of 12%.

## 3.1.4
### Alternative Approaches

A hallmark of all the syntheses summarized in this section is the fact that they have more than two retrosynthetic bond fissions within the pyrrolidine ring. These approaches use specialized key transformations, mostly developed in the respective research groups.

Deng and Overman [74] employed their aza-Cope rearrangement-Mannich cyclization reaction as the key step in an approach to both (+)- and (−)-preussin.

N-Cbz-protected phenylalanine (**132**) was transformed in four steps to the tandem reaction precursor **133** (Scheme 33). The N,O-acetal was subject-

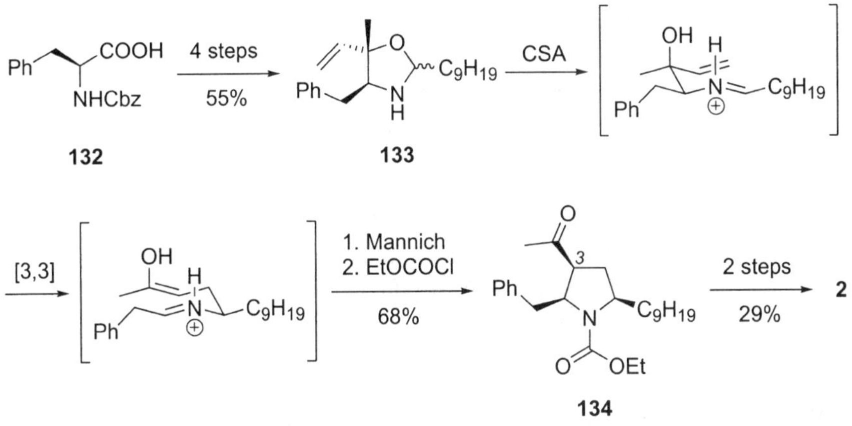

**Scheme 33**

ed to the acid-promoted aza-Cope rearrangement-Mannich cyclization, yielding the all-*cis* pyrrolidine as the major product. The crude product had to be N-protected again to prevent epimerization at carbon atom C-3. The enantiomeric purity of the intermediate **134** was determined to be 80% ee. In two additional steps ketone **134** was transformed into (+)-preussin (**2**), which was formed in seven steps with 11% overall yield. Overman and Deng also optimized a slightly different approach to (−)-preussin starting from **132** via an N-benzyl protected precursor for the domino reaction. The key reaction provided a (2R,3S,5S)-pyrrolidine related to **134**. After Baeyer-

Villiger oxidation the stereogenic center at C-3 was inverted by an oxidation-reduction sequence that yielded (−)-preussin with 18% yield over eleven steps.

An asymmetric 1,3-dipolar cycloaddition of decyl methyl nitrone (**135**) with (2R)-1-phenyl-but-3-en-2-ol (**136**) was the key reaction employed by Ohta and co-workers [75].

As the nitrone exists in the (E)-and (Z)-form in a ratio of approximately 50/50 four diastereoisomers were obtained, one of which was the precursor **137** to preussin (Scheme 34). It was transformed to (+)-preussin, yielding

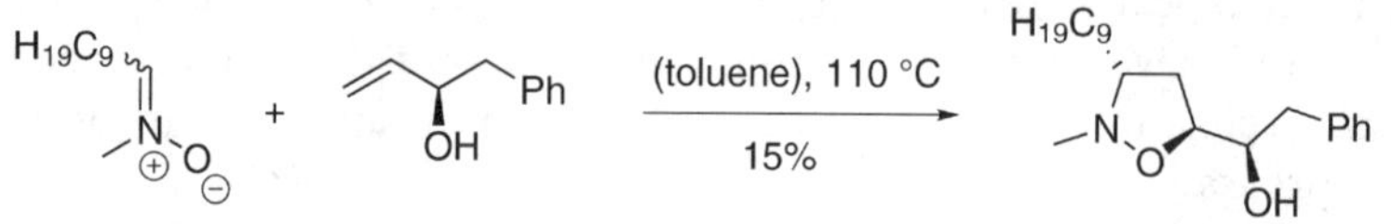

**Scheme 34**

the target compound in 4% over seven steps. Starting material of the sequence was (2R)-hydroxy-3-phenyl-propionic acid ethyl ester, which in turn can be synthesized from D-phenylalanine.

Greene and co-workers used a dichloroketene/enol ether cycloaddition and a Beckmann ring expansion as key reactions en route to (+)-preussin [76].

(R)-1-(2,4,6-Triisopropylphenyl)ethanol (**138**) as the source of chiral information was transformed to a chlorinated ynol by attack of the alkoxide on trichloroethylene (Scheme 35). Subsequent dehalogenation of the resulting ynol with butyl lithium and reaction with benzyl bromide provided the benzylated ynol which was partially reduced to the (Z)-enol ether **139** with Lindlar's catalyst. Cycloaddition with dichloroketene proceeded with high facial selectivity, yielding the major diastereoisomer **140** with a selectivity of 94/6. Using Tamura's Beckmann reagent, O-(mesitylenesulfonyl)hydroxylamine (MSH), the ring was expanded with complete regioselectivity. The dichlorosubstituted lactam so obtained was subsequently reduced using a zinc-copper couple to yield **141**. N-Boc-protection of the pyrrolidinone allowed for the Grignard addition of nonyl magnesium bromide. The reaction provided an α-hydroxy carbamate, which was in situ reduced with triethylsilane-boron trifluoride etherate to the all-cis pyrrolidine **142**. Trifluoracetic acid treatment caused cleavage of the carbamate and the ether. Subsequent reductive methylation using formaldehyde-sodium cyanoborohydride produced (+)-preussin (**2**), with an overall yield of 15%.

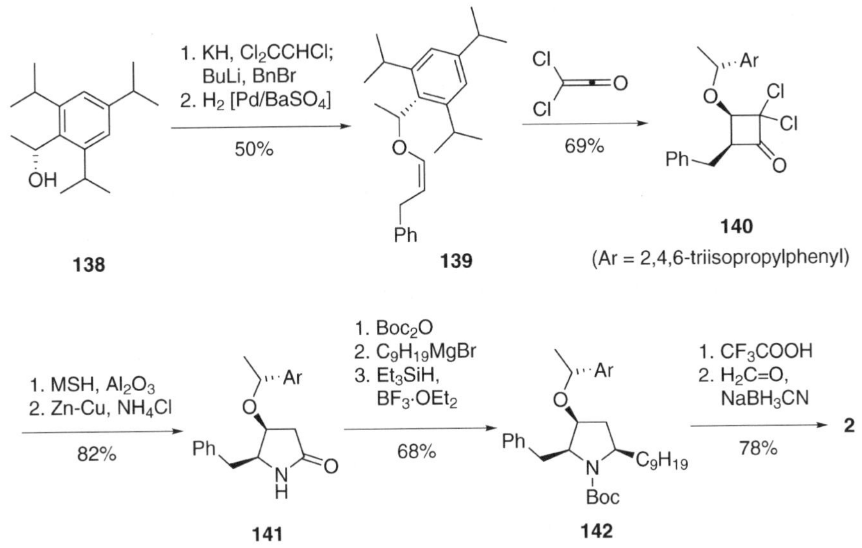

**Scheme 35**

Reißig and Hausherr synthesized (−)-preussin in ten steps starting from L-fructose. The chiral alkoxy allene **143** was formed as a mixture of diastereoisomers which differed in the configuration of the chiral axis [77].

Addition of the lithiated allene to an imine yielded a mixture of four diastereoisomers (ratio: 68/17/12/3; 68% *ds*) which was subjected to a silver nitrate induced cyclization. Product **145** derived from major diastereoisomer **144** was isolated and further converted to (−)-preussin (Scheme 36).

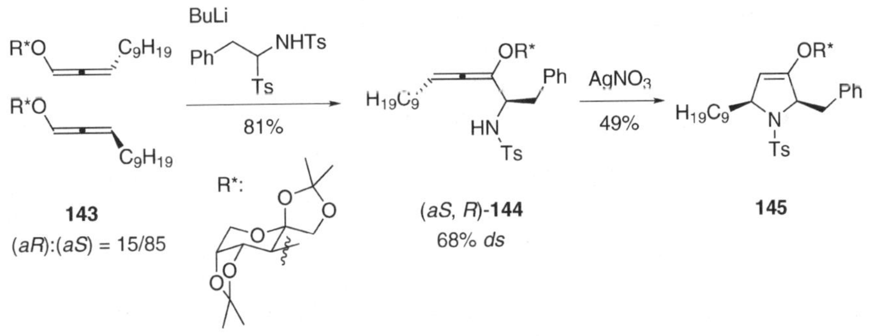

**Scheme 36**

## 3.2
## Paternò-Büchi Approach

### 3.2.1
### Background

The photocycloaddition of a carbonyl compound to an alkene was discovered as early as 1909 by Paternò and Chieffi [78] who employed sunlight as the irradiation source. In the 1950s the reaction was more intensively investigated by Büchi et al. [79] using artificial light sources. The Paternò-Büchi reaction has been studied mechanistically [80] and some important aspects are summarized in Scheme 37. Upon $n\pi^*$-excitation ($\lambda \cong 280$–350 nm), aldehydes

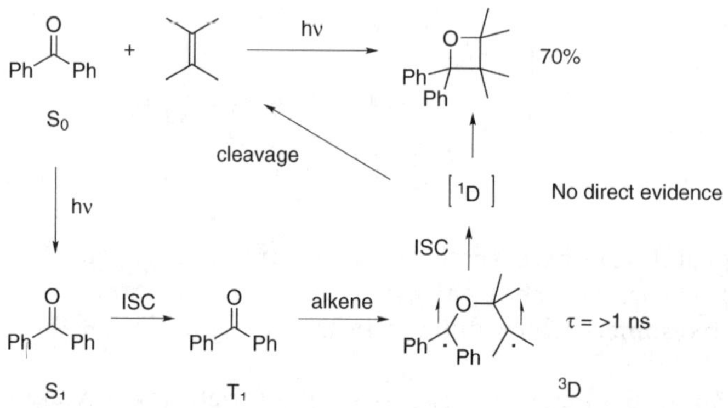

**Scheme 37**

and ketones such as benzophenone (Scheme 37) add to alkenes. Due to the high intersystem crossing (ISC) rate of most ketones and aldehydes ($S_1 \rightarrow T_1$) the addition occurs on the triplet hypersurface and is a step-wise process. The C-O-bond formation is the decisive parameter for the regio- and the facial diastereoselectivity whereas the simple diastereoselectivity is determined after ISC in the C-C-bond formation step. The cleavage of the biradical intermediate to the starting materials offers a reaction channel which may influence the regio- and the stereoselectivity of the photocycloaddition.

The application of regio- and stereoselective Paternò-Büchi reactions to synthetic problems was the first research topic pursued in our group and the project started in 1992. It became evident in early work [81, 82] that silyl enol ethers and aromatic aldehydes react with high selectivity to the corresponding 2-aryl-3-silyloxyoxetanes. The same key features were later established for N-acylenamines [83, 84] which form N-protected 3-amino-2-aryloxetanes. In combination with a hydrogenolytic ring opening the Paternò-Büchi reaction paves the way to a regioselective carbohydroxy-

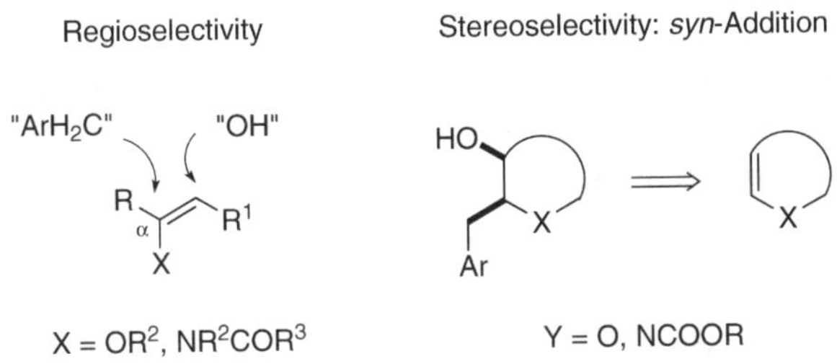

**Regioselectivity**

"ArH₂C"  "OH"

R\_α\_R¹
 X

X = OR², NR²COR³

**Stereoselectivity:** *syn*-Addition

HO  
 \_\_X  ⟹  X  
Ar

Y = O, NCOOR

**Scheme 38**

lation (Scheme 38) of noncyclic heteroatom substituted alkenes [85]. For obvious reasons it is restricted to a formal addition of an arylmethyl substitution to the α-position. If five- or six-membered cyclic enolethers or enamines are employed the Paternò-Büchi reaction proceeds in a *syn*-fashion. Upon hydrogenolysis it gives access to 2,3-*cis*-substituted heterocyclic rings.

Although the average yields (50–80%) in the Paternò-Büchi reaction are not outstandingly high, the combination [2+2]-photocycloaddition/hydrogenolysis provides a powerful tool for the synthesis of appropriately substituted heterocycles. With regard to chemo- and regioselectivity the restriction to 2-aryloxetanes is an advantage. Aromatic aldehydes and ketones are much better behaved in the Paternò-Büchi reaction than aliphatic carbonyl compounds. In Scheme 39 two typical carbohydroxylation reactions are

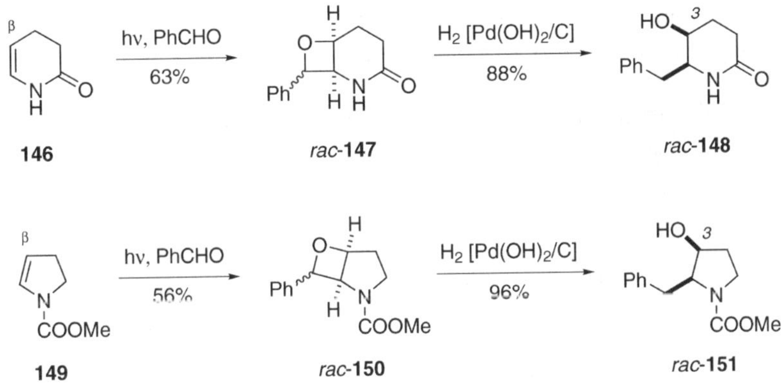

**146** → (hv, PhCHO, 63%) → ***rac*-147** → (H₂ [Pd(OH)₂/C], 88%) → ***rac*-148**

**149** → (hv, PhCHO, 56%) → ***rac*-150** → (H₂ [Pd(OH)₂/C], 96%) → ***rac*-151**

**Scheme 39**

summarized. Dihydropyridone (**146**) reacted cleanly with benzaldehyde upon irradiation at λ=300 nm (light source: RPR-3000 Å) to yield the oxetane **147** which could be hydrogenolytically cleaved with hydrogen in the presence of Pearlman's catalyst [86]. The resulting 3-hydroxypiperidine **148** was *cis*-configurated. The same sequence of events led to the *cis*-pyrrolidinol

151 starting from dihydropyrrole 149 via the intermediate oxetane 150. The regioselectivity of the reactions is determined by the high electrophilicity of the aldehyde oxygen atom in the $T_1$ state (Scheme 37). It attacks the alkene at the more electronrich position, $\beta$ to the heteroatom [83, 84, 86].

### 3.2.2
### Retrosynthesis

The reactions depicted in Scheme 39 were already conducted in view of a potential use in the synthesis of pyrrolidinols and piperidinols. The structural feature of a 2-arylmethyl-3-hydroxysubstitution is not only found in preussin but also in anisomycin (152) [87] or in the piperidine alkaloid FR 901483 (153) [88] (Fig. 5).

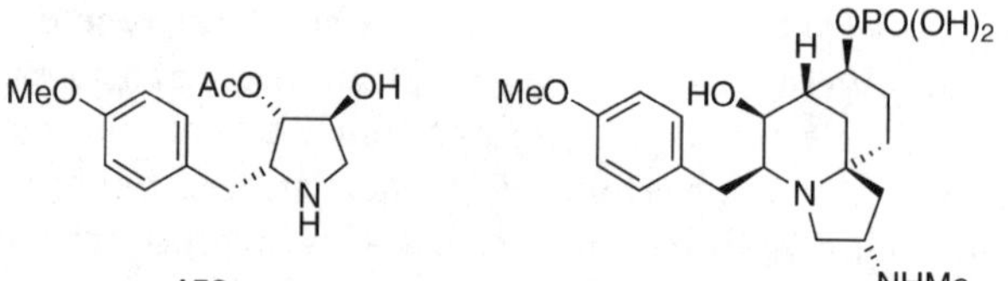

**Fig. 5** Chemical structures of anisomycin (152) and the piperidine alkaloid FR 901483 (153)

For the synthesis of (+)-preussin, commercially available L-pyroglutamic acid (154) could serve as an enantiomerically pure starting material. Given the key reaction we had in mind, the bond set for the synthesis of preussin was straightforward (Scheme 40). Attachment of the alkyl chain to the core

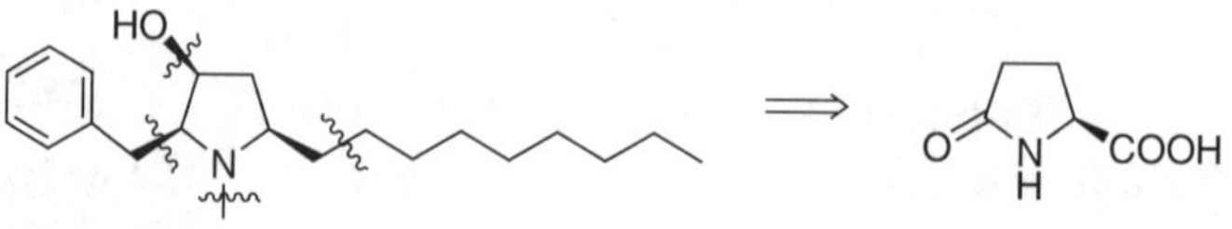

**Scheme 40**

fragment was envisaged to occur by a nucleophilic cuprate substitution. The reduction of the methoxycarbonyl group to the methyl group in the last step towards preussin was precedented [65].

In comparison to the other strategies employed for the synthesis of preussin the carbohydroxylation sequence we chose is unique. It has remained the only strategy which does not include the formation of the pyrrolidine ring. Despite its simplicity, the bond set depicted in Scheme 40 was associated with some uncertainty concerning the stereochemical outcome of the photocycloaddition. The key oxetane intermediate **155** with the right (2S,3S,5R)-configuration was to be formed from the dihydropyrrole **156** (Scheme 41). At first sight, it appears bewildering to postulate a cycloaddi-

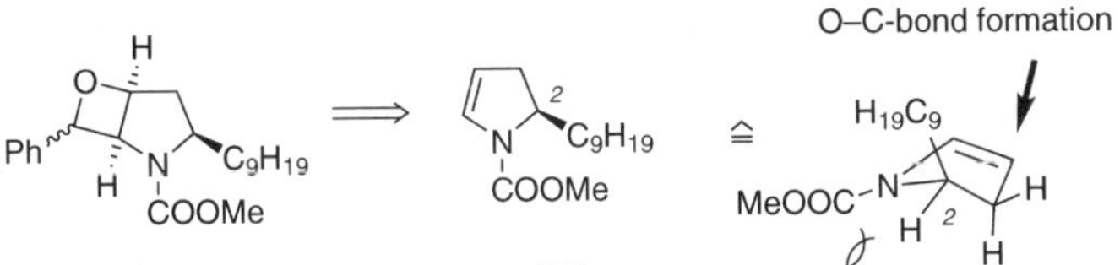

**Scheme 41**

tion from the apparently more shielded face. The skepticism about the facial diastereoselectivity is supported by the fact that ketene additions to a related 2,3-dihydropyrrole, indeed, occur from the face opposite to the substituent at C-2 [89]. A closer mechanistic consideration, however, reveals that the analogy between the ketene cycloaddition and the Paternò-Büchi reaction is not valid. Whereas in the former case the addition occurs both at C-4 and C-5 simultaneously in a [π2s+π2a]-mode the latter reaction proceeds by an initial O–C-bond formation (see above). Molecular models of the dihydropyrrole **156** suggested that the nonyl group was located in a pseudoaxial position to avoid 1,3-allylic strain with the N-methoxycarbonyl group. As a consequence there is a pseudoaxial and a pseudoequatorial hydrogen atom at C-3. The O–C-bond formation at C-4 as the first step in the Paternò-Büchi reaction is favored *trans* to the pseudoaxial hydrogen to avoid torsional strain. Model studies and the successful synthesis supported this notion.

### 3.2.3
### Synthesis

Our synthesis of preussin [90, 91] (Scheme 42) commenced with L-pyroglutaminol (**157**) which is commercially available or which can be prepared in 85% yield from L-pyroglutamic acid (**154**) [1. (MeO)$_2$CMe$_2$, HCl (cat.), 50°C (MeOH); 2. NaBH$_4$, r.t. (*i*-PrOH)]. There was literature precedence for improved yields in the nucleophilic displacement step upon using a pyrrolidinone tosylate and Lipshutz cuprates as compared to other methods [92, 93]. We therefore prepared the known tosylate [94] although the yield remained unsatisfactory. The analogous bromide [95] which is also a suitable electrophile was obtained in 65% yield [CBr$_4$, PPh$_3$, r.t. (CH$_2$Cl$_2$)]. The subsequent substitution reaction proceeded smoothly and gave lactam **158** which was acylated with methyl chloroformate. The conversion of pyrrolidinone **159** to

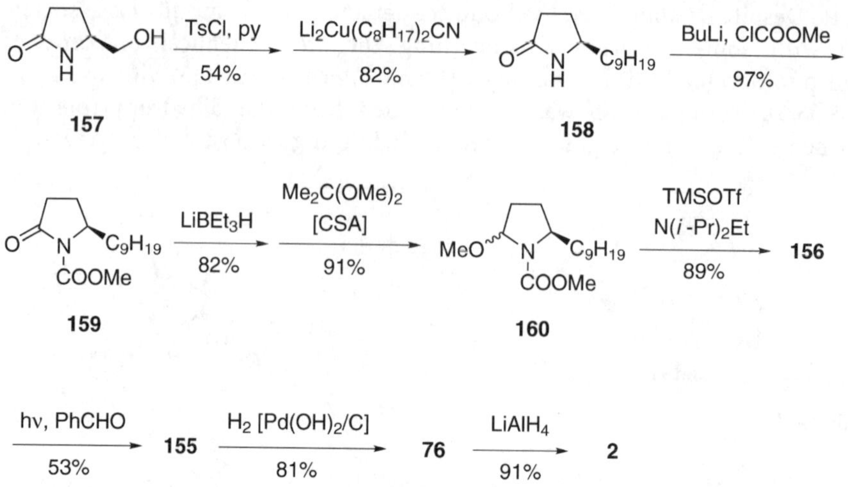

**Scheme 42**

dihydropyrrole **156** was based on a reduction/elimination sequence. Alternatively, the *N,O*-acetals required for the elimination can be prepared oxidatively [96]. Following a known protocol [97] we obtained the desired compound **160** in excellent yield. As the elimination according to reported procedures proceeded sluggishly we looked for possible alternatives. By applying a method which was originally used by Gassman et al. for the preparation of enolethers from *O,O*-acetals [98] we were able to convert the *N,O*-acetal **160** to dihydropyrrole **156** in high yields [99]. The reaction is applicable to a variety of *N,O*-acetals [99, 100].

Having the requisite alkene **156** in hand it was an unexpected and unpleasant surprise that the photochemical key step failed completely. Under the previously used irradiation conditions (MeCN as the solvent, Rayonet RPR 3000 Å lamps) an undefined but extremely fast decomposition of the starting material **156** occurred. For reasons we could not foresee the dihydropyrrole **156** was highly unstable upon UV irradiation at 300 nm. Fortunately the shock did not last very long. We noted quickly that an irradiation at longer wavelength (Rayonet RPR 3500 Å, $\lambda$=350 nm) was a viable solution to our problems. The longer wavelength irradiation was still effective for benzaldehyde excitation but did not interfere with the alkene. Under these conditions the photocycloaddition proceeded in decent yield providing two diastereoisomers in a ratio of 80/20. The desired diastereoisomer **155** could be isolated in 53% yield. The subsequent steps proceeded in line with our expectations. The hydrogenolysis provided the known pyrrolidine **76** which was reduced to preussin. Overall, our synthetic access to preussin comprised nine steps starting from L-pyroglutaminol (**157**) and eleven steps starting from the corresponding acid **154**. The total yield amounted to 11%.

Reactions which led to analogues of preussin were conducted subsequently [86]. A variation in the substituent at C-2 did not lead to major changes in the facial diastereoselectivity of the Paternò-Büchi reaction. The diastereomeric ratio of the two major products was consistently in the range of 70/30 to 80/20. The diastereoisomers result from the facial diastereoselectivity in the approach of the aldehyde to the diastereotopic faces of the dihydropyrrole but not from the simple diastereoselectivity of the photocycloaddition. After hydrogenolysis the diastereoisomeric alcohols were obtained in the same ratio as the oxetanes. The only major exception in the photochemical reaction was dihydropyrrole **161** which gave a single diastereoisomer upon benzaldehyde addition. A favorable $\pi\pi$- or van der Waals-interaction between the incoming benzaldehyde and the axial benzyl group might be responsible for the strong preference in favor of product **162** (Scheme 43).

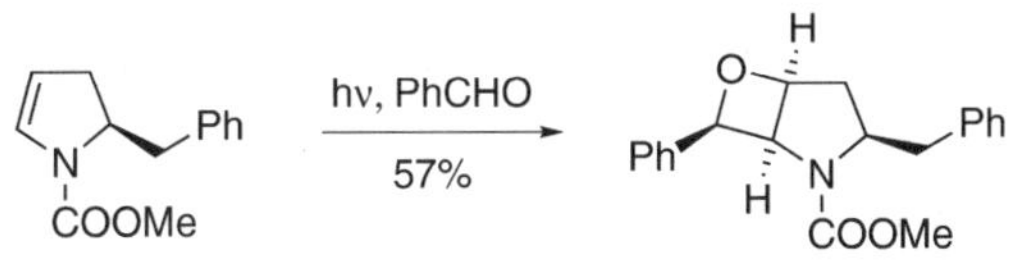

**Scheme 43**

# 4
# Conclusion

Both target compounds discussed in this review, kelsoene (**1**) and preussin (**2**), provide a fascinating playground for synthetic organic chemists. The construction of the cyclobutane in kelsoene limits the number of methods and invites the application of photochemical reactions as key steps. Indeed, three out of five completed syntheses are based on an intermolecular enone [2+2]-photocycloaddition and one—our own—is based on an intramolecular Cu-catalyzed [2+2]-photocycloaddition. A unique approach is based on a homo-Favorskii rearrangement as the key step. Contrary to that, the pyrrolidine core of preussin offers a plentitude of synthetic alternatives which is reflected by the large number of syntheses completed to date. The photochemical pathway to preussin has remained unique as it is the only route which does not retrosynthetically disconnect the five-membered heterocycle. The photochemical key step is employed for a stereo- and regioselective carbohydroxylation of a dihydropyrrole precursor.

**Note added in proof** After completion of this review the following papers, which deal with synthetic approaches to preussin, have appeared:
Raghavan S, Rasheed MA (2003) Tetrahedron 59:10307
Huang P-Q, Wu T-J, Ruan Y-P (2003) Org Lett 5:4341
Dikshit DK, Goswami LN, Singh VS (2003) Synlett 1737

# References

1. König GM, Wright AD (1997) J Org Chem 62:3837
2. Nabeta K, Yamamoto K, Hashimoto M, Koshino H, Funatsuki K, Katoh K (1998) J Chem Soc Chem Commun 1485
3. Warmers U, Wihstutz K, Bülow N, Fricke C, König WA (1998) Phytochemistry 49:1723
4. Warmers U, König WA (1999) Phytochemistry 52:1519
5. Nabeta K, Yamamoto M, Koshino H, Fukui H, Fukushi Y, Tahara S (1999) Biosci Biotechnol Biochem 63:1772
6. Nabeta K, Yamamoto M, Fukushima K, Katoh K (2000), J Chem Soc Perkin Trans 1:2703
7. Mehta G, Srinivas K (1999) Synlett 555
8. Mehta G, Srinivas K (1999) Tetrahedron Lett 40:4877
9. Fietz-Razavian S, Schulz S, Dix I, Jones PG (2001) J Chem Soc Chem Commun 2154
10. Fietz-Razavian S (2001) PhD thesis, Technische Universität Braunschweig
11. Piers E, Orellana A (2001) Synthesis 2138
12. Mehta G, Srinivas K (2001) Tetrahedron Lett 42:2855
13. Henry PM, Davies M, Ferguson G, Phillips S, Restivo R (1974) J Chem Soc Chem Commun 112
14. Ito Y, Hirao T, Saegusa T (1978) J Org Chem 43:1011
15. Dauben WG, Michno DM (1977) J Org Chem 42:682
16. Wolinsky J, Wolf H, Gibson T (1963) J Org Chem 28:274
17. Wolinsky J, Eustache EJ (1972) J Org Chem 37:3376
18. Magnus P, Roy G (1979) J Chem Soc Chem Comm 822
19. Magnus P, Roy G (1982) Organometallics 1:553
20. Taguchi H, Tanaka S, Yamamoto H, Nozaki H (1973) Tetrahedron Lett 2465
21. Lombardo L (1982) Tetrahedron Lett 23:4293
22. Lombardo L (1987) Org Synth 65:81
23. Zhang L, Koreeda M (2002) Org Lett 4:3755
24. Mann G, Muchall HM (1997) Cyclobutanes: synthesis by rearrangement of the carbon framework. In: de Meijere A (ed) Methods of organic chemistry (Houben-Weyl), 4th edn, vol E 17e. Thieme Verlag, Stuttgart, p 233
25. Wenkert E, Bakuzis P, Baumgarten RJ, Leicht CL, Schenk HP (1971) J Am Chem Soc 93:3208
26. Klinkmüller KD, Marschall H, Weyerstahl P (1975) Chem Ber 108:191
27. Erman WF, Treptow RS, Bakuzis P, Wenkert E (1971) J Am Chem Soc 93:657
28. Reetz MT, Kindler A (1995) J Organomet Chem 502:C5–C7
29. Koreeda M (2003) (personal communication)
30. Trost BM, Organ MG (1994) J Am Chem Soc 116:10320
31. Bach T, Pelkmann C, Harms K (1999) Tetrahedron Lett 40:2103
32. Bach T, Krüger C, Harms K (2000) Synthesis 305

33. Steiner G, Bach A, Bialojan S, Greger G, Hege HG, Höger T, Jochims K, Munschauer R, Neumann B Teschendorf HJ, Traut M, Unger L, Gross G (1998) Drugs Fut 23:191
34. Salomon RG, Coughlin DJ, Ghosh S, Zagorski MG (1982) J Am Chem Soc 104:998
35. Salomon RG, Ghosh S, Raychaudhuri SR, Miranti TS (1984) Tetrahedron Lett 25:3167
36. Steiner G, Munschauer R, Klebe G, Siggel L (1995) Heterocycles 40:319
37. Tietze LF, Beifuß U, Ruther M, Rühlmann A, Antel J, Sheldrick GM (1988) Angew Chem 100:1200; (1988) Angew Chem Int Ed 27:1186
38. Spiegel A (2002) PhD thesis, Technische Universität München
39. Salomon RG, Sachinvala ND, Roy S, Basu B, Raychaudhuri SR, Miller DB, Sharma RB (1991) J Am Chem Soc 113:3085
40. Langer K, Mattay J, Heidbreder A, Möller M (1992) Liebigs Ann Chem 257
41. Bach T, Spiegel A (2002) Eur J Org Chem 645
42. Shizuri Y, Suyama K, Yamamura S (1986) J Chem Soc Chem Commun 63
43. Chu A, Mander LN (1988) Tetrahedron Lett 29:2727
44. Corey EJ, Guzman-Perez A, Loh TP (1994) J Am Chem Soc 116:3611
45. Míčková R, Syhora K (1965) Coll Czech Chem Comun 30:2771
46. Desai MC, Singh J, Chawla HPS, Dev S (1981) Tetrahedron 37:2935
47. Bach T, Spiegel A (2002) Synlett 1305
48. Lovey RG, Saksena AK, Girijavallabhan VM (1994) Tetrahedron Lett 35:6047
49. Morgan B, Dodds DR, Zaks A, Andrews DR, Klesse R (1997) J Org Chem 62:7736
50. Schwartz RE, Liesch J, Hensens O, Zitano L, Honeycutt S, Garrity G, Fromtling RA, Onishi J, Monaghan R (1988) J Antibiot 41:1774
51. Johnson JH, Phillipson DW, Kahle AD (1989) J Antibiot 42:1184
52. Kasahara K, Yoshida M, Eishima J, Takesako K, Beppu T, Horinouchi S (1997) J Antibiot 50:267
53. Achenbach TV, Slater EP, Brummerhop H, Bach T, Müller R (2000) Antimicrob Agents Chemother 44:2794
54. Okue M, Watanabe H, Kitahara T (2001) Tetrahedron 57:4107
55. Okue M, Watanabe H, Kasahara K, Yoshida M, Horinouchi S, Kitahara T (2002) Biosci Biotechnol Biochem 66:1093
56. Overhand M, Hecht SM (1994) J Org Chem 59:4721
57. McGrane PL, Livinghouse T (1993) J Am Chem Soc 115:11485
58. Veeresa G, Datta A (1998) Tetrahedron 54:15673
59. Krasiński A, Gruza H, Jurczak J (2001) Heterocycles 54:581
60. Lee K-Y, Kim Y-H, Oh C-Y, Ham W-H (2000) Org Lett 2:4041
61. Beier C, Schaumann E (1997) Synthesis 1296
62. Luly JR, Dellaria JF, Plattner JJ, Soderquist JL, Yi N (1987) J Org Chem 52:1487
63. Caldwell JJ, Craig D, East SP (2001) Synlett 1602
64. Kadota I, Saya S, Yamamoto Y (1997) Heterocycles 46:335
65. Pak CS, Lee GH (1991) J Org Chem 56:1128
66. Yoda H, Yamazaki H, Takabe K (1996) Tetrahedron Asymmetry 7:373
67. de Armas P, García-Tellado F, Marrero-Tellado JJ, Robles J (1998) Tetrahedron Lett 39:131
68. Dong H-Q, Lin G-Q (1997) Chin Chem Lett 8:693
69. Dong H-Q, Lin G-Q (1998) Chin J Chem 16:458
70. Hatakeyama S, Sakurai K, Takano S (1985) Chem Commun 1759
71. Hoffmann RW (2003) Angew Chem 115:1128; (2003) Angew Chem Int Ed 42:1096
72. Verma R, Ghosh SK (1997) Chem Commun 1601
73. Verma R, Ghosh SK (1999) J Chem Soc Perkin Trans 1:265

74. Deng W, Overman LE (1994) J Am Chem Soc 116:11241
75. Shimazaki M, Okazaki F, Nakajima F, Ishikawa T, Ohta A (1993) Heterocycles 36:1823
76. Kanazawa A, Gillet S, Delair P, Greene AE (1998) J Org Chem 63:4660
77. Hausherr A (2001) PhD thesis, Freie Universität Berlin
78. Paternò E, Chieffi G (1909) Gazz Chim Ital 39(1):341
79. Büchi G, Inman CG, Lipinsky ES (1954) J Am Chem Soc 76:4327
80. Freilich SC, Peters KS (1985) J Am Chem Soc 107:3819
81. Bach T, Jödicke K (1993) Chem Ber 126:2457
82. Bach T (1995) Liebigs Ann 855
83. Bach T (1996) Angew Chem 108:976; (1996) Angew Chem Int Ed 35:884
84. Bach T, Schröder J (1999) J Org Chem 64:1265
85. Bach T (1995) Liebigs Ann 1045
86. Bach T, Bergmann H, Brummerhop H, Lewis W, Harms K (2001) Chem Eur J 7:4512
87. Sobin BA, Tanner FW (1954) J Am Chem Soc 76:4053
88. Sakamoto K, Tsuji E, Abe F, Nakanishi T, Yamashita M, Shigematsu N, Izumi S, Okuhara M (1996) J Antibiot 49:37
89. Carpes MJS, Miranda PCML, Correia CRD (1997) Tetrahedron Lett 38:1869
90. Bach T, Brummerhop H (1998) Angew Chem 110:3577; (1998) Angew Chem Int Ed 37:3400
91. Bach T, Brummerhop H, Harms K (2000) Chem Eur J 6:3838
92. Ackermann J, Matthes M, Tamm C (1990) Helv Chim Acta 73:122
93. Matthes M, Tamm C (1991) Helv Chim Acta 74:1585
94. Hardegger E, Ott H (1955) Helv Chim Acta 38:312
95. Silverman RB, Levy MA (1980) J Org Chem 45:815
96. Shono T, Matsumura Y, Tsubata K, Sugihara Y, Yamane S-I, Kanazawa T, Aoki T (1982) J Am Chem Soc 104:6697
97. Dieter RK, Sharma RR (1996) J Org Chem 61:4180
98. Gassman PG, Burns SJ (1988) J Org Chem 53:5574
99. Bach T, Brummerhop H (1999) J Prakt Chem 341:312
100. Gaulon C, Dhal R, Dujardin G (2003) Synthesis 2269

# Paraconic Acids –
# The Natural Products from *Lichen* Symbiont

Rakeshwar Bandichhor · Bernd Nosse · Oliver Reiser (✉)

Institut für Organische Chemie, Universität Regensburg, Universitätsstrasse 31, 93053 Regensburg, Germany
*Oliver.Reiser@chemie.uni-regensburg.de*

| | | |
|---|---|---|
| 1 | Introduction | 44 |
| 2 | Stereoselective Syntheses of Paraconic Acids | 45 |
| 2.1 | Diastereoselective Syntheses of Paraconic Acids | 45 |
| 2.2 | Enantioselective, Chiral Pool-Based Syntheses of Paraconic Acids | 49 |
| 2.2.1 | Starting from Isopropylidine Glyceraldehyde | 49 |
| 2.2.2 | Starting from Diacetone-D-glucose | 52 |
| 2.2.3 | Starting from D-Mannitol | 53 |
| 2.2.4 | Starting from Bicyclic Oxaenones | 54 |
| 2.2.5 | Starting from Chiral Butenolide Synthons | 56 |
| 2.3 | Enantioselective, Chiral Auxiliary Based Syntheses of Paraconic Acids | 57 |
| 2.3.1 | [2+2]-Cycloaddition | 57 |
| 2.3.2 | Sequential Aldol-Lactonization | 58 |
| 2.3.3 | Carbolithiation Towards (−)-Roccellaric Acid | 59 |
| 2.3.4 | Nicholas-Schreiber Reaction | 60 |
| 2.4 | Catalytic Asymmetric Syntheses of Paraconic Acids | 61 |
| 2.4.1 | Nickel Catalyzed Intramolecular Carbozincation | 61 |
| 2.4.2 | Palladium(II) Catalyzed Construction of $\gamma$-Butyrolactones | 62 |
| 2.4.3 | Base-Induced Intramolecular Michael Addition | 63 |
| 2.4.4 | Chemoenzymatic Transformations Toward Phaseolinic Acid | 64 |
| 2.4.5 | Asymmetric Cyclopropanation | 65 |
| 2.4.6 | Ireland-Claisen Rearrangement | 67 |
| 2.4.7 | Tungsten-$\pi$-allyl Complexes | 68 |
| 3 | Conclusion | 69 |
| | References | 69 |

**Abstract** Paraconic acids, belonging to the class of $\gamma$-butyrolactone natural products, display a broad range of biological activities such as antibiotic and antitumor properties. Consequently a great number of synthetic strategies have been devised for them, ranging from diastereoselective and chiral pool approaches to the application of asymmetric catalysis. This review gives a critical account on the different methods developed that lead to paraconic acids of great structural variety.

**Keywords** Paraconic acid · Chiral pool · Chiral auxiliary · Asymmetric synthesis · $\gamma$-Butyrolactones

**List of Abbreviations**

| | |
|---|---|
| Ac | Acetyl |
| AIBN | 2,2'-Azobisisobutyronitrile |
| 9-BBN | 9-Borabicyclo[3.3.1]nonane |
| Bn | Benzyl |
| Bu | Butyl |
| t-Bu | tert-Butyl |
| Bz | Benzoyl |
| CAN | Ceric ammonium nitrate |
| cat | Catalyst |
| Cp | Cyclopentadienyl |
| DABCO | 1,4-Diazabicyclo[2.2.2]octane |
| DET | Diethyl tartrate |
| DIBALH | Diisobutylaluminum hydride |
| DMAP | 4-(Dimethylamino)pyridine |
| DBU | 1,8-Diazabicyclo[4.3.0]undec-7-ene |
| DEAD | Diethyl azodicarboxylate |
| DME | 1,2-Dimethoxyethane |
| DMF | Dimethylformamide |
| DMSO | Dimethyl sulfoxide |
| ee | Enantiomer excess |
| equiv | Equivalent(s) |
| Et | Ethyl |
| h | Hour(s) |
| HMPA | Hexamethylphosphoric triamide |
| LDA | Lithium diisopropylamide |
| m-CPBA | m-Chloroperoxybenzoic acid |
| Me | Methyl |
| Mes | Mesityl, 2,4,6-trimethylphenyl |
| min | Minute(s) |
| NaHMDS | Sodium hexamethyldisilazide |
| NBS | N-Bromosuccinimide |
| Nu | Nucleophile |
| PDC | Pyridinium dichromate |
| Ph | Phenyl |
| PPTS | Pyridinium p-toluenesulfonate |
| rt | Room temperature |
| Tf | Trifluoromethanesulfonyl (triflyl) |
| THF | Tetrahydrofuran |
| THP | Tetrahydropyran-2-yl |
| TMEDA | $N,N,N',N'$-Tetramethyl-1,2-ethylenediamine |
| TMS | Trimethylsilyl |
| Ts | Tosyl, 4-Toluenesulfonyl |

# 1
# Introduction

*Lichens*, the natural source of paraconic acids [1], are a successful alliance between fungi and algae. Each is doing what they do best, and thriving as a

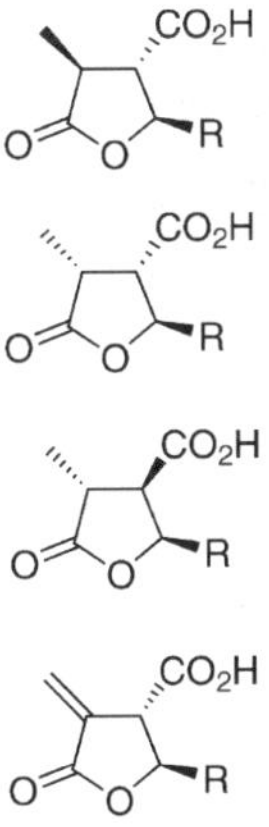

R = $n$-C$_{13}$H$_{27}$: (+)-Roccellaric acid (**1**)
R = $n$-C$_{11}$H$_{23}$: (+)-Nephrosteranic acid (**2**)
R = (CH$_2$)$_{13}$CH(OH)CH$_3$: (+)-Neodihydromurolic acid (**3**)

R = $n$-C$_{13}$H$_{27}$: (+)-Dihydroprotolichesterinic acid (**4**)
R = (CH$_2$)$_{13}$COCH$_3$: (+)-Dihydropertusaric acid (**5**)

R = $n$-C$_5$H$_{11}$: (+)-Phaseolinic acid (**6**)
R = C$_{13}$H$_{27}$: (+)-Nephromopsinic acid (**7**)
R = (CH$_2$)$_{13}$COCH$_3$: (+)-Pertusarinic acids (**8**)

R = $n$-C$_{12}$H$_{24}$CO$_2$H: (+)-Protopraesorediosic acid (**9**)
R = $n$-C$_{13}$H$_{27}$: (+)-Protolichesterinic acid (**10**)
R = $n$-C$_5$H$_{11}$: (+)-Methylenolactocin (**11**)

**Fig. 1** Most prominent paraconic acids

result of a natural cooperation. They live as one organism, both inhabiting the same body, and produce a variety of substances friendly to environment and human health except few poisonous examples. Screening and isolation of molecules from suitable *Lichen* symbionts or medicinal plants has resulted in a great number of paraconic acids with antineoplastic and antibiotic properties [2], which prompted many researchers to devise synthetic routes towards these natural products. Paraconic acids are naturally occurring $\gamma$-butyrolactones [3], having a carboxylic acid group in the 3-position as their characteristic functionality (Fig. 1).

Consequently, a number of stereoselective syntheses have been developed leading to a variety of paraconic acids either in racemic or in enantiopure form, using starting materials from the chiral pool, chiral auxiliaries or applying catalytic asymmetric methodology. Moreover, a number of strategies leading to paraconic acids in a non-stereoselective way have been reported, which will not be described in detail in this review [4, 5].

# 2
# Stereoselective Syntheses of Paraconic Acids

## 2.1
## Diastereoselective Syntheses of Paraconic Acids

$\gamma$-Butyrolactones are readily obtained from 4-hydroxy carboxylic esters and can therefore be build up by a homoaldol reaction [6]. The principle difficulties of stability and self-condensation encountered with generating homoenolates is not as severe in the case of paraconic acids due to the carboxy functionality being present in the 3-position required for paraconic acids, thus, resulting in enolate stabilization. Consequently, if a mixed enolate/homoenolate of type **14** can be *regioselectively* formed, aldol reaction with **15** and subsequent cyclization results in the target compounds **12** (Scheme 1).

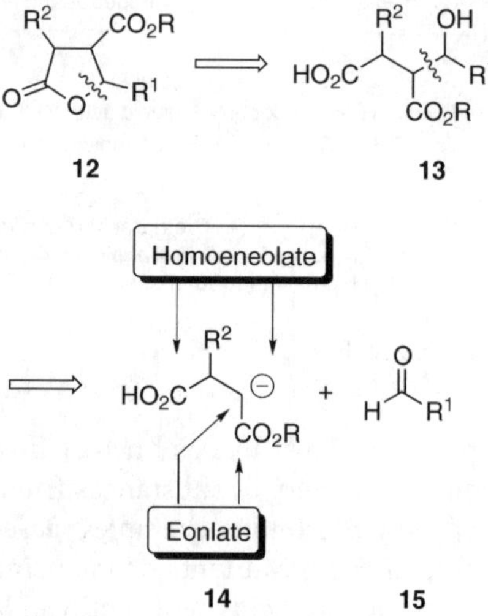


**Scheme 1**

The norbornene derivative **16**, obtained exclusively as the *exo* adduct via a Diels-Alder reaction of itaconic anhydride with cyclopentadiene followed by hydrolysis and esterification [7], was found to be a suitable precursor for an enolate of type **14** (Scheme 2). Due to the quaternary center at C-3 enolization with base proceeded unambiguously, giving rise to a diastereomeric mixture of lactones **17/18** after reaction with hexanal. Retro-Diels-Alder reaction led to the monocyclic lactones **19/20** (2:1), elegantly unmasking the *exo*-methylene group found in so many paraconic acids [8]. Hydrolysis of this mixture in refluxing butanone with 6 N HCl [9] effected epimerization

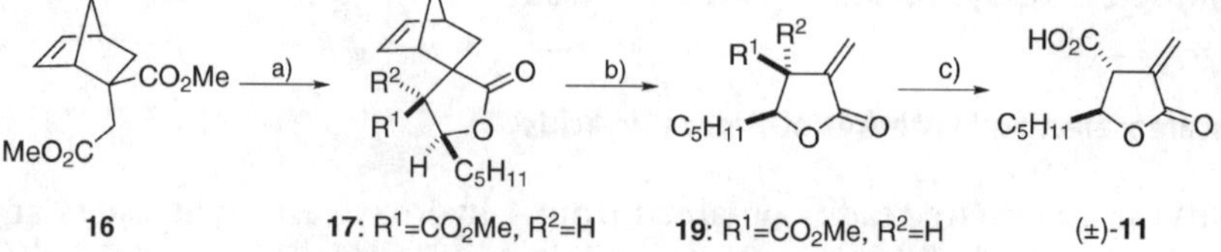


**16** **17**: $R^1$=$CO_2Me$, $R^2$=H; **18**: $R^1$=H, $R^2$=$CO_2Me$ **19**: $R^1$=$CO_2Me$, $R^2$=H; **20**: $R^1$=H, $R^2$=$CO_2Me$ (±)-**11**

*cis:trans* = 2:1

**Scheme 2** Synthesis of $\gamma$-butyrolactones from ester enolates and aldehydes. Reagents and conditions: (a) LDA, THF, −78 °C, $C_5H_{11}CHO$, 79%; (b) FVT at 500 °C (0.005 mm), 92%; (c) 6 N HCl, butanone, reflux, 3 h, 71%

[10] of the *cis* isomer **19** to afford exclusively the *trans*-substituted methylenolactocin (±)-(**11**).

Vicinally donor-acceptor-substituted cyclopropanol carboxylic esters have been proven to be versatile synthetic building blocks in organic synthesis [11]. They readily undergo a retroaldol reaction, thus creating a stable enolate that at the same time can be considered as a homoenolate in relation to the newly formed carbonyl function. Shimada et al. applied this strategy to the preparation of γ-substituted lactones starting from cyclopropane **21** (Scheme 3) [12].

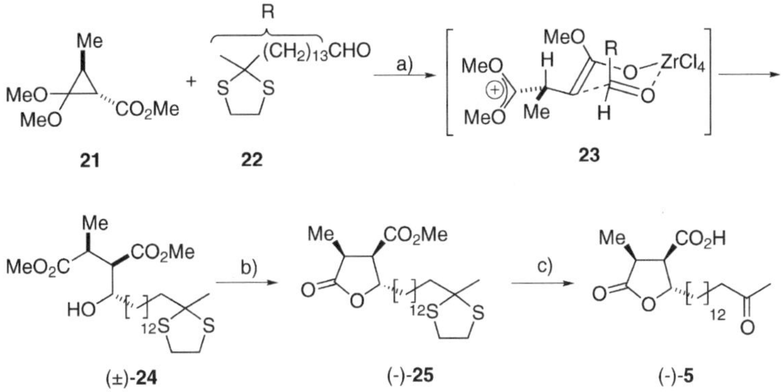

**Scheme 3** Reagents and conditions: (a) ZrCl$_4$, CH$_2$Cl$_2$, −94 °C to −78 °C; (b) (i) cat. *p*-TsOH, 64% (two steps), (ii) optical resolution (*ee*=98%); (c) (i) NBS, acetone, (ii) dioxane, 6 mol/l HCl, reflux, 97%

A retroaldol/aldol cascade is achieved by Lewis acid catalysis, triggering first the ring opening reaction of **21** followed by addition of the aldehyde **22** to the resulting enolate. The relative stereochemistry at the newly formed stereocenter in 3- and 4-position is dictated by the orientation of the aldehyde **22** and the stereogenic side chain generated from **21** after ring opening in the transition state: The outcome of the reaction is rationalized by a Zimmerman Traxler model, placing R in an axial position in order to avoid 1,3-diaxial-like repulsion with the methyl group of the equatorially positioned side chain as depicted in **23**. The choice of the Lewis acid turned out to be important for achieving high diastereoselectivity: ZrCl$_4$ (diastereoselectivity 2,3-position *cis/trans* 92:8; 3,4-position *cis/trans* 15:85) compared to TiBr$_4$ (diastereoselectivity 2,3-position *cis/trans* 88:12; 3,4-position *cis/trans* 37:63) as a Lewis acid minimizes better the interactions between R and an axial halogen due to longer bonds between the metal and the oxygens. (±)-**24** was readily lactonized and the racemic mixture (±)-**25** was resolved via chiral HPLC to obtain both enantiomers in optically pure form. Subsequent cleavage of the thioketal in (−)-**25** with NBS and acid induced hydrolysis of the ester finally yielded (−)-dihydropertusaric acid (−)-(**5**).

Highly functionalized γ-butyrolactones can also be prepared by homolytic cleavage of epoxides followed by intramolecular cyclization of the resulting radical with an alkyne moiety. This methodology was elegantly applied by Roy et al. for the synthesis of two racemic paraconic acids [13], but this strategy should be easily applicable to enantiomerically pure starting materials, being readily available by kinetic resolution of secondary allylic alcohols [14]. The reaction sequence started with the epoxidation of **26** (Scheme 4), followed by

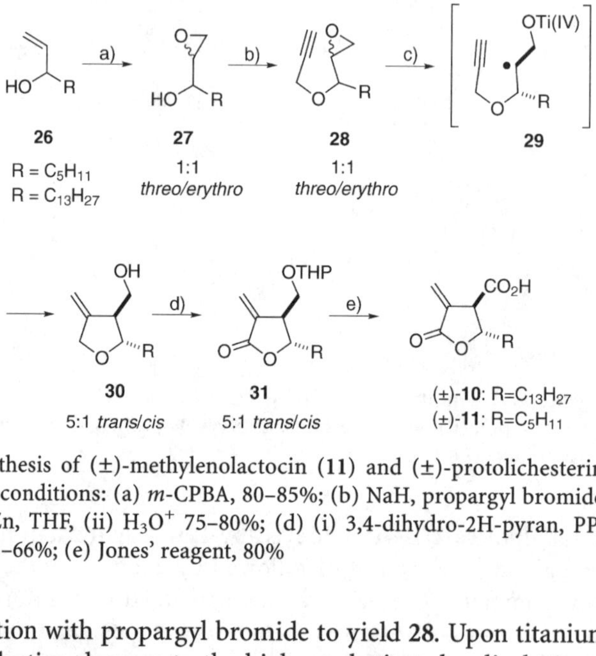

**Scheme 4** Synthesis of (±)-methylenolactocin (**11**) and (±)-protolichesterinic acid (**10**). Reagents and conditions: (a) *m*-CPBA, 80–85%; (b) NaH, propargyl bromide 75–80%; (c) (i) Cp$_2$TiCl, Zn, THF, (ii) H$_3$O$^+$ 75–80%; (d) (i) 3,4-dihydro-2H-pyran, PPTS, 98%, (ii) PDC, DMF, 63–66%; (e) Jones' reagent, 80%

ether formation with propargyl bromide to yield **28**. Upon titanium(III) catalysis, regioselective cleavage to the higher substituted radical **29** occurs, which undergoes 5-*exo-dig* cyclization to **30**. Preferentially, the sterically less hindered *trans*-substituted tetrahydrofurans are formed [15, 16]; moreover, this strategy establishes an *exo*-methyleno group, being found in many paraconic acids, in a straightforward way. After protection of the hydroxy functionality, allylic oxidation to the γ-butyrolactone **31** becomes possible [17], which was converted to protolichesterinic acid: (±)-**10** or methylenolactocin (±)-**11**.

The same strategy was also employed in the synthesis of the saturated analogue of protolichesterinic acid: (±)-dihydroprotolichesterinic acid (**4**) can be synthesized using epoxide **32** as a starting material for the intramolecular radical cyclization to yield **33** as the major diastereomer (Scheme 5). Subsequent oxidation gave finally rise to (±)-**4**, which could be obtained in diastereomerically pure form by fractional crystallization.

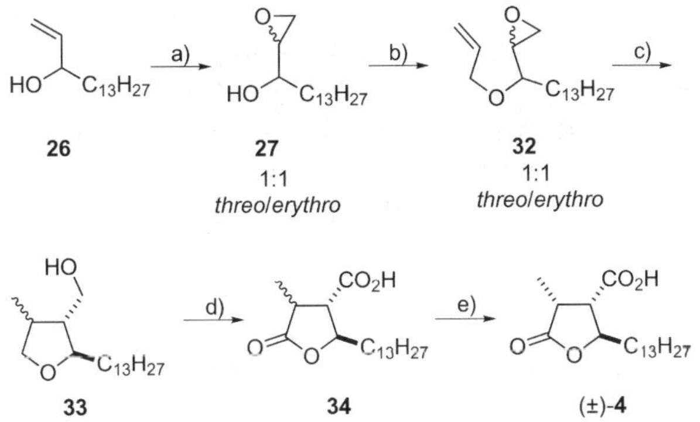

**Scheme 5** Synthesis of (±)-dihydroprotolichsterinic acid (**4**). Reagents and conditions: (a) *m*-CPBA, 80–85%; (b) NaH, allyl bromide, THF/DMSO, rt, 6 h, 81%; (c) (i) Cp$_2$TiCl, Zn, THF, (ii) 10% H$_2$SO$_4$, 76% (two steps); (d) RuCl$_3$ (cat.), NaIO$_4$, H$_2$O/CCl$_4$/CH$_3$CN, 2 h, 90%; (e) fractional crystallization, 78%

## 2.2
## Enantioselective, Chiral Pool-Based Syntheses of Paraconic Acids

### 2.2.1
### Starting from Isopropylidine Glyceraldehyde

(R)-2,3-O-Isopropylidene glyceraldehyde (**35**) has been used as the starting point for a number of different approaches towards paraconic acids. For the synthesis of (−)-nephromopsinic acid (**7**), the great preparative value of chromium(III)-mediated allylations of aldehydes (Hiyama reaction) was

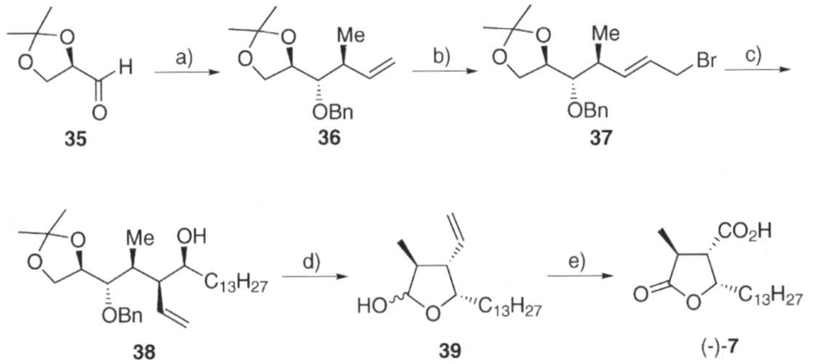

**Scheme 6** Synthesis of (−)-nephromopsinic acid (**7**) via a Hiyama reaction. Reagents and conditions: (a) (i) CrCl$_3$, LiAlH$_4$, crotyl bromide, THF, 38%, (ii) NaH, benzyl chloride, DMF, 77%; (b) (i) O$_3$ then PPh$_3$, CH$_2$Cl$_2$, −78 °C to 22 °C, 84%, (ii) (EtO)$_2$P(O)CH$_2$CO$_2$Et, NaH, THF, 0 °C–22 °C, 24 h, 70%, (iii) DIBALH, toluene/Et$_2$O, −40 °C to −5 °C, 4 h, 78%, (iv) PBr$_3$, Et$_2$O, −25 °C to 22 °C, 79%; (c) CrCl$_3$, LiAlH$_4$, THF, tetradecanal, −5 °C, 36 h, 60%; (d) (i) *p*-TsOH cat., MeOH, 22 °C, 24 h, 84%, (ii) Na/NH$_3$, THF, −40 °C, 30 min, 92%, (iii) H$_5$IO$_6$, Et$_2$O, 22 °C, 1 h then K$_2$CO$_3$, MeOH, 15 min, 73%; (e) (i) PCC, CH$_2$Cl$_2$, 2 h, 92%, (ii) RuCl$_3$ cat., NaIO$_4$, CCl$_4$/H$_2$O/CH$_3$CN, 22 °C, 2 h, 62%

demonstrated with the stereoselective assembly of **36** and **38**, which could be lactonized and further oxidized to (−)-**7** (Scheme 6) [18].

An instructive strategy to cyclize acyclic precursors to lactones was developed by Mulzer et al. (Scheme 7) [19]. The lactone **46** was formed via a halolactonization-dehalogenation [20] sequence from precursor **45**, which in turn was build up by a Claisen rearrangement from the chiral allylic alcohol

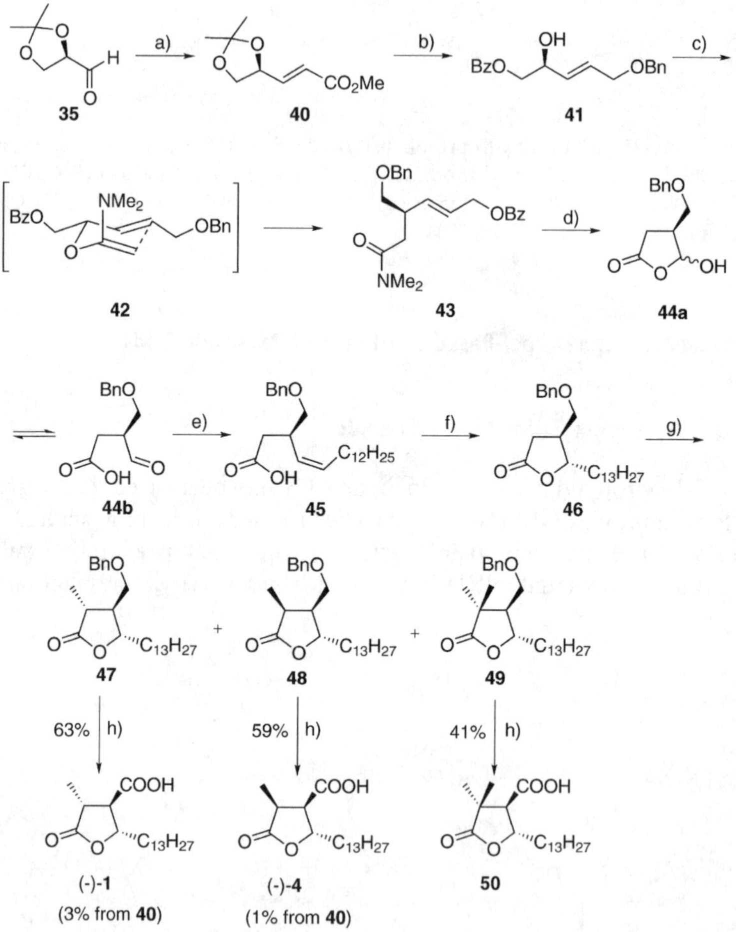

**Scheme 7** Reagents and conditions: (a) $Ph_3PCHCO_2Me$ [22b]; (b) (i) DIBALH, toluene, −78 °C, 3 h, (ii) NaH, DMF, 40 °C, BnCl, (iii) MeOH, p-TsOH, 22 °C, 24 h, (iv) BzCl, pyridine, 0 °C, (+10% di-benzoate), 47% (four steps); (c) N,N-dimethylacetamide dimethyl acetal, toluene, reflux, 5 h, 92%; (d) (i) MeOH, NaOH, 4 h, 22 °C, (ii) $O_3$, MeOH, −78 °C, $PPh_3$-workup, 71% (two steps); (e) $C_{12}H_{25}$-CH=$PPh_3$, THF, −78 °C, 2 h, 52%; (f) (i) $I_2$·KI, THF-$H_2O$, $K_2CO_3$, 22 °C, 48 h, (ii) $Bu_3SnH$, AIBN, toluene, reflux, 4 h, 59% (two steps); (g) LDA, −78 °C, MeI, 92% (**47**:**48**:**49**=3:1:1); (h) (i) $H_2$/Pd, MeOH, 3 bar, 22 °C, (ii) PDC, DMF, 22 °C, 15 h

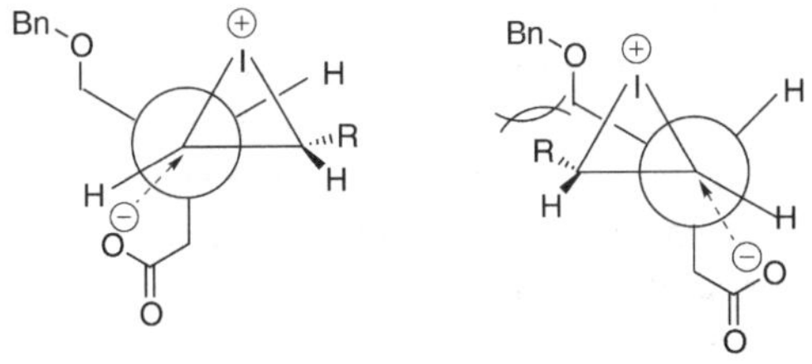

Paraconic Acids – The Natural Products from *Lichen* Symbiont    51

**Fig. 2** Stereochemical model for the iodolactonization of **45** to **46**

**41** [21]. The reaction sequence started with the elongation of readily available (R)-2,3-isopropylidene glyceraldehyde (**35**) to the acrylate **40** [22]. After reduction and protecting group manipulation, the alcohol **41** has been subjected to a Eschenmoser-Claisen rearrangement [23] in the presence of N,N-dimethylacetamide dimethyl acetal to give **43**.

Hydrolysis of the amide, followed by reductive ozonolysis furnished **44** (99.5:0.5 mixture of **44a** and **44b**), which can be alkenylated to **45** via the aldehyde **44b**. Halolactonization could be achieved to the *trans*-disubstituted lactone **46** via the iodonium ion **51** (Fig. 2), being preferred to **52** because of lesser allylic 1,3-strain [24]. Methylation of **46** gave rise to a 3:1:1 mixture of **47–49**, which can be easily separated by chromatography. Deprotection followed by oxidation furnished (−)-roccellaric acid (**1**) [25], (−)-dihydroprotolichesterinic acid (**4**) and the dimethylated acid **50** in pure form.

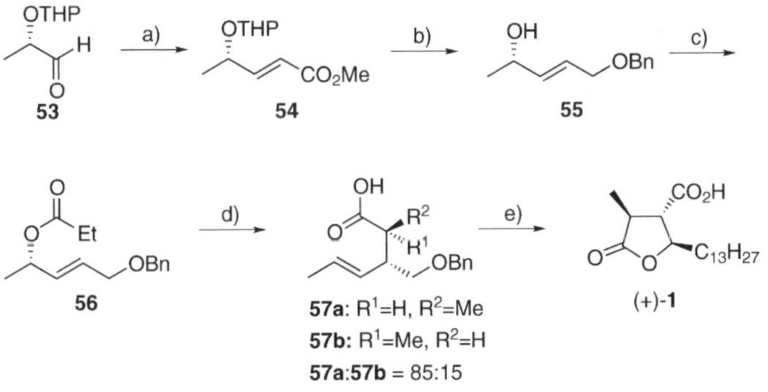

**Scheme 8** Synthesis of (+)-roccellaric acid (**1**) from **53**. Reagents and conditions: (a) (i) NaH, carbomethoxymethane diethylphosphonate, THF, −20 °C to 22 °C, 12 h, 78%; (b) (i) DIBALH, toluene, −78 °C, 3 h, (ii) NaH, DMF, 40 °C, BnCl, (iii) MeOH, p-TsOH, 22 °C, 24 h, 52% (three steps); (c) propionic anhydride, DMAP, pyridine, 97%; (d) LDA, TMSCl, −78 °C, then 22 °C, aqueous workup, 73%; (e) (i) O$_3$, MeOH, −78 °C, PPh$_3$-workup, (ii) C$_{12}$H$_{25}$-CH=PPh$_3$, THF, −78 °C, 41% (two steps) (iii) I$_2$·KI, THF-H$_2$O, K$_2$CO$_3$, 22 °C, 48 h, (iv) Bu$_3$SnH, AIBN, toluene, reflux, 4 h, (v) H$_2$/Pd, MeOH, 3 bar, 22 °C, (vi) PDC, DMF, 22 °C, 15 h, 41% (four steps)

Moreover, the authors developed an alternative sequence which is shorter and higher yielding, leading to the naturally occurring enantiomeric series (Scheme 8). In contrast to 41, alcohol 55 is lacking a base labile functionality and could therefore be subjected to an Ireland-Claisen rearrangement. After acylation to 56, treatment with LDA/TMSCl yielded 57 as a 85:15 mixture of diastereomers. The analogous strategy as described for the conversion of 45 in (−)-1 and 4 (Scheme 7h) gave rise to (+)-1 [19].

## 2.2.2
### Starting from Diacetone-D-glucose

Synthesis of optically pure (+)-1 and (+)-4 was also achieved from diacetone-D-glucose 58 (Scheme 9) [19]. Oxidation of the free hydroxy group at C-3 followed by Wittig methylation, selective hydroboration from the convex face and final benzylation led to 59 [26], having the side chain installed for later conversion to the carboxylic acid required in all paraconic acids. Selective cleavage of the sterically less hindered acetonide followed by oxidative degradation of the resulting diol to an aldehyde set the stage for introduction of the C13-alkyl side chain by Wittig alkenylation found in roccellaric acid (1) and dihydroprotolichesterinic acid (4). Subsequent hydrogenation of the double bond in 60 followed by deprotection of the remaining acetonide gave rise to the furanoside 61 as mixture of two anomers ($\beta/\alpha$=15:1). Deoxygenation to 62 using the method of Barton and McCombie [27] was found to be effective for the reductive removal of the 2-OH group. Oxidation

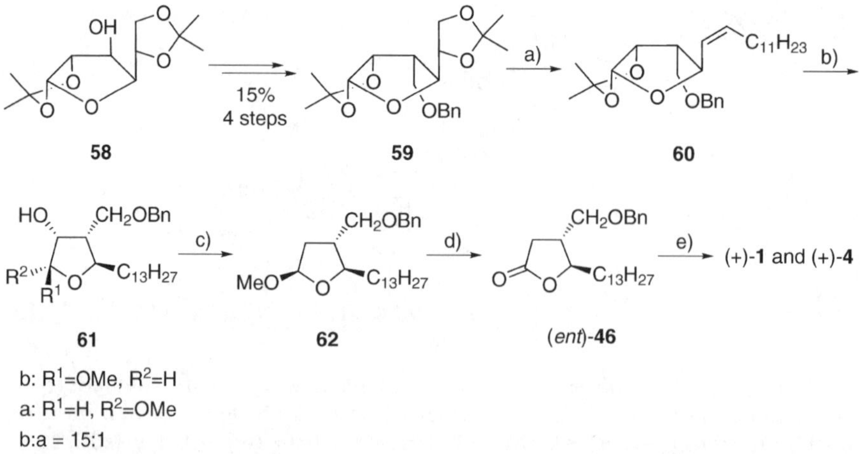

b: $R^1$=OMe, $R^2$=H
a: $R^1$=H, $R^2$=OMe
b:a = 15:1

**Scheme 9** Reagents and conditions: (a) (i) 50% HOAc, (ii) Pb(OAc)$_4$, CH$_2$Cl$_2$, 22 °C, (iii) H$_{23}$C$_{11}$CH=PPh$_3$, THF, 42% (three steps); (b) (i) H$_2$/Pd/BaSO$_4$, MeOH, 1 bar, 22 °C, (ii) p-TsOH, MeOH, 65% (two steps); (c) NaH, CS$_2$, MeI, THF; then Bu$_3$SnH, AIBN, reflux, 52%; (d) Na$_2$Cr$_2$O$_7$, acetone, 10% H$_2$SO$_4$, 0 °C, 70%; (e) see Scheme 7

of the acetal functionality in **62** with Jones reagent gave rise to (*ent*)-**46**, converging at this point with the synthesis of **1** and **4** already described in Scheme 7.

## 2.2.3
### Starting from D-Mannitol

D-Mannitol (**63**) has been used for the synthesis of γ-butyrolactones, making again use of a Claisen rearrangement as the key step (Scheme 10). The $C_2$-symmetrical 1,4-diol **65**, obtained from **63** via the alkene **64** [28], could be converted to **67** by applying the Eschenmoser-Claisen variant. Cyclization to **68** was readily achieved upon heating **67** in xylene, establishing two differ-

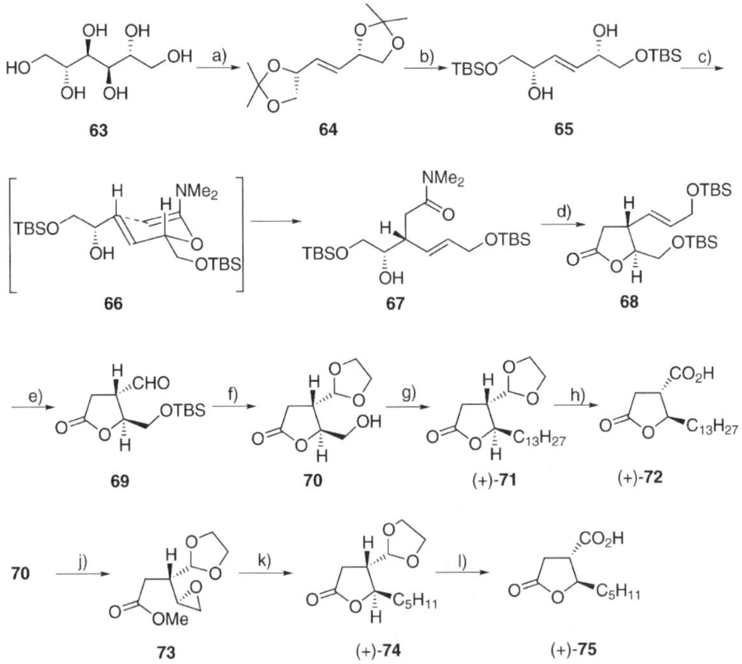

**Scheme 10** Synthesis of precursors [(+)-**72** and (+)-**75**] of various paraconic acids. Reagents and conditions: (a) (i) $ZnCl_2$, acetone, (ii) $HC(OMe)_2$, $CH_2Cl_2$, (ii) MeI, toluene, reflux, 41% (three steps); (b) (i) 75% AcOH, (ii) TBSCl, imidazole, 72% (two steps); (c) $MeC(OMe)_2NMe_2$, benzene, reflux, 2 h 98%; (d) xylene, reflux, 15 h, 80%; (e) $OsO_4$(cat.), $NaIO_4$, *t*-BuOH-$H_2O$, rt, 1 h (74%) or $O_3$, DMS, $CH_2Cl_2$, -78 °C (71%); (f) (i) 2-butanone, ethylene acetal, TsOH, benzene, rt, 8 h, (ii) $Bu_4NF$, THF, rt, 0.5 h, 55% (two steps); (g) (i) $Tf_2O$, $Et_3N$, $CH_2Cl_2$, 0 °C, 4 h, (ii) *n*-$C_{12}H_{25}MgBr$, $CuBr\cdot Me_2S$, THF, 0 °C, 25% (two steps); (h) (i) 75% AcOH-$H_2SO_4$, (ii) Jones oxidation, 76% (two steps); (j) (i) TsCl, pyridine, rt, 7.5 h, 77%, (ii) $K_2CO_3$, MeOH, rt, 6 h, 92%; (k) *n*-$Bu_2CuLi$, $Et_2O$, -60 °C, 1 h, 65%; (l) (i) 75% AcOH, $H_2SO_4$, (ii) Jones' reagent, 77% (two steps)

entiated side chains in the lactone ring being positioned *trans* to each other. Their further conversion to either (+)-**72** or (+)-**75**, being precursors to various paraconic acids [47, 49], could be accomplished in a straightforward way.

Intriguingly, the enantiomeric *trans*-disubstituted γ-butyrolactone (−)-**75**, completing a formal synthesis of (−)-methylenolactocin (**11**) [47], can be obtained by cyclizing **67** through an iodolactonization after protection of the free OH group (Scheme 11).

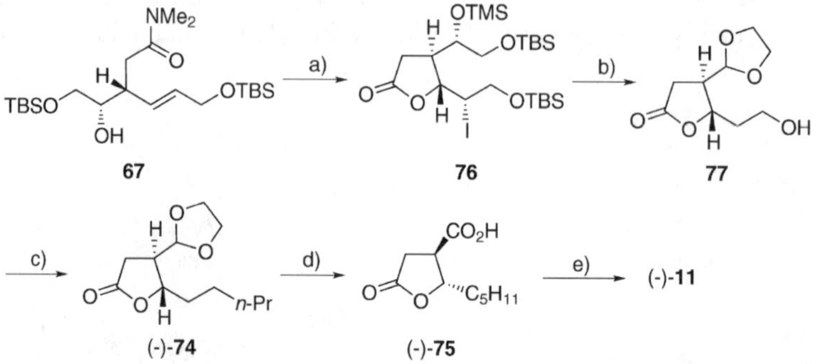

**Scheme 11** Synthesis of (−)-**75**. Reagents and conditions: (a) (i) TMSCl, Et$_3$N, THF, rt, 9 h, (ii) I$_2$, NaHCO$_3$, THF-H$_2$O, −10 °C, 7 h, 74% (two steps); (b) (i) CF$_3$CO$_2$H, MeOH, rt, 6 h, 86%, (ii) NaIO$_4$, THF-H$_2$O, rt, 20 min, (iii) 2-butanone, ethylene acetal, TsOH, benzene, rt, 16 h, 58% (two steps); (c) (i) *n*-Bu$_3$SnH, AIBN, toluene, reflux, 0.5 h, 72% (ii) TsCl, pyridine, 0 °C, 90%, (iii) *n*-Pr$_2$CuLi, Et$_2$O, toluene, −60 °C, 25 min, 62%; (d) (i) 75% AcOH, H$_2$SO$_4$, 60 °C, 1 h, (ii) Jones' reagent, 60 °C, 5 min, 80% (two steps); (e) [46]

## 2.2.4
### Starting from Bicyclic Oxaenones

Both enantiomers of the bicyclic enone **78** and their derivatives have been proved to be useful chiral building blocks for the synthesis of natural products [29], among them γ-butyrolactones. **78** is readily available in either enantiomeric form by a Diels-Alder reaction of furan with α-acetoxyacrylonitrile and subsequent hydrolysis, followed by a resolution of the racemate [30]. Strategies for the direct enantiopure preparation of **78** are also known [31], but suffer from low stereoselectivity or are tedious.

Renaud and co-workers used **78** for the synthesis of (−)-phaseolinic acid (**6**) and (−)-pertusarinic acid (**8**) (Scheme 12) [32, 33]. Radical addition of dimethyl phenylselenomalonate to **78** proceeded with rearrangement of the bicyclics to yield the seleno-acetal **79** [34]. After reductive deselenylation and Baeyer-Villiger oxidation treatment of **80** with Bu$_4$NI and BBr$_3$ led to a simultaneous cleavage of the ether, the lactone, and the methyl ester func-

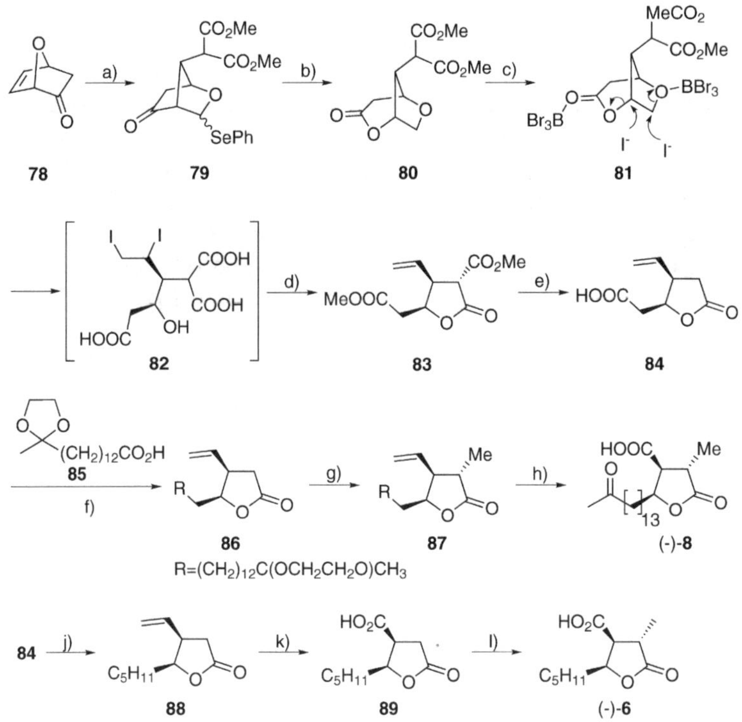

**Scheme 12** Reagents and conditions: (a) dimethyl 2-(phenylselenyl)propanedioate, benzene, $h\nu$ (300 W), 30 °C, 83% (crude); (b) (i) Bu$_3$SnH, AIBN, $h\nu$, 10 °C, 90%, (ii) m-CPBA, 50%; (c) Bu$_4$NI, BBr$_3$; (d) CH$_2$N$_2$, 75% (two steps); (e) (i) DMF, H$_2$O, Δ, 85%, (ii) 1 N HCl, 100 °C, 81%; (f) **85** (8.0 equiv.), Et$_3$N, MeOH, electrolysis, 40%; (g) LDA, MeI, 62%; (h) (i) 1 N HCl, 80%, (ii) O$_3$, DMS, −78 °C, (iii) NaClO$_2$, NaH$_2$PO$_4$, 70% (two steps); (j) pentanoic acid (8.0 equiv.), Et$_3$N, MeOH, electrolysis, 40%; (k) RuCl$_3$, NaIO$_4$, CCl$_4$/H$_2$O/CH$_3$CN, 3 h, 80%; (l) [35]

tionalities to yield the lactone **83** upon esterification with diazomethane. Decarboxylation and ester hydrolysis afforded **84**, which can be envisioned as a precursor for a variety of paraconic acids having different γ-chains.

The required side-chain for (−)-pertusarinic acid (**8**) was introduced by a mixed Kolbe electrolysis with **85** [36], elegantly demonstrating the value of this process for natural product synthesis in the presence of various functional groups. In an analogous way (−)-phaseolinic acid (**6**) was synthesized by reacting pentanoic acid in a mixed Kolbe electrolysis with **84** to **88**.

## 2.2.5
## Starting from Chiral Butenolide Synthons

Chiral butenolides are valuable synthons towards γ-butyrolactone natural products [37] and have also been successfully applied to the synthesis of paraconic acids. The lactone 91, readily available from the hydroxyamide (*rac*)-90 by enzymatic resolution [38] followed by iodolactonization, proved to be an especially versatile key intermediate. Copper(I)-catalyzed cross coupling reactions with Grignard reagents allowed the direct introduction of alkyl side chains, as depicted in 92a and 92b (Scheme 13) [39, 40]. Further

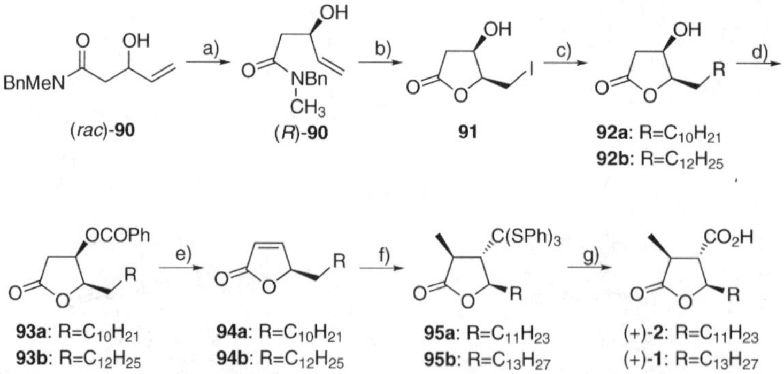

**Scheme 13** Reagents and conditions: (a) immobilized Amano PS, vinylacetate, pentane, 44%; (b) $I_2$, DME/$H_2O$, 63% (*cis/trans*=7.7:1); (c) RMgBr, CuBr·$Me_2S$, 61% for 92a, 48% for 92b; (d) PhCOCl, pyridine, 84% for 93a, 80% for 93b; (e) $NH_3$, MeOH, 81% for 94a, 72% for 94b; (f) (i) (PhS)$_3$CLi, THF, −78 °C, (ii) MeI, HMPA, THF, −78 °C to rt, 24 h, 78% for 95a, 75% for 95b; (g) HgO, $BF_3·OEt_2$, THF/$H_2O$, 89% for (+)-2, 86% for (+)-1

conversion of 92 to the *lichen* components (+)-roccellaric acid (1) and (+)-nephrosteranic acid (2) [41] was achieved via the butenolide 94, which was found to be an excellent Michael acceptor [42]: Conjugate addition of lithiated tris(phenylthio)methane to 94 at −78 °C followed by quenching of the resulting lactone enolate with methyl iodide yielded the all-*trans*-trisubstituted γ-butyrolactones 95. Hydrolysis of the tris(phenylthio)methyl moiety finally gave rise to nephrosteranic acid (+)-2 and roccellaric acid (+)-1.

Similarly (−)-methylenolactocin (11) was accessible using the same strategy as mentioned above, but employing (*ent*)-90 as starting material. The authors also achieved the synthesis of the flavor components (+)-*trans*-whisky lactone [43] and (+)-*trans*-cognac lactone [44] by applying the same strategy.

## 2.3
## Enantioselective, Chiral Auxiliary Based Syntheses of Paraconic Acids

### 2.3.1
### [2+2]-Cycloaddition

The stereoselective [2+2] cycloaddition between ketenes and enolethers can be used as a key step in the construction of $\gamma$-butyrolactones (Scheme 14) [45], if the resulting cyclobutanones can subsequently undergo ring enlargement by a regioselective Baeyer-Villiger oxidation.

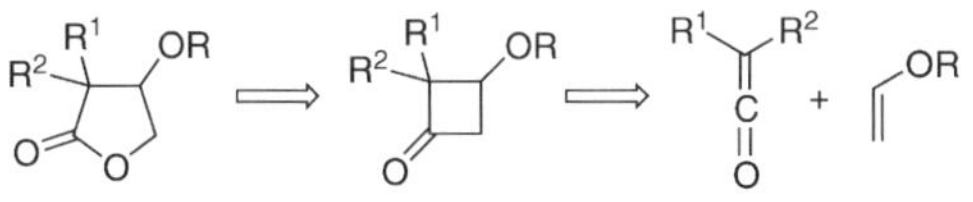

**Scheme 14** Chiral olefin-ketene [2+2] cycloaddition

Using (−)-100 [46] as a chiral auxiliary tethered to the enolether, one face of the alkene can be specifically blocked by a $\pi$-$\pi$ interaction of the phenyl rest for the $[2\pi_s+2\pi_a]$ cycloaddition with a ketene [47], resulting in the highly diastereoselective formation of the cyclobutanone 102 (Scheme 15). The observed regio- and stereoselectivity is in accord with the stereochemical predictions made on the basis of the Woodward-Hoffmann

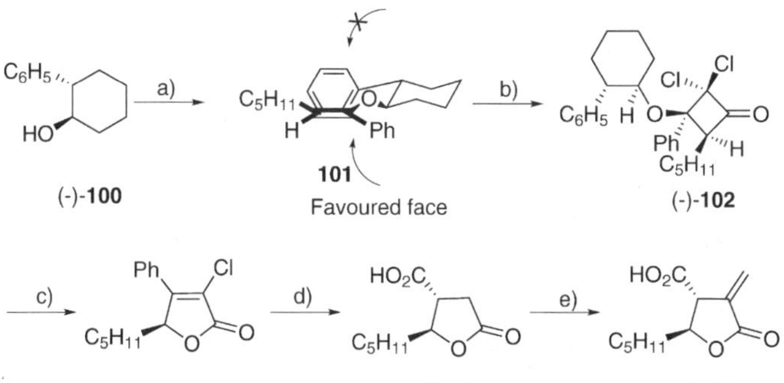

**Scheme 15** Reagents and conditions: (a) (i) $C_6H_5COCl$, DMAP, $C_5H_5N$, 20 °C, 12 h, 99%, (ii) $C_6H_{12}Br_2$, Zn, $TiCl_4$, TMEDA, $THF-CH_2Cl_2$, 0 to 20 °C, 3 h, quant. (E/Z=9:1); (b) $Cl_3CCOCl$, Zn-Cu, $Et_2O$, 20 °C, 6 h, 79%; (c) (i) m-CPBA, $NaHCO_3$, $CH_2Cl_2$, (ii) $Cr(ClO_4)_2$, acetone, 0 °C, 1 h, 73% (two steps); (d) (i) $H_2$, Pd/C, MeOH, (ii) $RuCl_3$, $NaIO_4$, $CCl_4$-$CH_3CN$-$H_2O$, 38 °C, 72 h, (iii) $CH_2N_2$, DBU, $Et_2O$, 20 °C, 72 h; 6 N HCl, dioxane, reflux, 2 h, 61% (three steps); (e) $MeOMgOCO_2Me$, aq. HCHO, $C_6H_5NHCH_3$, NaOAc, HOAc, 20 °C, 2 h, 66%

rules. Regioselective Baeyer-Villiger oxidation involving the more electron rich carbon-carbon bond followed by reductive elimination with chromous perchlorate provided the $\alpha$-chlorobutenolide (+)-**103** in 73% yield with concurrent recovery of the chiral auxiliary **100**. After reduction of the chloroalkene, oxidative degradation of the phenyl to a carboxy group was achieved with $RuCl_3$-$NaIO_4$ to yield a *trans*/*cis* mixture of $\gamma$-butyrolactones, which could be conveniently equilibrated with DBU to give in 61% overall yield the enantiopure *trans*-lactone (−)-**75**. Treatment of (−)-**75** with Stiles reagent and formaldehyde [48] afforded finally (−)-methylenolactocin (**11**).

## 2.3.2 Sequential Aldol-Lactonization

Acyloxazolidinones of acyclic dicarboxylic acids can be used in the stereoselective synthesis of $\gamma$-butyrolactones, which has been applied to one of the shortest syntheses of (−)-roccellaric acid (**1**) reported to date (Scheme 16) [49].

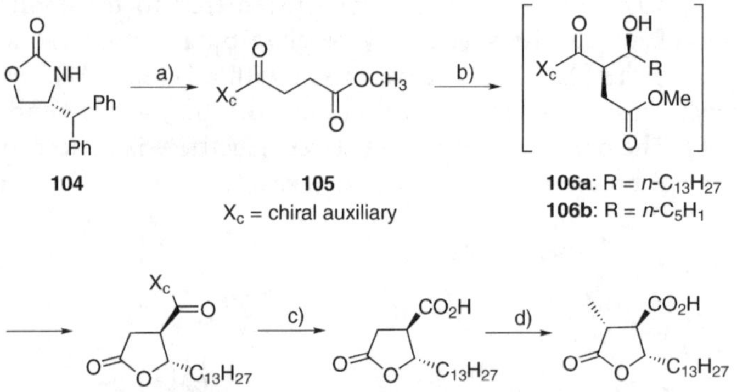


**104** **105** $X_c$ = chiral auxiliary **106a**: R = $n$-$C_{13}H_{27}$ **106b**: R = $n$-$C_5H_1$

**107a**: R = $n$-$C_{13}H_{27}$ **107b**: R = $n$-$C_5H_{11}$ (-)-**72**: R = $n$-$C_{13}H_{27}$ (-)-**75**: R = $n$-$C_5H_{11}$ (-)-**1**

**Scheme 16** Synthesis of (−)-**1** via a oxazolidinone-mediated sequential aldol-lactonization reaction. Reagents and conditions: (a) BuLi, $ClCOCH_2CH_2CO_2OCH_3$, THF, −78 °C, 97%; (b) (i) $Bu_2BOTf$, $Et_3N$, $CH_2Cl_2$, 0 °C, (ii) RCHO, −78 °C to 0 °C, 55% for **107a**, 82% for **107b**; (c) LiOH, $H_2O_2$, THF, $H_2O$, 85% for (−)-**72**, 81% for (−)-**75**; (d) NaHMDS (2.2 equiv.), THF, −78 to 0 °C, 3 h, $CH_3I$, 55%

Aldol addition of aliphatic aldehydes to **105** [50] yielded the expected *syn*-aldol adducts **106**, which spontaneously lactonized upon warming to 0 °C to give the $\gamma$-butyrolactones **107** in good yields and excellent optical purities. The chemoselective deprotonation of **105** in $\alpha$-position to the imide rather than to the ester reflects the selective activation of the imide carbonyl group by a Lewis acid due to its possibility of chelation with the chiral auxiliary [51]. The synthesis of roccellaric acid (−)-1 was accomplished by cleavage of the chiral auxiliary and methylation at C-3 in a straightforward manner. Moreover, the successful synthesis of (−)-72 and (−)-75 constitutes a formal synthesis of (−)-protolichesterinic acid (10) and (−)-methylenolactocin (11) respectively [47b]. This strategy could be further extended to the fumarate 108 (Scheme 17), which underwent a che-

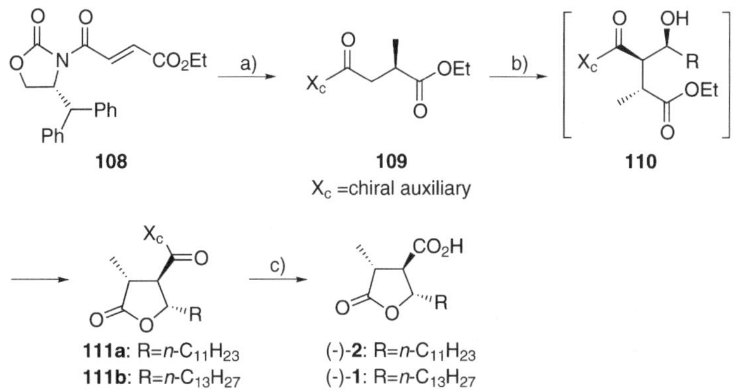

**Scheme 17** Synthesis of (−)-1 and (−)-2 via radical addition. Reagents and conditions: (a) (i) Sm(OTf)$_3$, ClCH$_2$I, Bu$_3$SnH, Et$_3$B/O$_2$, CH$_2$Cl$_2$, −78 °C, 1 h, 91% (>100:1), (ii) Bu$_3$SnH, AIBN, toluene, reflux, 12 h, 76%; (b) Bu$_2$BOTf, CH$_2$Cl$_2$, Et$_3$N, −78 to 0 °C, RCHO, 12 h, 84% for 111a, 65% 111b; (c) LiOH, H$_2$O$_2$, THF/H$_2$O, rt, 92% for (−)-2, 94% for (−)-1

mo- and stereoselective alkylation to give rise to 109, installing a methyl group required for the synthesis of many paraconic acids. The methylation was achieved by a radical process, using the combination of ClCH$_2$I/Bu$_3$SnH to introduce a chloromethylene group and Sm(OTf)$_3$ to ensure the activation of the imide carbonyl through chelation with the auxiliary. Subsequently, following the strategy outlined in Scheme 16, the synthesis of (−) roccellaric acid (1) and (−)-nephrosteranic acid (2) could be completed [52].

## 2.3.3
### Carbolithiation Towards (−)-Roccellaric Acid

The application of mixed enolates/homoenolates of type 14 for the racemic synthesis of γ-butyrolactones has been already discussed (cf. Scheme 1). An ingenious way to render this strategy asymmetric was demonstrated with the regio- and stereoselective carbolithiation of 114, generating the organolithium intermediate 115, which could be reacted with the appropriate aldehyde to the lactones **117**. The minor **117b** could be equilibrated to the more stable *all-trans* substituted **117a**, which was transformed to (−)-roccellaric acid (**1**) by cleavage of the chiral auxiliary [53] (Scheme 18).

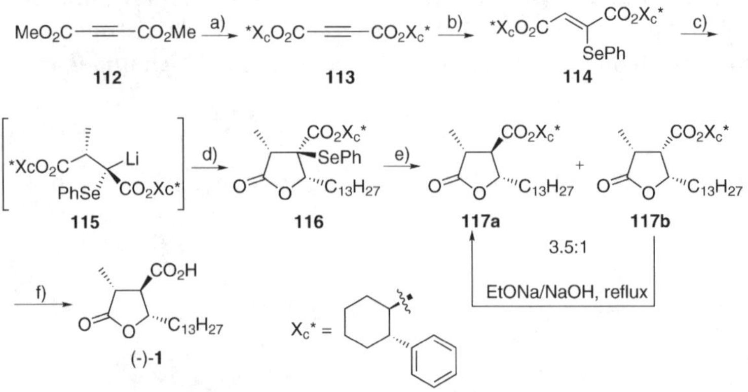

**Scheme 18** Synthesis of (−)-**1** through carbolithiation reaction. Reagents and conditions: (a) Ti(OEt)$_4$, X$_c$OH, 0.1 torr, 75 °C, 1 h, 65%; (b) *n*-Bu$_3$P, PhSeSePh, H$_2$O, CH$_3$CN, 0 °C, 1 h, 83%; (c) MeLi, cumene/THF; (d) C$_{13}$H$_{27}$CHO, HMPA, CH$_2$Cl$_2$, −78 °C, 62% (two steps); (e) NiCl$_2$, NaBH$_4$, THF/MeOH, 91% (**117a**/**117b**=3.5:1); (f) K$_2$CO$_3$, MeOH, rt, 82%

## 2.3.4
### Nicholas-Schreiber Reaction

The nucleophilic displacement of propargyl alcohols or ethers can be affected by complexation of the alkyne moiety with dicobaltoctacarbonyl, which facilitates the heterolytic cleavage of neighboring alcohols or ethers and subsequently allows nucleophilic attack at this position [54, 55]. Based on this reaction, an efficient synthesis of (+)-nephrosteranic acid (**2**) was developed starting from enantiopure alkoxyaldehyde **118** (Scheme 19). After alkynylation to **119**, the propargylic methoxy group can be displaced by the enolate **121** upon cobalt activation to give rise to **123** with high stereoselectivity regarding the two newly formed stereocenters. A series of deprotection events and oxidation of the alkynyl group to create a carboxy function led to **126**, which was readily equilibrated to the target molecule (+)-**2** under basic conditions.

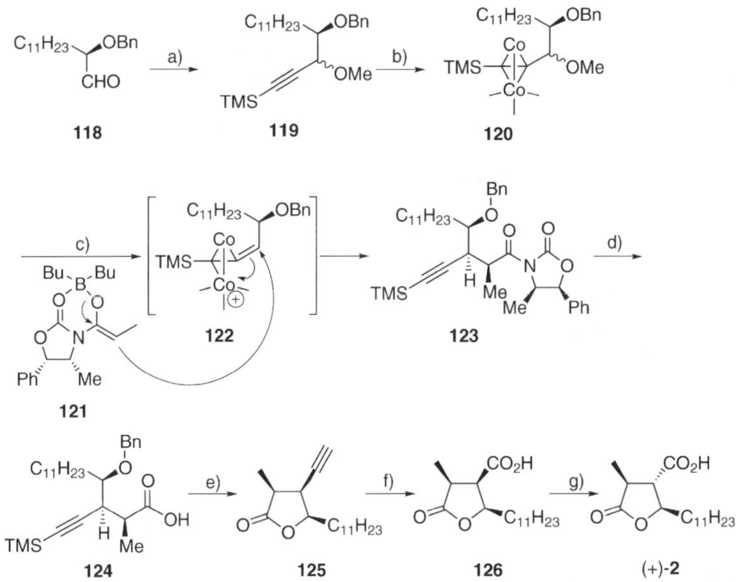

**Scheme 19** Synthesis of (+)-2 through a Nicholas-Schreiber reaction. Reagents and conditions: (a) (i) *n*-BuLi, trimethylsilylacetylene, THF, −78 °C, (ii) dimethylsulfate, rt, 48 h, 83% (two steps); (b) $Co_2(CO)_8$, 93%; (c) (i) **121** (2.0 equiv.), $CH_2Cl_2$, −78 °C to rt, (ii) CAN, acetone, rt, 71% (two steps); (d) LiOH, $H_2O_2$, THF/DMF/$H_2O$, 85%; (e) $P_4S_{10}$, $CH_2Cl_2$, rt, 24 h, 90%; (f) $RuCl_3$, $NaIO_4$, $CCl_4$/$CH_3CN$/$H_2O$, 92%; (g) DBU, toluene, reflux, 3 days, 86%

## 2.4
## Catalytic Asymmetric Syntheses of Paraconic Acids

### 2.4.1
### Nickel Catalyzed Intramolecular Carbozincation

Radical cyclization of polyfunctional 5-hexenyl halides mediated by $Et_2Zn$ and catalyzed by nickel or palladium salts has been demonstrated to produce stereoselectively polyfunctional 5-membered carbo- and heterocycles [56, 57]. Based on this strategy a formal synthesis of methylenolactocin (**11**) was achieved (Scheme 20). The acetal **130**, readily being built up by asymmetric alkylation of aldehyde **127** followed by reaction with butyl vinyl ether and NBS, served as the key intermediate for the construction of the lactone ring. Nickel(II)-catalyzed carbometallation was initiated with diethylzinc to yield exclusively the *trans*-disubstituted lactol **132**, which could be oxidized directly by air to **134**. Final oxidation under more forcing conditions then yielded the lactone (−)-**75** as a known intermediate in the synthesis of (−)-methylenolactocin (**11**) [47a].

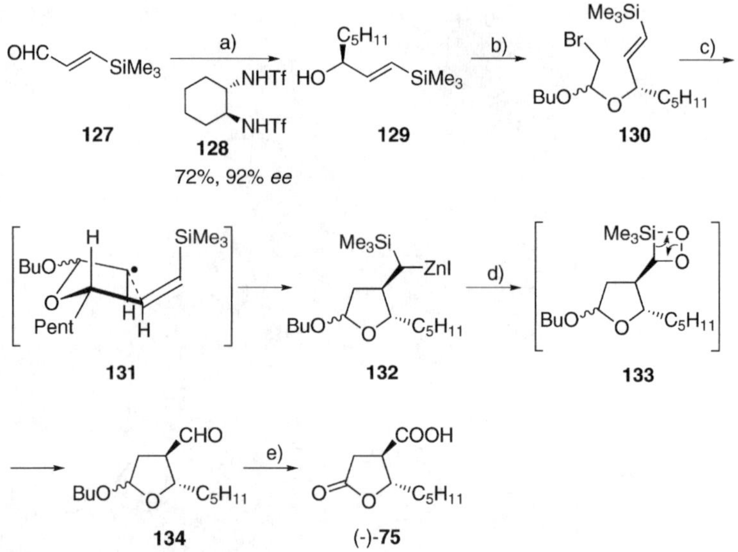

**Scheme 20** Stereospecific construction of *trans*-γ-butyrolactole (−)-75 via intramolecular carbozincation. Reagents and conditions: (a) Pent$_2$Zn, Ti(O*i*-Pr)$_4$, **128** (8 mol%), 70%, 92% *ee*; (b) butyl vinyl ether, NBS, CH$_2$Cl$_2$, 88%; (c) Et$_2$Zn, LiI, Ni(acac)$_2$ (5 mol%), THF, 40 °C; (d) O$_2$, TMSCl, THF, −5 °C, 4 h, 55% (two steps); (e) Jones oxidation, acetone, 0 °C, 15 min, 90%.

## 2.4.2
### Palladium(II) Catalyzed Construction of γ-Butyrolactones

Palladium (II)-catalyzed ring closure of acyclic allylic 2-alkynoates of type **137** provides an efficient entry to stereodefined α-methylene-γ-butyro-

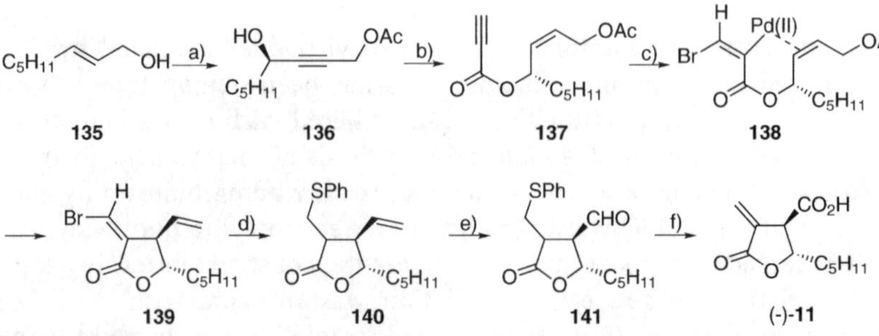

**Scheme 21** Stereoselective Pd(II)-catalyzed cyclization as a key step for the synthesis of (−)-**11**. Reagents and conditions: (a) [59]; (b) (i) P$_2$-Ni, NH$_2$CH$_2$CH$_2$NH$_2$, EtOH, H$_2$ (1 atm), rt, 95%, (ii) propynoic acid, DEAD, PPh$_3$, THF, rt, 85%; (c) LiBr, Pd(OAc)$_2$ (0.05 equiv.), HOAc, rt, 65%; (d) (i) Zn-Ag, MeOH, rt, (ii) NEt$_3$, PhSH, THF, rt, 93% (two steps); (e) O$_3$, MeOH/CH$_2$Cl$_2$, −78 °C; (f) (i) PDC, DMF, 0 °C to rt, (ii) NaIO$_4$, MeOH/benzene/H$_2$O, rt, (iii) toluene, reflux, 60% (four steps)

lactones. This strategy was applied to the enantioselective synthesis of (−)-methylenolactocin (**11**) (Scheme 21) [58], which started from the enantiomerically pure alcohol **136**, being prepared by asymmetric epoxidation of hept-2-en-1-ol (**135**) following a method reported by Kibayashi and coworkers [59]. After conversion to **137**, the combination of Pd(OAc)$_2$/LiBr initially transformed the alkyne group to the vinylpalladium intermediate **138**, which cyclized to the sterically favored *trans*-lactone **139** as the key intermediate towards the target compound.

### 2.4.3
### Base-Induced Intramolecular Michael Addition

Functionalized γ-acyloxy-α,β-unsaturated esters provided yet another versatile strategy towards paraconic acids as exemplified with the synthesis of protolichesterinic acid ((+)-**10**) (Scheme 22) [60]. The prerequisite precur-

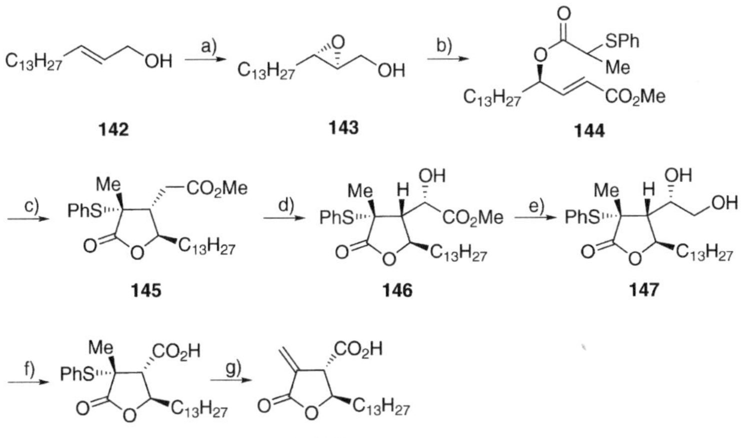

**Scheme 22** Reagents and conditions: (a) Ti(OiPr)$_4$, (+)-DET, *t*-BuOOH, CH$_2$Cl$_2$, 78%, >95% ee; (b) (i) 2-(phenylthio)propionic acid, Ti(OiPr)$_4$, CH$_2$Cl$_2$, (ii) NaIO$_4$, THF/H$_2$O, (iii) NaH, Ph$_3$P=CHCO$_2$CH$_3$, benzene, 74% (three steps), E/Z=20:1; (c) LiN(TMS)$_2$, THF/HMPA, 91%; (d) LiN(TMS)$_2$, oxodiperoxomolybdenum-pyridine-HMPA complex (MoOPH), THF, −78 to −50 °C, 85%; (e) NaBH$_4$ (cat.), BH$_3$·SMe$_2$, THF, 92%; (f) NaIO$_4$, KMnO$_4$, dioxane, H$_2$O, 83%; (g) (i) NaIO$_4$, MeOH/benzene/H$_2$O, 50 h, 55% (isomeric mixture of sulfoxides), (ii) toluene, reflux, 1 h, 85%

sor **143** is obtained from allylic alcohol **142**, reliably introducing the stereogenic hydroxy function by a Sharpless epoxidation. Upon deprotonation of **144** with LiHMDS, a stereoselective intramolecular 1,4-addition to **145** took place. Subsequent transformation to (+)-**10** was initiated by a remarkably chemo- and stereoselective hydroxylation in the α-position to give rise to the highly functionalized lactone **146**, in which after reduction of the ester the resulting diol was cleaved to form the acid **148**. The thiophenyl group can be utilized in a twofold way: oxidation to the corresponding sulfoxide allows its elimination, ultimately arriving at the target molecule (+)-**10**, having installed an *exo*-methyleno group found in many paraconic acids.

Alternatively, the thiophenyl group could be reduced in **147** under retention of configuration to **149**, which was further converted to roccellaric acid (**1**) in very high yield (Scheme 23).

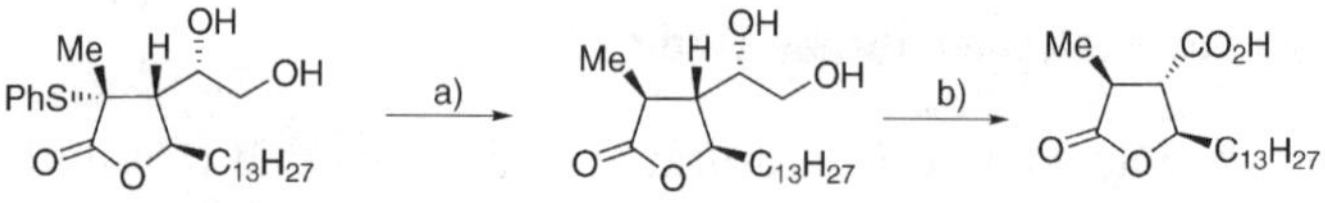

**Scheme 23** Reagents and conditions: (a) NiCl$_2$, NaBH$_4$, H$_2$ (1 atm), EtOH, 82%; (b) RuCl$_3$, NaIO$_4$, CH$_3$CN/CCl$_4$/H$_2$O, 2 h, 90%

## 2.4.4
### Chemoenzymatic Transformations Toward Phaseolinic Acid

The synthesis of chiral carboxylic acids by enzymatic resolution of the corresponding racemate is a widely established method, and for this purpose a broad variety of esterases are commercially available. Consequently, this

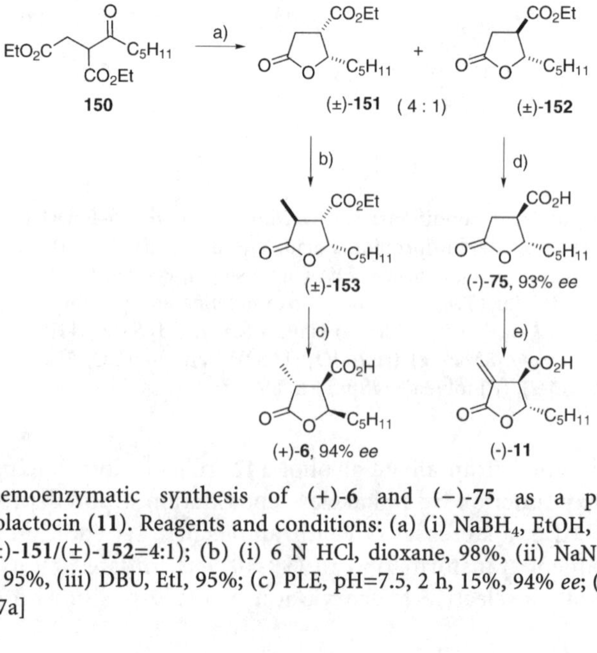

**Scheme 24** Chemoenzymatic synthesis of (+)-**6** and (−)-**75** as a precursor for (−)-methylenolactocin (**11**). Reagents and conditions: (a) (i) NaBH$_4$, EtOH, (ii) HCl, 91% (two steps, (±)-**151**/(±)-**152**=4:1); (b) (i) 6 N HCl, dioxane, 98%, (ii) NaN(TMS)$_2$, MeI, then 1 N HCl, 95%, (iii) DBU, EtI, 95%; (c) PLE, pH=7.5, 2 h, 15%, 94% *ee*; (d) PPL, 20%, 93% *ee*; (e) [47a]

strategy should also be applicable to paraconic acids, and indeed, the synthesis of (+)-phaseolinic acid (**6**) as well as (−)-methylenolactocin (**11**) has successfully been achieved this way (Scheme 24) [61]. This approach is especially advantageous if the precursors are readily available on large scale to compensate for the principle loss of at least 50% caused by the presence of the undesired enantiomer. Thus, a mixture of the lactones (±)-**151** and (±)-**152** is available in a single step from the ketodiester **150** [5c, 62] via a reduction-lactonization sequence. After separation, (±)-**151** was transformed into racemic (±)-**153**, which was hydrolyzed by pig liver esterase (PLE) to yield (+)-phaseolinic acid ((+)-**6**) in up to 94% *ee*. Alternatively, the remaining (−)-**153** could be obtained in up to 96% *ee*, which could be hydrolyzed to (−)-phaseolinic acid ((−)-**6**).

Analogously, (±)-**152** was enzymatically hydrolyzed by porcine pancreatic lipase (PPL) to give (−)-**75** in up to 93% *ee*, which could be further converted to (−)-methylenolactocin (**11**).

## 2.4.5
### Asymmetric Cyclopropanation

The cyclopropane aldehyde **156** was identified as a versatile chiral building block for the enantioselective synthesis of 4,5 disubstituted γ-butyrolactones of type **158** or **159**. Both enantiomers of **156** can be easily obtained in a highly diastereo- and enantioselective manner from furan-2-carboxylic ester **154** using an asymmetric copper-catalyzed cyclopropanation as the key step followed by an ozonolysis of the remaining double bond (Scheme 25) [63]. Addition of

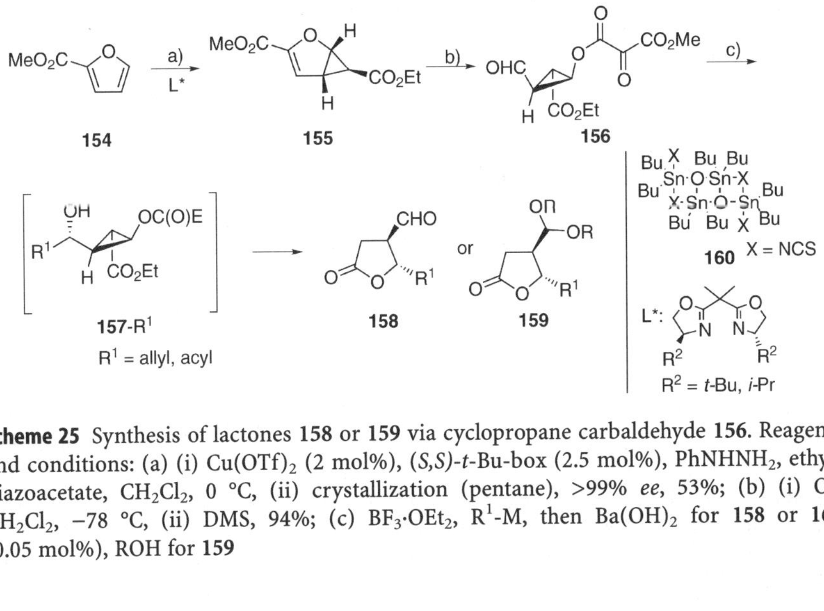

**Scheme 25** Synthesis of lactones **158** or **159** via cyclopropane carbaldehyde **156**. Reagents and conditions: (a) (i) Cu(OTf)$_2$ (2 mol%), (*S,S*)-*t*-Bu-box (2.5 mol%), PhNHNH$_2$, ethyldiazoacetate, CH$_2$Cl$_2$, 0 °C, (ii) crystallization (pentane), >99% *ee*, 53%; (b) (i) O$_3$, CH$_2$Cl$_2$, −78 °C, (ii) DMS, 94%; (c) BF$_3$·OEt$_2$, R$^1$-M, then Ba(OH)$_2$ for **158** or **160** (0.05 mol%), ROH for **159**

allylsilanes or silyl enol ethers to **156** catalyzed by $BF_3·Et_2O$ gave rise to the Felkin-Anh [64] adduct **157** with diastereoselectivities of up to ≥95:5, which can be directly converted without the need of isolation by a retroaldol/lactonization cascade to either **158** or **159** depending on the reaction conditions.

Based on this approach, several paraconic acids became available [65]. A formal synthesis of (−)-methylenolactocin (**11**) was accomplished by the addition of 1,3-pentadienyltrimethylsilane (**161**) to the aldehyde **156** (Scheme 26).

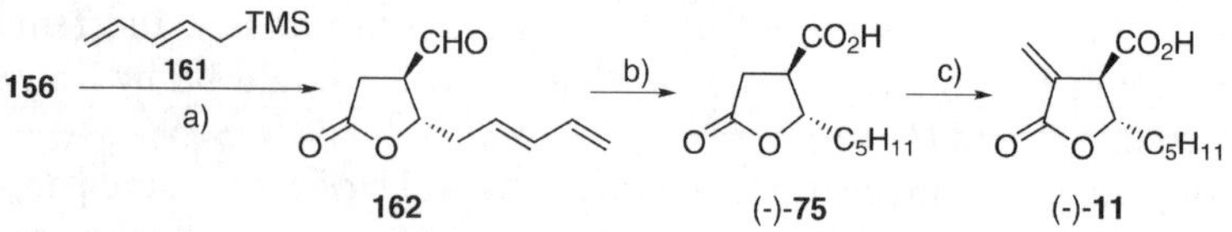

**Scheme 26** Formal synthesis of (−)-methylenolactocin (**11**). Reagents and conditions: (a) (i) $BF_3·OEt_2$, **161** (1.25 equiv.), −78 °C, 12 h, quant, (ii) $Ba(OH)_2$, MeOH, 0 °C, 66%, *trans/cis*=97:3; (b) (i) Pd/C, $H_2$ (1 atm), MeOH, quant., (ii) $NaClO_2$, $H_2O_2$, 86%; (c) [47a]

The resulting lactone **162** was hydrogenated to (−)-**75**, being a direct precursor for (−)-**11** as previously reported [47a].

For the synthesis of paraconic acids being substituted with alkyl chains of various lengths the lactones **163** were employed (Scheme 27). It was envi-

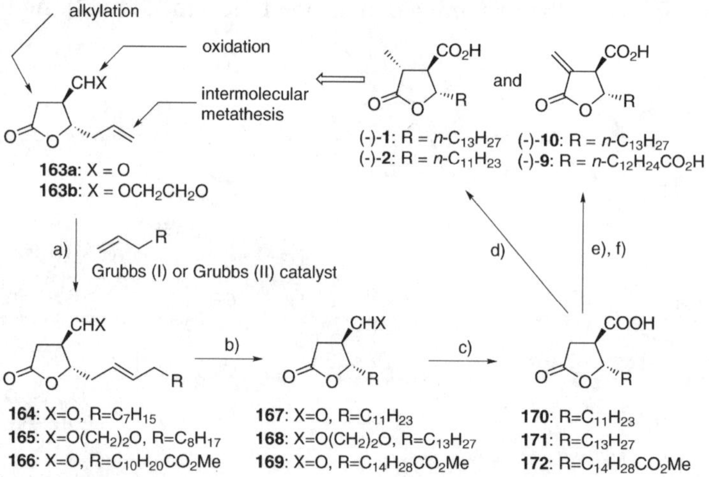

**Scheme 27** General access towards paraconic acids employing the lactones **169** as starting material. Reagents and conditions: (a) alkene, Grubbs (I) [$(PCy_3)_2$-$Cl_2Ru$=CHPh] for **165** (57%) or Grubbs (II) [(4,5-dihydroIMES)($PCy_3$)$Cl_2Ru$=CHPh] for **164** (53%) and **166** (38%); (b) Pd/C, $H_2$ (1 bar), 90% (**167**), 99% (**168**), 88% (**169**); (c) $NaClO_2$, $Na_2HPO_4$ for **170** (72%) or $CrO_3$, $H_2SO_4$, for **171** (88%) and **172** (96%); (d) NaHMDS then MeI, 90% (−)-(**2**), 96% (−)-(**1**); (e) (i) $MeOMgOCO_2Me$, (ii) *N*-methylaniline, $CH_2O$, 62% (−)-**9**; (f) [47b] for (−)-**10**

sioned that the allyl group could be converted directly to the desired side chains by intermolecular metathesis reactions.

Indeed, the lactones **163** could be reacted individually with an excess of a terminal alkene, using the Grubbs-I catalyst ((PCy$_3$)$_2$Cl$_2$Ru=CHPh) for the metathesis of the protected aldehyde **163b**, while the unprotected aldehyde **163a** dictated the use of the more active Grubbs-II catalyst (4,5-dihydroIMES)(PCy$_3$)Cl$_2$Ru=CHPh [66]. The resulting alkenes (E/Z=7:1 to 3.5:1) **164–166** were subsequently hydrogenated, followed by oxidation of the aldehyde or the acetal. Alkylation with MeI gave rise to (−)-roccellaric acid (**1**) from **171** and to (−)-nephrosteranic acid (**2**) from **170**, while introducing a *exo*-methylene group according to a method described by Greene et al. [47b] yielded (−)-protopraesorediosic acid (**9**) and (−)-protolichesterinic acid (**10**).

## 2.4.6
## Ireland-Claisen Rearrangement

The strategy for the synthesis of γ-butyrolactones by Claisen rearrangement of acylated 1,4-diols has been already discussed (Schemes 6–11). A stereodi-

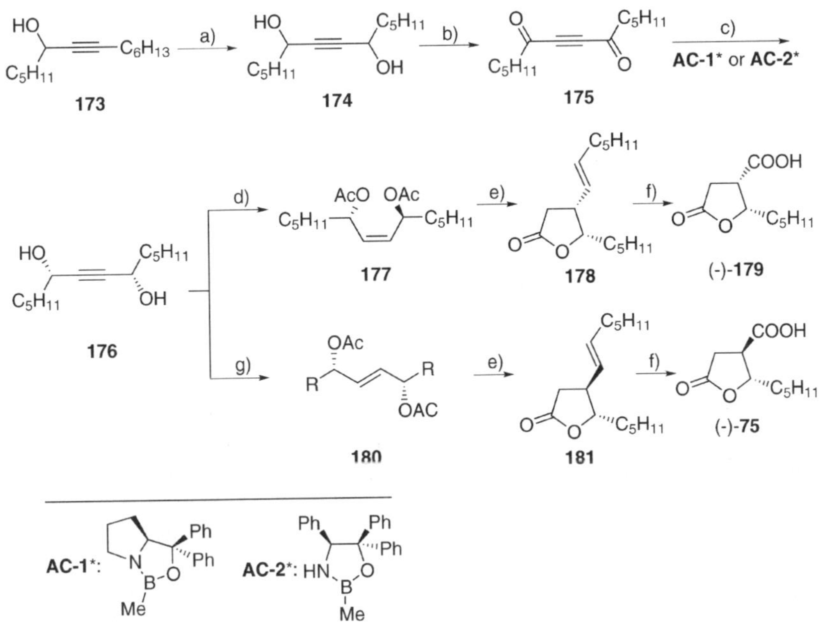

**Scheme 28** Synthesis of methylenolactocin (−)-**11**. Reagents and conditions: (a) (i) *t*-BuOOH, SeO$_2$, CH$_2$Cl$_2$, rt, 24 h, (ii) NaBH$_4$, MeOH, 0 °C, 55% (two steps); (b) CrO$_3$, aq. H$_2$SO$_4$, 0 °C, 75%; (c) BMS (2.0 equiv.), **AC-1\*** or **AC-2\*** (0.4 equiv.), 70%, >99% ee (*dl/meso* 4.6:1); (d) (i) Lindlar catalyst, H$_2$, EtOAc, 75%, (ii) acetylation; (e) (i) KHMDS, *t*-BuMe$_2$SiCl, THF, −78 °C, then toluene, Δ, (ii) LiOH, H$_2$O/THF, Δ, (iii) aq. HCl/THF, Δ, 65% for **178** (three steps), 70% for **181** (three steps); (f) (i) cat. RuCl$_3$, NaIO$_4$, CCl$_4$/MeCN/H$_2$O, 68% for (−)-**179**, 70% for (−)-**75**

vergent variant to paraconic acids from $C_2$-symmetric *trans-* and *cis*-alk-2-ene-1,4-diols was developed by Garcia et al. (Scheme 28) [67]. Starting from the diketone **175**, oxaborolidine-mediated reduction yielded the diol **176** with >99% ee. Hydrogenation followed by acetylation can be carried out selectively to either the *cis* or the *trans* alkenes **177** and **180**, respectively, which could be converted to the lactones **178** or **181** by rearrangement under Ireland-Claisen conditions followed by hydrolysis of the esters. Oxidation to the lactones (−)-**179** or (−)-**75** completed the formal synthesis of (−)-phaseolinic acid (**6**) and (−)-methylenolactocin (**11**).

## 2.4.7
### Tungsten-π-allyl Complexes

The application of tungsten-π-allyl complexes provided yet another elegant and efficient access to paraconic acids (Scheme 29) [68]. The propargylic al-

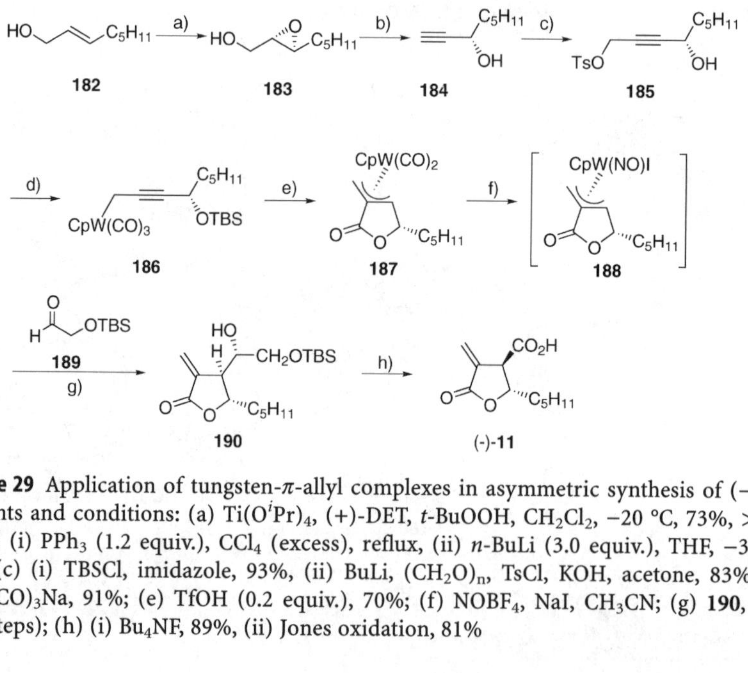

**Scheme 29** Application of tungsten-π-allyl complexes in asymmetric synthesis of (−)-**11**. Reagents and conditions: (a) Ti(O$^i$Pr)$_4$, (+)-DET, *t*-BuOOH, CH$_2$Cl$_2$, −20 °C, 73%, >98% ee; (b) (i) PPh$_3$ (1.2 equiv.), CCl$_4$ (excess), reflux, (ii) *n*-BuLi (3.0 equiv.), THF, −35 °C, 56%; (c) (i) TBSCl, imidazole, 93%, (ii) BuLi, (CH$_2$O)$_n$, TsCl, KOH, acetone, 83%; (d) CpW(CO)$_3$Na, 91%; (e) TfOH (0.2 equiv.), 70%; (f) NOBF$_4$, NaI, CH$_3$CN; (g) **190**, 67% (two steps); (h) (i) Bu$_4$NF, 89%, (ii) Jones oxidation, 81%

cohol **185**, readily obtained from **182** by making use of an asymmetric Sharpless epoxidation, was metallated to yield the tungsten-$\eta^1$-propargyl complex **186** as the key intermediate. Acid induced cyclization yielded the tungsten-π-allyl complex **187** with exclusive *syn* stereochemistry between the alkyl side chain and the metal fragment. Treatment of **187** with NOBF$_4$ and NaI generated the corresponding CpW(NO)I(π-allyl) derivative **188**, which belongs to a class of complexes being known to be an allyl anion equivalent that reacts at its more substituted allyl carbon with electrophiles [69]. Hence, reaction with TBS-protected lactic aldehyde yielded the α-methylene butyrolactone **190** as a single stereoisomer. Deprotection and oxidative cleavage of **190** completed the synthesis of (−)-methylenolactocin (**11**).

# 3
# Conclusion

Biologically important γ-butyrolactone natural products, among them the paraconic acids, display broad structural diversity and continue to be of great interest. With the absence of a general solution for the synthesis of chiral γ-butyrolactones, many useful approaches have been devised towards them that greatly differ in their overall strategy. Paraconic acids being *trans*-substituted in the lactone ring are now readily available, while the stereoselective synthesis of their *cis*-substituted analogs still remains a challenge.

## References

1. David F, Elix JA, Samsudin MW (1990) Aust J Chem 43:1297
2. (a) Maier DI, Marimon G, Stortz CA, Adler MT (1999) J Nat Prod 62:1565; (b) Park BK, Nakagawa M, Hirota A, Nakayama M (1988) J Antibiot 41:751; (c) Bérdy J, Aszalos A, Bostian M, McNitt KL (1982) In: CRC handbook of antibiotic compounds. CRC Boca Raton Fla, p 39
3. (a) Miyabe H, Fujii K, Goto T, Naito T (2000) Org Lett 2:4071; (b) Fernandez AM, Plaquevent JC, Duhamel L (1997) J Org Chem 62:4007; (c) Sibi MP, Lu J, Talbacka CL (1996) J Org Chem 61:7848; (d) Peng Z-H, Woerpel KA (2001) Org Lett 3:675
4. For non-stereoselective syntheses see: (a) Pohmakotr M, Reutrakul V, Phongpradit T, Chansri A (1982) Chem Lett 687; (b) Brückner C, Reissig HU (1988) J Org Chem 53:2440; (c) Lawlor JM, McNamee MB (1983) Tetrahedron Lett 24:2211; (d) van Tamelen EE, Bach SR (1958) J Am Chem Soc 80:3079; (e) Pohmakotr M, Harnying W, Tucinda P, Reutrakul V (2002) Helv Chim Acta 85:3792
5. For stereoselective syntheses before 1991 see: (a) Damon RE, Schlessinger RH (1976) Tetrahedron Lett 1561; (b) Carlson RM, Oyler AR (1976) J Org Chem 41:4065; (c) Martin J, Watts PC, Johnson F (1974) J Org Chem 39:1676; (d) Löffler A, Pratt RD, Pucknat J, Gelbard G, Dreiding AS (1969) Chimica 23:413
6. (a) Hoppe D (1984) Angew Chem Int Ed Engl 23:932; (b) Roder H, Helmchen G, Peters EM, Peters K, Schnering HG (1984) Angew Chem Int Ed Engl 23:898; (c) Hoppe D, Zschage O (1989) Angew Chem Int Ed Engl 28:69; (d) Hoppe D, Hense T (1997) Angew Chem Int Ed Engl 36:2282; (e) Ahlbrecht H, Beyer U (1999) Synthesis 365; (f) Whisler MC, Vaillancourt L, Beak P (2000) Org Lett 2:2655; (g) Johnson TA, Curtis MD, Beak P (2001) J Am Chem Soc 123:1004; (h) Lim SH, Curtis MD, Beak P (2001) Org Lett 3:711; (i) Schlappbach A, Hoffmann RW (2001) Eur J Org Chem 323; (j) Gaul C, Seebach D (2002) Helv Chim Acta 85:963; (k) Özlügedik M, Kristensen J,

Wibbeling B, Fröhlich R, Hoppe D (2002) Eur J Org Chem 414 and references cited therein
7. Fotiadu F, Michel F, Buono G (1990) Terahedron Lett 31:4863
8. (a) Sarkar S, Ghosh S, (1996) Tetrahedron Lett 37:4809; (b) For the synthesis of (±)-protolichesterinic acid see: Ghatak A, Sarkar S, Ghosh S (1997) Tetrahedron 53:17335
9. Mawson SD, Weavers RT (1995) Terahedron 51:11257
10. Maiti G, Roy SC (1996) J Chem Soc Perkin Trans 1 5:403
11. Reissig H-U (1988) Top Curr Chem 144:73
12. Shimada S, Hashimoto Y, Saigo K (1993) J Org Chem 58:5226
13. (a) Mandal PK, Maiti G, Roy SC (1998) J Org Chem 63:2829; (b) Mandal PK, Roy SC (1999) Tetrahedron 55:11395
14. cf Martin VS, Woodard SS, Katsuki T, Yamada Y, Ikeda M, Sharpless KB (1981) J Am Chem Soc 103:6237
15. Beckwith ALJ, Easton CJ, Lawrence T, Serelis AK (1983) Aust J Chem 36:545
16. Rajanbabu TV (1991) Acc Chem Res 24:139
17. Corey EJ, Schmidt G (1979) Tetrahedron Lett 20:399
18. Mulzer J, Kattner L, Strecker AR, Schröder C, Buschmann J, Lehmann C, Luger P (1991) J Am Chem Soc 113:4218
19. Mulzer J, Salimi N, Hartl H (1993) Tetrahedron Asym 4:457
20. Review: Mulzer J, Altenbach J, Braun M, Krohn K, Reissig HU (1990) In: Organic synthesis highlights. VCH, Weinheim New York, p 158
21. Ziegler GE (1988) Chem Rev 88:1423
22. (a) Mulzer H, Kappert M (1982) Angew Chem Suppl 23; (b) Minami M, Ko SS, Kishi Y (1982) J Am Chem Soc 104:1109; (c) Katsuki T, Lee AWM, Ma P, Martin VS, Masamune S, Sharpless KB, Tuddenham D, Walker FJ (1982) J Org Chem 47:1373
23. (a) Wick AE, Felix D, Steen K, Eschenmoser A (1964) Helv Chim Acta 74:2425; (b) Welch JT, Eswarakrishnan S (1985) J Org Chem 50:5907
24. Hoffman RW (1989) Chem Rev 89:1841
25. Ashahina Y, Yanagita M (1937) Ber Dtsch Chem Ges 70:227
26. Ulrich Steffen (1989) PhD thesis, FU Berlin
27. Barton DHR, McCombie SW (1975) J Chem Soc Perkin Trans I 1574
28. (a) Morpain C, Nasser B, Laude B, Latruffe N (1990) Org Prep Proced Intern 22:540; (b) Eastwood FW, Harrington KJ, Josan JS, Pura JL (1970) Tetrahedron Lett 11:5223
29. Forster A, Kovac T, Mosimann H, Renaud P, Vogel P (1999) Tetrahedron Asymmetry 10:567 and references cited therein
30. Black KA, Vogel P (1984) Helv Chim Acta 67:1612
31. (a) Vieira E, Vogel P (1983) Helv Chim Acta 66:1865; (b) Reymond JL, Vogel P (1990) Tetrahedron Asymmetry 1:729; (c) Ronan B, Kagan HB (1991) Tetrahedron Asymmetry 2:75; (d) Aggarwal VK, Lightowler M, Lindell SD (1992) Synlett 730; (e) Corey EJ, Loh TP (1993) Tetrahedron Lett 34:3979
32. Brecht-Forster A, Fitremann J, Renaud P (2002) Helv Chim Acta 85:3965
33. According to Huneck S, Toensberg T, Bohlmann F (1986) Phytochemistry 25:453 and [12] the product is named as pertusarinic acid and not as dihydropertusarinic acid as Renaud et al. proposed
34. (a) Vogel P, Fattori D, Gasparini F, Le Drian C (1990) Synlett 173; (b) Renaud P, Vionnet JP (1993) J Org Chem 58:5895
35. Zhang Z, Lu X (1996) Tetrahedron Asymmmetry 7:1923
36. Stucky G (1988) GIT Fachz Lab 32:535

37. For reviews concerning the synthesis of butenolides and saturated $\gamma$-lactones see: (a) Nagao Y, Ochiai M, Shiro M (1989) J Org Chem 54:5211 and references cited therein; (b) Corey EJ, Cheng XM (1989) In: The logic of chemical synthesis. Wiley, New York
38. Takahata H, Uchida Y, Momose T (1994) J Org Chem 59:7201
39. Takahata H, Uchida Y, Momose T (1995) J Org Chem 60:5628
40. For a review on electrophile-mediated heterocyclization see: (a) Harding KE, Timer TH (1991) In: Trost BM (ed) Comprehensive organic synthesis. Pergamon, Oxford, p 353; (b) Cardillo G, Orena M (1990) Tetrahedron 46:3321; (c) Takahata H (1993) Yakugaku Zasshi 113:737
41. (a) Hesse O (1898) J Prakt Chem 57:232; (b) Huneck S, Follmann GJ (1967) Naturforsch B 22:666; (c) cf. [19]
42. (a) Perkmutter P (1992) In: Baldwin JE (ed) Conjugation addition reactions in organic synthesis. Pergamon, Oxford, p 283; (b) Stork G, Rychnovsky SD (1987) J Am Chem Soc 109:1564; (c) Vigneron JP (1984) Tetrahedron 40:6521; (d) von Oeveren A, Jansen JFGA, Feringa BL (1994) J Org Chem 59:5999
43. (a) Masuda M, Nishimura K (1971) Phytochemistry 10:401; (b) Masuda M, Nishimura K (1981) Chem Lett 1333; (c) Pai Y-C, Fang J-M, Wu S-H (1994) J Org Chem 59:6018; (d) Ebata T, Matsmoto K, Yoshikoshi H, Koseki K, Kawakami H, Okano K, Matsushita H (1993) Heterocycles 36:1017
44. (a) ter Heide R, deValois PJ, Visser J, Jaegers PP, Timar R (1978) In: Charalambous (ed) Analysis of food and beverages. Academic Press, New York, p 275; (b) Nishimura K (1987) Chem Today 189:30; (c) Orutuno RM, Merce R, Font J (1987) Tetrahedron 43:4497
45. (a) Greene AE, Charbonnier F (1985) Tetrahedron Lett 26:5525; (b) Greene AE, Charbonnier F, Luche M-J, Moyano A (1987) J Am Chem Soc 109:4752; (c) Frater G, Müller U, Günther W (1986) Helv Chim Acta 69:1858
46. (a) Whitesell JK, Chen HH, Lawrence RM (1985) J Org Chem 50:4663; (b) Whitesell JK, Lawrence RM (1986) Chimica 40:318; (c) Schwartz A, Madan P, Whitesell JK, Lawrence RM (1990) Org Synth 69:1
47. (a) Azevedo MBM, Murta MM, Greene AE (1992) J Org Chem 57:4567; (b) Murta MM, Azevedo MBM, Greene AE (1993) J Org Chem 58:7537
48. Gras J-L (1978) Tetrahedron Lett 19:2111
49. Sibi MP, Deshpande PK, La Loggia AJ (1996) Synlett 343
50. The choice of the 4-diphenylmethyl-2-oxazolidinone as a chiral auxiliary was based on its known superiority in radical reactions; see: (a) Sibi MP, Despande PK, La Loggia AJ, Christensen JW (1995) Tetrahedron Lett 36:8961; (b) Sibi MP, Jasperse CP, Ji J (1995) J Am Chem Soc 117:10779; (c) Sibi MP, Ji J (1997) Angew Chem Int Ed Engl 36:274
51. (a) Evans DA, Bartoli J, Shih TL (1981) J Am Chem Soc 103:2127; (b) Evans DA, Urpi F, Somer TC, Clark JS, Bilodeau MT (1990) J Am Chem Soc 112:8215; (c) Evans DA, Bilodeau MT, Somer TC, Clardy JS, Cherry D, Kato Y (1991) J Org Chem 56:5750; (d) Oppolzer W, Cintas-Moreno P, Tamura O, Cardinaux F (1993) Helv Chim Acta 76:187
52. Sibi MP, Liu P, Ji J, Hajra S, Chen J (2002) J Org Chem 67:1738
53. Bella M, Margarita R, Orlando C, Orsini M, Parlanti L, Piancatelli G (2000) Tetrahedron Lett 41:561
54. (a) Lockwood RF, Nicholas KM (1977) Tetrahedron Lett 18:4163; (b) Nicholas KM, Nestle MO, Deyferth D (1978) In: Halper (ed) Transition metal organometallics. Academic Press, New York, p 1; (c) Schreiber SL, Sammakia T, Crowe WE (1986) J Am Chem Soc 108:3128; (d) Nicholas KM, Mulvaney M, Bayer M (1980) J Am Chem Soc 102:2508; (e) Hodes HD, Nicholas KM (1978) Tetrahedron Lett 19:4350; (f) Bramwell

AF, Combie L, Knight MH (1965) Chem Ind (London) 1265 and references cited therein; (g) Saha M, Bogby B, Nicholas KM (1986) Tetrahedron Lett 27:915; (h) Schreiber SL, Kalimas MT, Sammakia T (1987) J Am Chem Soc 109:5749
55. Jacobi PA, Herradura P (2001) Can J Chem 79:1727
56. (a) Stadtmüller H, Lentz R, Tucker CE, Stüdemamm T, Dörner W, Knochel P (1993) J Am Chem Soc 115:7027; (b) Stadtmüller H, Tucker CE, Vaupel A, Knochel P (1993) Tetrahedron Lett 34:7911; (c) Vaupel A, Knochel P (1994) Tetrahedron Lett 35:8349
57. (a) Vaupel A, Knochel P (1995) Tetrahedron Lett 36:231; (b) Vaupel A, Knochel P (1996) J Org Chem 61:5743
58. Zhu G, Lu X (1995) J Org Chem 60:1087
59. Aoyagi S, Wang TC, Kibayashi C (1993) J Am Chem Soc 115:11393
60. Martin T, Rodriguez CM, Martin VS (1996) J Org Chem 61:6450
61. Drioli S, Felluga F, Forzato C, Nitti P, Pitacco G, Valentin E (1998) J Org Chem 63:2385
62. Patrick TM Jr (1952) J Org Chem 17:1009
63. Böhm C, Schninnerl M, Bubert C, Zabel M, Labahn T, Parisini E, Reiser O (2000) Eur J Org Chem 2955
64. Review: Mengel A, Reiser O (1999) Chem Rev 99:1191
65. (a) Chhor RB, Nosse B, Sörgel S, Böhm C, Seitz M, Reiser O (2003) Chem Eur J 9:260; (b) Böhm C, Reiser O (2001) Org Lett 3:1315
66. Chatterjee AK, Grubbs RH (1999) Org Lett 1:1751
67. Ariza X, Garcia J, Lopez M, Montserrat L (2001) Synlett 120
68. Chandrasekharam M, Liu R-S (1998) J Org Chem 63:9122
69. (a) Faller JW, Linebarrier DL (1989) J Am Chem Soc 111:1939; (b) Faller JW, John JA, Mazzier MR (1989) Tetrahedron Lett 30:1769

# Recent Progress in the Total Synthesis of Dolabellane and Dolastane Diterpenes

Martin Hiersemann (✉) · Hannes Helmboldt

Institut Für Organische Chemie, Technische Universität Dresden,
01062 Dresden, Germany
*martin.hiersemann@chemie.tu-dresden.de*

| 1 | Introduction | 74 |
|---|---|---|
| 2 | Biosynthesis | 75 |
| 3 | Dolabellane and Neodolabellane Diterpenes | 78 |
| 3.1 | Isolation, Structure and Biological Activities | 78 |
| 3.2 | Total Synthesis of Dolabellanes and Neodolabellanes | 82 |
| 3.2.1 | Yamada's Strategy | 83 |
| 3.2.1.1 | (−)-Claenone (1998) | 83 |
| 3.2.1.2 | (−)-Stolonidiol (2001) | 88 |
| 3.2.1.3 | (−)-Palominol and (+)-Dolabellatrienone (2003) | 93 |
| 3.2.2 | Williams' Total Syntheses of Neodolabellanes | 93 |
| 3.2.2.1 | (+)-4,5-Deoxyneodolabelline | 93 |
| 3.2.2.2 | (±)-Neodolabellenol | 98 |
| 4 | Dolastane Diterpenes | 99 |
| 4.1 | Isolation, Structure and Biological Activities | 99 |
| 4.2 | Total Syntheses of Dolastanes | 101 |
| 4.2.1 | Williams' Total Synthesis | 106 |
| 5 | Neodolastane Diterpenes | 112 |
| 5.1 | Isolation, Structure and Biological Activities | 112 |
| 5.2 | Completed and Formal Total Syntheses | 115 |
| 5.2.1 | Danishefsky's Total Synthesis | 115 |
| 5.2.2 | Snider's Formal Total Synthesis | 121 |
| 5.3 | Approaches Toward Guanacastepene | 127 |
| 5.3.1 | Strategies Based on Metathesis | 127 |
| 5.3.2 | Strategies Based on Cycloadditions | 129 |
| 5.3.3 | Miscellaneous | 130 |
| 6 | Conclusion | 132 |
| References | | 132 |

**Abstract** The isolation, structure and total synthesis of members of four classes of diterpenes has been summarized. Dolabellanes, neodolabellanes, dolastanes and neodolastanes are structurally related bi- or tricyclic diterpenes. Dolabellanes belong to a continuously growing class of diterpenes being isolated from marine and terrestrial sources. The published work on isolation and synthesis since 1998 has been summarized. Neodolabellanes represent a scarce class of diterpenes that have been isolated exclusively from marine sources. 5-7-6-tricyclic diterpenes classified as dolastanes have been isolated mainly from marine sources. The intense research devoted to the total synthesis of these diterpenes between 1986 and 1993 has been reviewed. Guanacastepenes have been isolated from fungi. These diterpenes are classified as neodolastanes and recently completed syntheses as well as synthetic approaches have been summarized.

**Keywords** Diterpene · Total Synthesis · Natural Product · Dolabellane · Dolastane · Guanacastepene

**List of Abbreviations**
| | |
|---|---|
| DCE | Dichloroethane |
| DIBAH | Diisobutylaluminium hydride |
| DPPB | Bis(diphenylphosphino)butane |
| Cy | Cyclohexyl |
| DBA | *trans,trans*-Dibenzylideneacetone |
| DMDO | Dimethyldioxirane |
| DMP | 2,2-Dimethoxypropane |
| $ED_{50}$ | Effective dose 50% |
| HWE | Horner-Wadsworth-Emmons |
| LAH | Lithium aluminium hydride |
| NMO | 4-Methylmorpholine *N*-oxide |
| NOE | Nuclear Overhauser effect |
| RCM | Ring closing metathesis |
| SAE | Sharpless asymmetric epoxidation |
| TBS | *tert*-Butyldimethylsilyl |
| TEMPO | 2,2,6,6-Tetramethyl-1-piperidinyloxy |
| TES | Triethylsilyl |
| TFP | Tri-2-furylphospine |
| TLC | Thin layer chromatography |
| TPAP | Tetrapropylammonium perruthenate |
| TPS | *tert*-Butyldiphenylsilyl |

# 1
# Introduction

Through the eyes of a synthetic organic chemist, monocyclic and polycyclic diterpenes amalgamate the beauty of structure with the importance of biological activity to provide challenging target molecules for total synthesis. From this perspective, it is not surprising that the report of a structurally novel diterpene exerting a promising biological activity can trigger off a synthetic gold rush. The story of guanacastepene A represents a case study for such an event. First reported in 2000, at least a dozen different research groups have embarked on the total synthesis of these neodolastane diterpenes and published conceptually different approaches. One total and one formal synthesis have been completed so far and some more will follow. Dolabellanes on the other hand have been know for almost 30 years and according to Rodríguez's leading review article, the structures of about 140 different dolabellanes having been published up until June 1998 [1]. Consequently, one would expect that numerous ingenious total syntheses of dolabellane diterpenes would have been realized. The objective of this review is to provide a non-comprehensive and subjective presentation of the total syntheses of dolabellane and dolastane diterpenes. If appropriate, details to isolation and structural elucidation are provided. Six syntheses are covered in-depth. Whenever possible, details regarding the reaction conditions are given and important mechanisms will be discussed within the schemes. It was my intention to enable the less experienced reader to follow the syntheses without consultation of the original literature. Approaches toward and older completed total syntheses are summarized in single schemes and equations. Therefore, the reader will be able to compare different synthetic strategies which have evolved over time.

# 2
# Biosynthesis

The biosynthesis of diterpenes was created to produce molecular diversity. Literally hundreds of different macrocyclic, bicyclic and polycyclic diterpenes

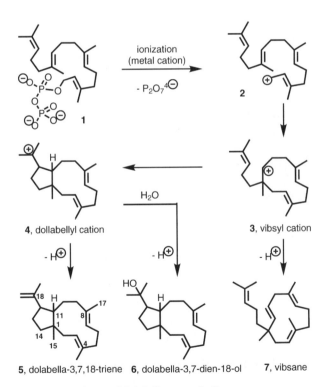

**Scheme 1** Proposed biosynthesis of dolabellanes and vibsanes

can be isolated from microorganisms, plants and marine animals and algae. They are all biosynthesised by an anabolic pathway that utilizes a single achiral substrate, geranylgeranyl diphosphate. The structure of diterpenes is often characteristic for a certain plant or animal family, genus or species. Significant work is currently devoted to study the molecular biology of diterpene synthases (cyclases) [2–9]. Though a detailed investigation of the cyclases which produce dolabellane and dolastane diterpenes is not yet available, it is reasonable to assume that geranylgeranyl diphosphate is first ionised by an enzymatic, metal ion-initiated process (Scheme 1). The first cyclization generates the vibsyl cation (3) which undergoes a second cyclization to afford the dolabellyl cation (4). The loss of a proton or the nucleophilic attack by water provides the dolabellane scaffold (5, 6) that can be further chemically diversified for example by enzymatic oxidations. Vibsanes represent a rare class of diterpenes which have been isolated along with dolabellanes from the liverwort *Odontoschisma denudatum* [10–13]. Their structure fits nicely into the proposed biosynthetic scheme assuming that the initial cyclization is followed by deprotonation and not by a second cyclization reaction.

The dolabellyl cation (4) may be transformed into the neodolabellyl cation (10) by a series of [1,2]-sigmatropic rearrangements (Scheme 2). Subse-

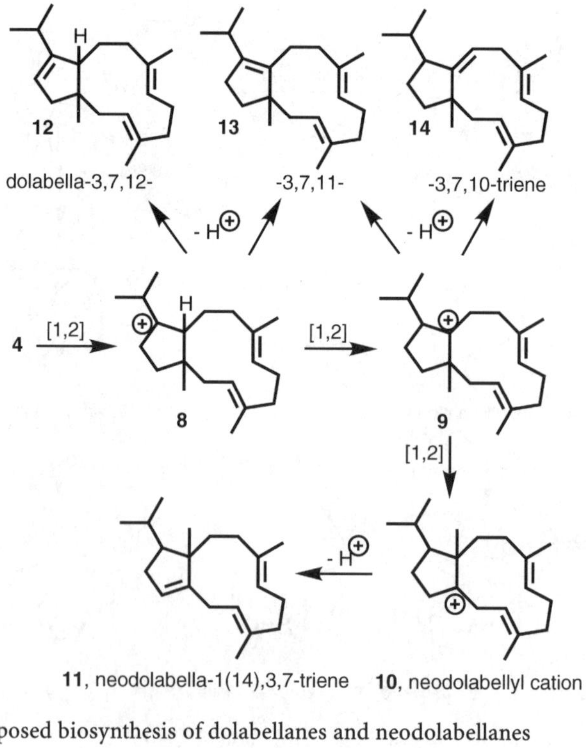

**Scheme 2** Proposed biosynthesis of dolabellanes and neodolabellanes

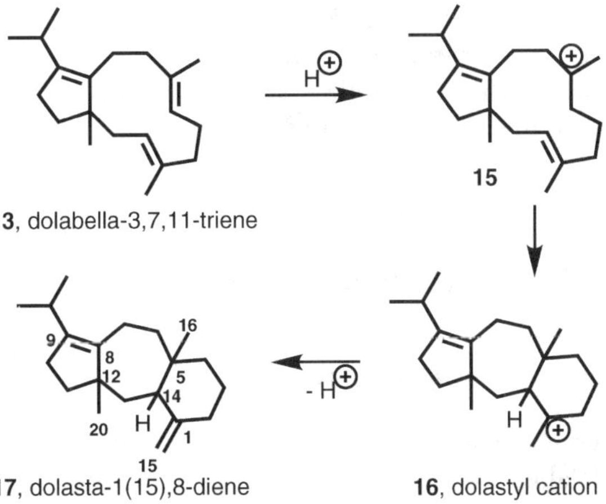

**Scheme 3** Proposed biosynthesis of dolastanes

quent deprotonation of the neodolabellyl cation (**10**) affords the neodolabellane framework (**11**). Loss of a proton from the intermediate cation (**8, 9**) could lead to the formation of dolabellanes (**12–14**) featuring a double bond within the cyclopentane moiety.

The ionisation-induced cyclization sequence could be followed by a proton-induced cyclization (Scheme 3). Protonation of the $C^{7/8}$ double bond of **13** generates a dolabellyl cation (**15**) that is transformed into a dolastyl cation (**16**) by a transannular cyclization between $C^3$ and $C^8$ (dolabellane numbering). Subsequent deprotonation at $C^{15}$ (dolastane numbering) leads to the dolasta-1(15),8-diene which is further functionalized by enzymatic oxidation.

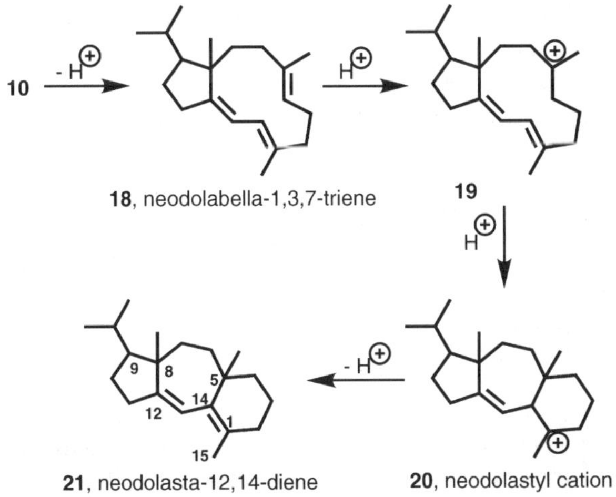

**Scheme 4** Proposed biosynthesis of neodolastanes

It is tempting to speculate that the neodolastane carbon skeleton (**21**) is biosynthesized by a transannular cyclization of a neodolabellyl cation (**10**) (Scheme 4). However, this assumption is only based on structural analogy. A detailed analysis of the biosynthesis has yet to be published.

# 3
# Dolabellane and Neodolabellane Diterpenes

## 3.1
## Isolation, Structure and Biological Activities

The leading reference that covers the isolation and total synthesis of dolabellanes and neodolabellanes from 1975 through June 1998 was published by Rodríguez and coworkers [1]. Our chapter was written to provide an update on the chemistry of dolabellanes and neodolabellanes from 1999 through October 2003. Dolabellanes are being isolated from marine and terrestrial sources. The review of Rodríguez covers 140 dolabellanes and neodolabellanes from which 44% were isolated from algae, 23% from coelenterates (soft corals and gorgonians), 11% from marine molluscs and 22% from moulds, liverworts and higher plants (Fig. 1). Interestingly, dolabellanes that are isolated from marine animals possess the opposite absolute configuration at $C^1$ and $C^{11}$ compared to dolabellanes found in algae, liverworts or higher plants. Neodolabellanes are rare compounds isolated exclusively from corals.

Diverse biological activities have been reported for dolastanes. Depending on structural features, the diterpenes exhibit cytotoxicity against various cancer cell lines, antimicrobial activity against Gram-positive, Gram-negative bacteria, fungi and viruses. They possess ichthyotoxic and phytotoxic activities [1].

A collection of new dolabellanes whose isolation was published after the review of Rodrígez had appeared will be outlined below.

Three new dolabellanes (**24–26**) were isolated from the wood of *Trichilia trifolia* (family *Meliaceae*, "mahogany family") collected in Yucatán, México

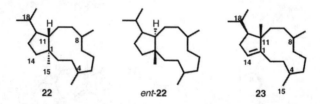

**Fig. 1** Carbon skeleton of dolabellanes isolated from higher plants, liverworts, algae, molluscs, sponge (**22**), corals, fungi (mould) (*ent*-**22**) and neodolabellanes isolated from corals (**23**)

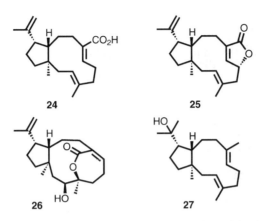

**Fig. 2** Dolabellanes (24–26) isolated from the plant *Trichilia trifolia* (2000). Dolabellane (27) from the liverwort *Pleurozia gigantea*

(Fig. 2) [14]. The relative configuration was assigned based on an X-ray crystal structure analysis of 26. The absolute configuration was deduced from the Cotton effects of 26. Significantly, this work defined for the first time the absolute configuration of a dolabellane isolated from a higher plant. The *cis* arrangement between the isopropyl group at $C^{12}$ and the methyl group at $C^1$ is unusual and was found previously in the dolastane 27 that was isolated from the liverwort *Pleurozia gigantea* [1].

Two new dolabellanes (28, 29) were isolated from the mould *Stachybotrys chartarum* cultivated on Uncle Ben's rice (Fig. 3) [15]. The relative configuration was assigned based on 1D and 2D NMR spectroscopy. Though not ascertained, the absolute configuration depicted in Fig. 3 was suggested in analogy to the absolute configuration of previously isolated dolabellanes from mould [1].

The structure and relative configuration of six new dolabellanes (30–35) isolated from the soft coral *Clavularia inflata* collected at Orchid Island located 44 nautical miles off Taiwan's southeast coast is depicted in Fig. 4 [16]. The relative configuration was assigned based on 1D and 2D NMR spectroscopy and supports the *cis* arrangement between the isopropyl group at $C^{12}$

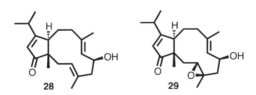

**Fig. 3** Two dolabellanes (28, 29) from the mould *Stachybotrys chartarum* (2000). Absolute configuration not verified

**Fig. 4** Six dolabellanes (30–35) isolated from the soft coral *Clavularia inflata* (2001)

and the methyl group at $C^1$. The dolabellanes **30–35** are cytotoxic against the cell lines A549 (human, lung carcinoma), HT-29 (human, colon carcinoma) and P388 (mouse, leukaemia). Hydroperoxide **35** was found to be the most active compound with $ED_{50}$ values of 0.56, 0.31 and 0.052 µg/mL respectively.

The investigation of the essential oil of the plant *Cyperus alopecuroides* (order *Poales*, family *Ceperaceae*) revealed the existence of the dolabellane diterpene **36** (Fig. 5) [17]. The relative configuration was deduced from NOESY correlations but the absolute configuration remains undisclosed.

The dolabellane **37** (2-acetoxystolonidiol) was isolated from the soft coral *Clavularia koellikeri* (octocorallia: order *Alcyonacea*, family *Clavulariidae*) collected from a coral reef of Ishigaki Island (Yaeyama archipelago, prefec-

**Fig. 5** Dolabellane (36) from the plant *Cyperus alopecuroides* (2001)

**Fig. 6** 2-Acetoxystolonidiol (37) and stolonidiol (38) isolated from *Clavularia sp.* (2002 and 1987)

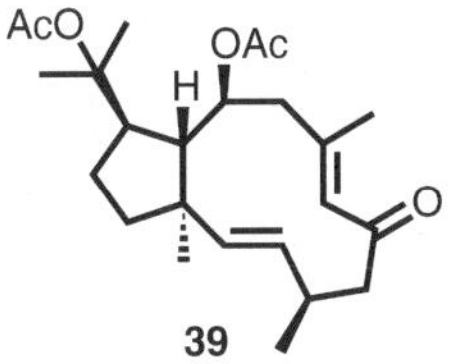

**Fig. 7** The first dolabellane (**39**) isolated from a marine sponge (1999)

ture Okinawa, Japan) (Fig. 6) [18]. The relative configuration was assigned based on NOE spectroscopy and the absolute configuration was determined by the improved version of Mosher's empirical method based on $^1$H NMR data. 2-Acetoxystolonidiol (**37**) is structurally closely related to stolonidiol (**38**) whose absolute configuration has been proved by an enantioselective total synthesis (see below). 2-Acetoxystolinidiol (**37**) is cytotoxic against human colorectal adenocarcinoma cells (DLD-1) with a moderate IC$_{50}$ of 5.0 µg/mL.

The marine sponge *Sigmosceptrella quadrilobata* (family *Latrunculiidae*, order *Hadromerida*) has been identified as a source for the dolabellane **39** (Fig. 7) [19]. The sponge was collected along the coast of the Island Mayotte (Comorian archipelago) located 200 miles off the East African coast approximately halfway between the island of Madagascar and northern Mozambique. The relative configuration was assigned by NOE spectroscopy and the absolute configuration is suggested based on the Cotton effect in the CD spectrum. The dolabellane **39** is cytotoxic against four cancer cell lines with an IC$_{50}$ between 7.7 and 17.2 µg/mL.

Two new dolabellanes (**40, 41**) were isolated from an unspecified *Clavularia* species that was also collected from a coral reef of Ishigaki Island (Fig. 8) [20]. The relative configuration of **40** and **41** was assigned from NMR spectroscopy and X-ray crystallography. The absolute configuration of the dolabellane **40** was suggested to be same as the configuration of claenone (**42**) which was isolated before from *Clavularia viridis* and whose absolute configuration was corroborated by total synthesis (see below). The absolute configuration of the hydroperoxide (**41**) was proved through chemical corre-

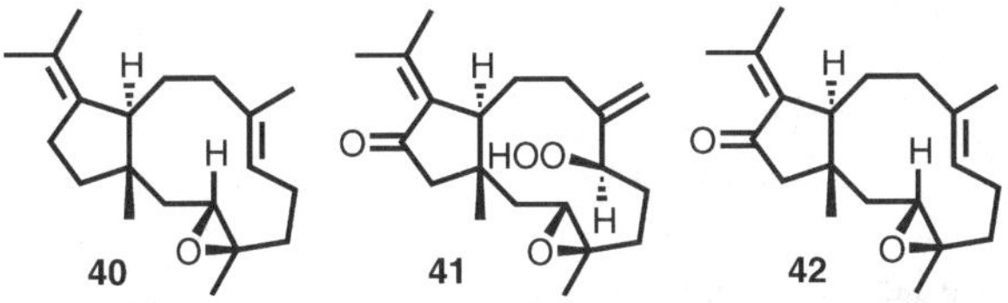

**Fig. 8** Two new dolabellanes (**40, 41**) from a *Clavularia* sp. (2002) and the known (−)-claenone (**42**)

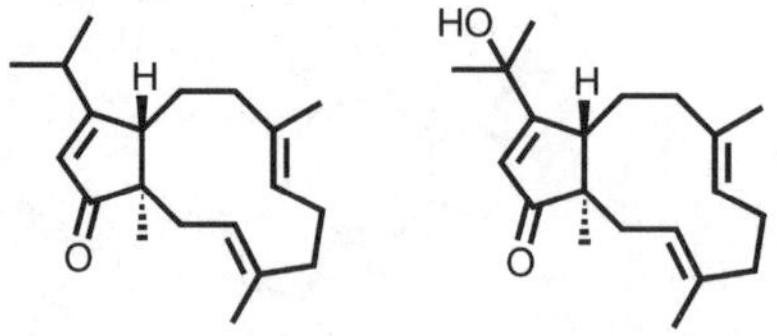

Fig. 9 Two new dolabellanes (43, 44) from a *Casearia membranacea* (2003)

lation with (−)-claenone (42). The observation that claenone is easily oxidized by air led the conclusion that the hydroperoxide (41) might be an artefact of the isolation process. This revelation raises the question whether the hydroperoxide (35, Fig. 4) is formed from the dolabellane 33 during the isolation process. The hydroperoxide (41) showed moderate cytotoxicity against various cancer cell lines according to the Japanese Foundation for Cancer Research 39 cell line assay.

The two new dolabellanes Casearimene A (43) and B (44) have been isolated from dried stems of the Asian *Casearia membranacea* (Family: *Flacourtiaceae*) (Fig. 9) [21]. *C. membranacea* is a small evergreen tree indigenous to Hainan Island (off the southeast coast of China) and the southern part of Taiwan. The structure and relative configuration was assigned by NMR and X-ray crystal structure analysis. Casearimene A (43) and B (44) did not show a significant cytotoxicity against P-388, A549 and HT-29 cancer cell lines in vitro.

## 3.2
## Total Synthesis of Dolabellanes and Neodolabellanes

The dolabellanes are a prominent family of natural products isolated from different marine and terrestrial sources. They possess a broad spectrum of biological activities. One expects that especially the dolabellanes from marine animals should be worthwhile targets for enantioselective total synthesis due to their biological activity and the unsolved supply issue. However, only a limited number of completed total syntheses have been reported since the structure of the first dolabellane (isolated from the sea hare *Dolabella californica* by Faulkner and Clardy) was published in 1976 [22]. Confined on completed total syntheses of natural dolabellanes, it appears that only five dolabellanes (38, 42, 46–49) and two neodolabellanes (45, 50) have been synthesized (Fig. 10). Approaches toward and completed total syntheses of artificial and natural dolabellanes that have been summarized in the leading review article of Rodrígez will not be covered [1, 23–34]. The progress in the field during the last five years will be summarized in the following sub-sections.

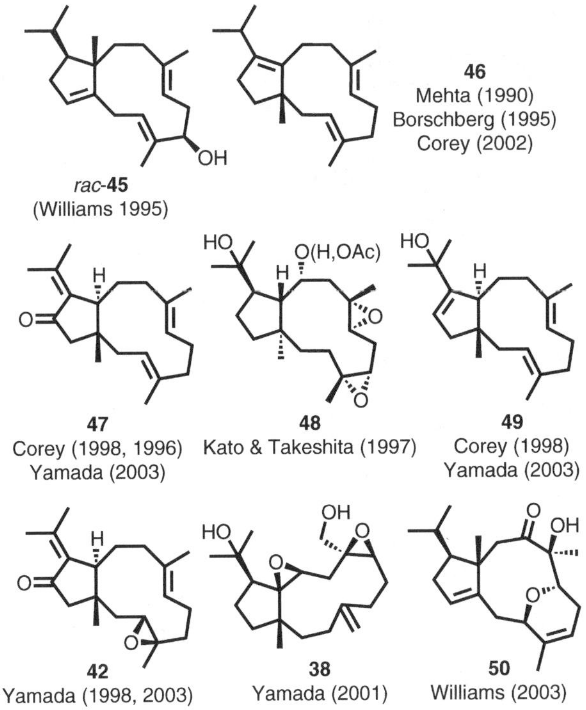

**Fig. 10** Naturally occurring dolabellanes for which total syntheses have been reported

## 3.2.1
## Yamada's Strategy

### 3.2.1.1
### (−)-Claenone (1998)

The research group of Yasuji Yamada utilized an effective sequence of two sequential Michael additions between the enolate of 3-methoxymethoxy-cyclopent-2-enone (**53**) and the easily accessible chiral $\alpha,\beta$-unsaturated ester **54** followed by a retro-aldol addition for the synthesis of the highly substituted cyclopentanone **52** (Fig. 11) [35, 36]. However, the efficiency of the strategy is hampered by a multitude of functional and protecting group transformations that were required for the conversion of a tricyclic Michael-Michael-addition product (**56**) into the desired cyclopentanone **52**. A significant synthetic effort was also required to realize the seemingly straightforward transformation of the cyclopentanone **52** into the sulfone **51**. An intramolecular alkylation of the sulfone **51** served as pivotal ring closing reaction for the synthesis of the 11-membered macrocycle.

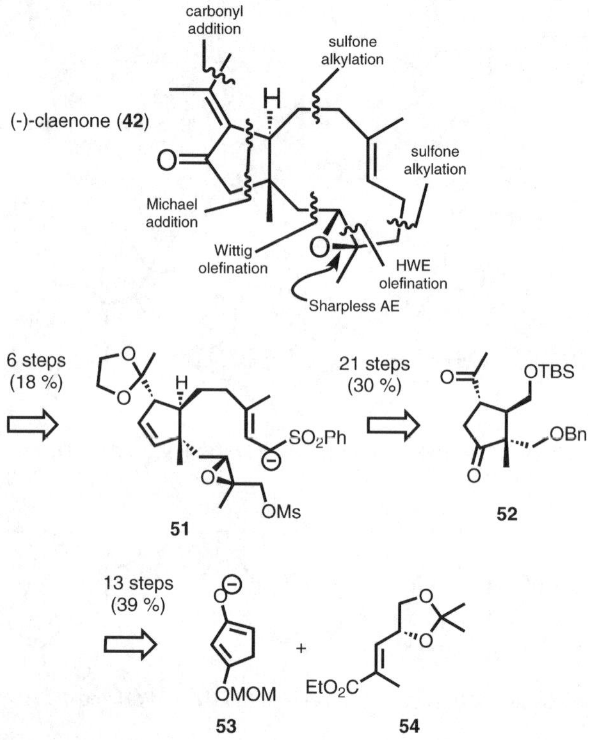

**Fig. 11** Retrosynthetic overview of Yamada's ex-chiral-pool total synthesis of (−)-claenone (**42**) (1998)

The detailed course of the sequential Michael/Michael addition between the enolate **53** and the α,β-unsaturated ester **54** is outlined in Scheme 5. Regioselective deprotonation of 3-methoxymethoxy-cyclopent-2-enone generated a cyclic enolate (**53**) that approached the α,β-unsaturated ester **54** preferentially following a (10S,11Si,12Re) topicity (natural product numbering is used throughout the review). The (11Si) attack could be favoured due to steric reasons based on a conformation of the α,β-unsaturated ester **54** that is dictated by 1,3-allylic strain. The first Michael addition generated an ester enolate (**55**) that underwent the second (intramolecular) Michael addition following a (1Re,14Si) topicity. The stereochemical course of the second Michael addition was determined by the intramolecularity (14Si) of the reaction and by a reactive conformation of the ester enolate moiety (1Re) that could be favoured due the chelation of the lithium cation between the enolate oxygen atom and the enone carbonyl oxygen atom and/or due to minimized 1,3-allylic strain. The sequence of two Michael additions afforded a 4/1 mixture of two diastereomers (**56**, **57**) that were separated by chromatography.

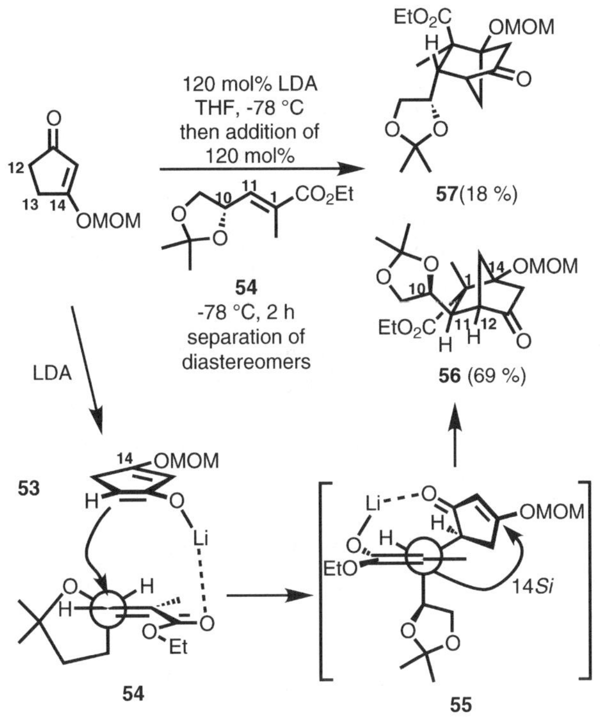

**Scheme 5** Stereochemical course of the sequential Michael/Michael addition

Continuing with the diastereomerically pure tricycle **56**, an 11-step sequence consisting of redox and protective group chemistry was necessary to generate a β-hydroxy keton (**58**) suitable for a retro-aldol addition via an intermediate alkoxide to the highly substituted cyclopentanone **52** (Scheme 6).

A further eight steps were required to convert the cyclopentanone **52** into the sulfone **59** that was deprotonated and treated with an allylic bromide (**60**) to afford the alkylated sulfone **61** (Scheme 7). The sulfone moiety and the benzyl ether protecting group were reductively removed in a one-pot procedure to afford a mono-protected diol (**62**).

Scheme 8 summarizes the introduction of the missing carbon atoms and the diastereoselective epoxidation of the $C^3/C^4$ double bond using a Sharpless asymmetric epoxidation (SAE) of the allylic alcohol **64**. The primary alcohol **62** was converted into the aldehyde **63** which served as the starting material for a Horner-Wadsworth-Emmons (HWE) reaction to afford an E-configured tri-substituted double bond. The next steps introduced the sulfone moiety via a Mukaiyama redox condensation and a subsequent sulfide to sulfone oxidation. The sequence toward the allylic alcohol **64** was com-

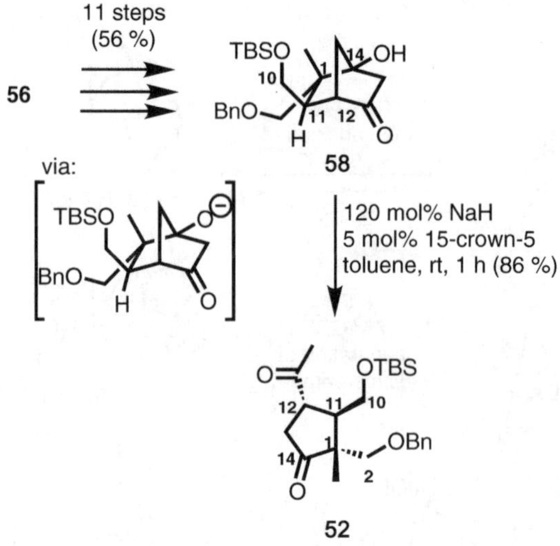

**Scheme 6** Set-up and realization of the retro-aldol addition

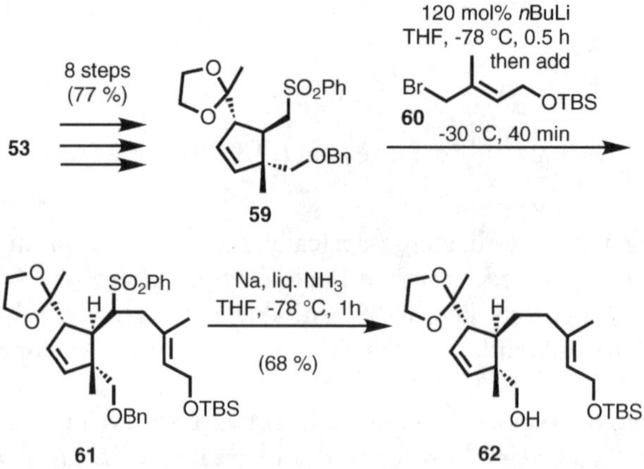

**Scheme 7** Stepwise construction of the macrocycle, part I: sulfone alkylation

pleted by DIBAH reduction of an intermediate α,β-unsaturated ester. The allylic alcohol **64** was then epoxidized following the Sharpless protocol to provide the oxirane **65** as a single diastereomer [37, 38].

Mesylation of the alcohol **65** followed by deprotonation afforded the sulfone-stabilized carbanion **66** that underwent a macrocyclization to afford the artificial dolabellane **67** in moderate yield (Scheme 9). Hydrolytic cleavage of the ketal (**67**) followed by a base-mediated double bond isomerization (into conjugation) afforded an enone containing an exocyclic carbonyl group. Nucleophilic 1,2-addition of methyl lithium introduced the missing

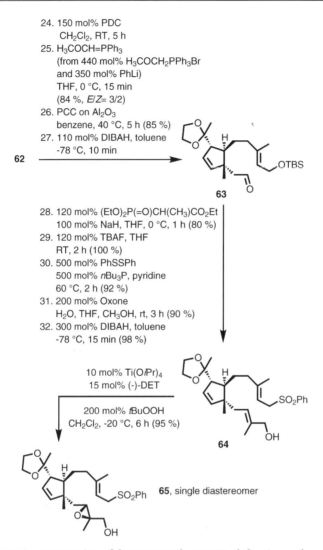

**Scheme 8** Stepwise construction of the macrocycle, part II: olefination and epoxidation

$C^{20}$ methyl group. The reductive removal of the sulfone moiety and subsequent $Cr^{VI}$-mediated oxidative transposition of the tertiary hydroxyl function concluded the enantioselective total synthesis of (−)-claenone (**42**) [39].

In conclusion, the longest linear sequence of Yamada's (−)-claenone (**42**) synthesis consist of 40 steps (6 C/C connecting transformation) with an overall yield of 2.1%. The centrepiece of Yamada's synthetic strategy is the sequence of two Michael additions and a retro-aldol addition to provide a highly substituted cyclopentanone building block (**52**).

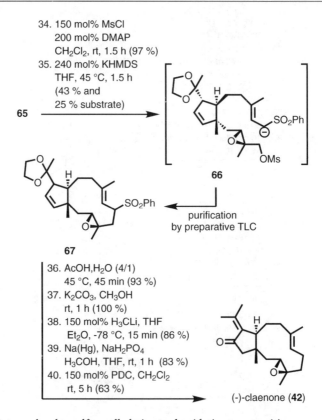

**Scheme 9** Intramolecular sulfone alkylation and oxidative transposition

## 3.2.1.2
### (−)-Stolonidiol (2001)

The structural similarity between claenone (**42**) and stolonidiol (**38**) enabled Yamada to exploit an almost identical strategy for the total synthesis of (−)-stolonidiol (**38**) [40]. A short retrosynthetic analysis is depicted in Fig. 12. An intramolecular HWE reaction of **68** was successfully applied for the macrocyclization. The highly substituted cyclopentanone **69** was made available by a sequence that is highlighted by the sequential Michael-Michael addition between the enolate **53** and the $\alpha,\beta$-unsaturated ester **70** followed by a retro-aldol addition. However, as is the case for the claenone (**42**) synthesis, the synthesis of stolonidiol (**38**) is characterized by numerous functional and protecting group transformations that are a consequence of Yamada's synthetic strategy.

The sequential Michael/Michael addition between 3-methoxymethoxycyclopent-2-enone and the $\alpha,\beta$-unsaturated ester **70**, which is accessible from ascorbic acid [41], afforded the tricycle **73** as a mixture of diastereomers (Scheme 10). The (1R) configuration was found to be kinetically favoured

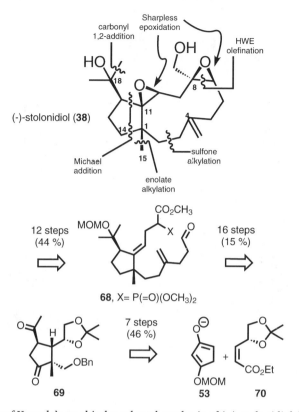

**Fig. 12** Survey of Yamada's ex-chiral-pool total synthesis of (−)-stolonidiol (**38**) (2001)

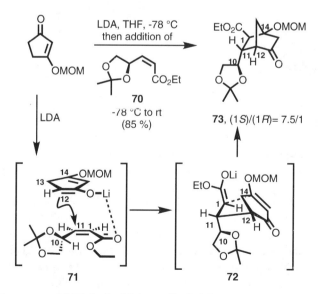

**Scheme 10** Two sequential Michael additions afforded the tricyclic building block **73**

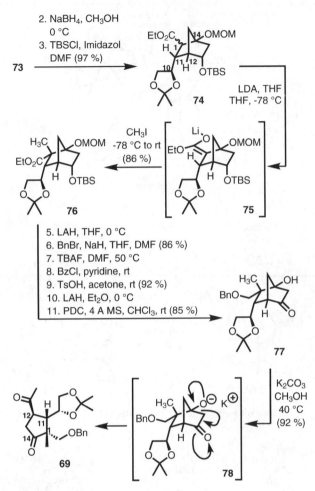

**Scheme 11** Enolate alkylation and retro-aldol addition as key steps toward the building block **69**

whereas the (1S) configuration was preferentially obtained under reaction conditions that should lead to the thermodynamically favoured product (warming the reaction mixture to room temperature).

The ketone **73** was reduced chemo- and diastereoselectively and protected to provide the silyl ether **74**. The ester function was then deprotonated to the corresponding ester enolate (**75**) that was alkylated with methyl iodide exclusively from the *Re* face of the enolate to afford the bicycle **76** (Scheme 11). The substrate for the retro-aldol reaction (**77**) was prepared by a sequence that consists of seven functional and protecting group transformations. The retro-aldol reaction converted the bicyclic β-hydroxy ketone **77** into the 1,3-diketone **69** via the alkoxide (**78**) in very good yield.

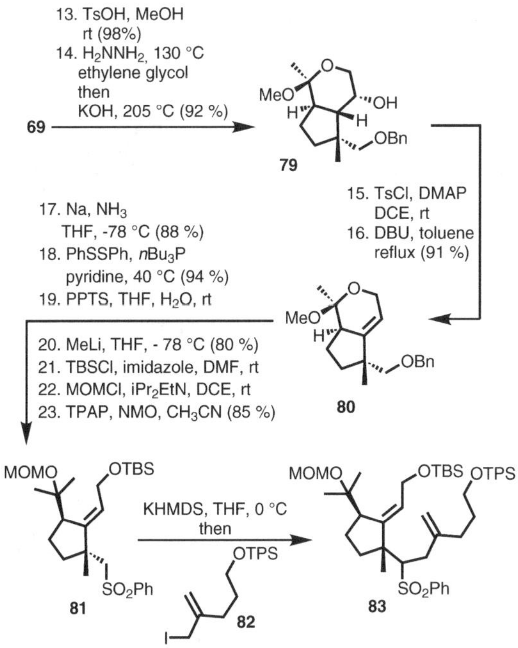

**Scheme 12** Functional and protecting group transformations and a sulfone alkylation for chain elongation

The exocyclic carbonyl function was protected as an intramolecular acetal (**79**) and the superfluous endocyclic carbonyl oxygen atom was reductively removed using the Huang-Minlon modification of the Wolff-Kishner reduction (Scheme 12). Tosylation and elimination afforded the allylic ether **80** with a remarkable regioselectivity. The next seven steps were devoted to convert the benzyl protected primary hydroxyl function into a sulfone moiety and to introduce the MOM as well as the TBS protecting group regioselectively (**81**). The sulfone (**81**) was then deprotonated and treated with an allylic iodide (**82**) to provide the corresponding alkylated sulfone (**83**).

The sulfone moiety was reductively removed and the TBS ether was cleaved chemoselectively in the presence of a TPS ether to afford a primary alcohol (Scheme 13). The alcohol was transformed into the corresponding bromide that served as alkylating agent for the deprotonated ethyl 2-(diethylphosphono)propionate. Bromination and phosphonate alkylation were performed in a one-pot procedure [33]. The TPS protecting group was removed and the alcohol was then oxidized to afford the aldehyde **68** [42]. An intramolecular HWE reaction under Masamune-Roush conditions provided a macrocycle as a mixture of double bond isomers [43]. The E/Z isomers were separated after the reduction of the $\alpha,\beta$-unsaturated ester to the allylic alcohol **84**. Deprotection of the tertiary alcohol and protection of the prima-

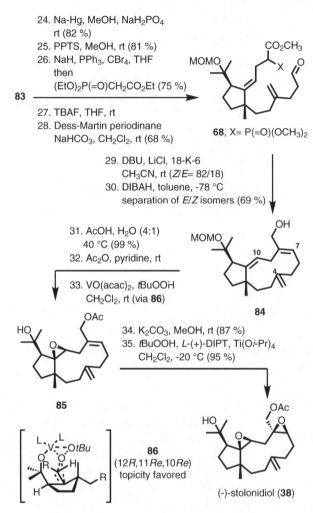

**Scheme 13** Intramolecular HWE reaction and Sharpless epoxidation as key transformations toward stolonidiol (**38**)

ry alcohol was accomplished next. A $V^{IV}$-catalysed substrate-directed epoxidation of the $C^{10}/C^{11}$ allylic alcohol double bond afforded the oxirane **85** [44]. Deprotection of the primary alcohol and subsequent SAE provided (−)-stolonidiol (**38**) [37, 38].

In conclusion, the ex-chiral-pool total synthesis of (−)-stolonidiol (**38**) required a longest linear sequence of 35 steps (6 C/C connecting transformations) with an overall yield of 3%.

### 3.2.1.3
### (−)-Palominol and (+)-Dolabellatrienone (2003)

Starting with a late intermediate (64) of the (−)-claenone (42) synthesis (Scheme 8), Yamada and coworker realized the total synthesis of (−)-palominol (49) and (+)-dolabellatrienone (47) as depicted in Scheme 14 [36]. Me-

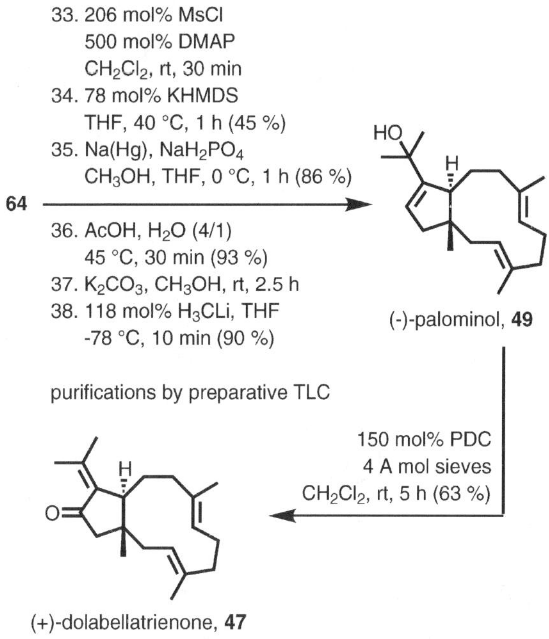

**Scheme 14** Final steps toward the total synthesis of (−)-palominol (49) and (+)-dolabellatrienone (47) (2003)

sylation of the allylic alcohol 64 followed by an intramolecular sulfone alkylation, reductive removal of the sulfone moiety, hydrolysis of the ketal, base-mediated double bond isomerization into conjugation and, finally, methyl lithium 1,2-addition to the ketone carbonyl group afforded (−)-palominol (49). The oxidative transposition as exercised during the claenone synthesis converted (−)-palominol (49) into (+)-dolabellatrienone (47) [39].

### 3.2.2
### Williams' Total Syntheses of Neodolabellanes

### 3.2.2.1
### (+)-4,5-Deoxyneodolabelline

The vast majority of total syntheses of dolastanes and dolabellanes rely on more or less linear strategies. The total synthesis of (+)-4,5-deoxyneodola-

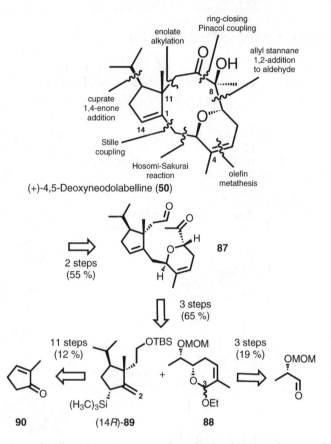

Fig. 13 Overview of Williams' convergent total synthesis of (+)-4,5-deoxyneodolabelline (50) (2003)

belline (50) represents a remarkable and convincing exception [45]. Figure 13 depicts a retrosynthetic survey. The longest linear sequence consists of 16 steps starting from 2-methyl-cyclopenten-2-one (90). The macrocyclization was realized by an intramolecular pinacol coupling of a keto aldehyde (87). The intermolecular Hosomi-Sakurai reaction between the allylic silane (14R)-89 and an oxocarbenium ion generated from the dihydropyran 88 is a particularly noteworthy transformation.

The synthesis of the non-racemic cyclopentanone (+)-93 is outlined in Scheme 15. Starting with 2-methyl-cyclopent-2-enone (90), sequential cuprate addition and enolate alkylation afforded the racemic cyclopentanone rac-92 as a single diastereomer. The double bond was cleaved by ozonolysis, the resulting aldehyde chemoselectively reduced in the presence of the keto function and the primary hydroxyl function was subsequently protected as a silyl ether to provide racemic rac-93. This sequence has been applied fre-

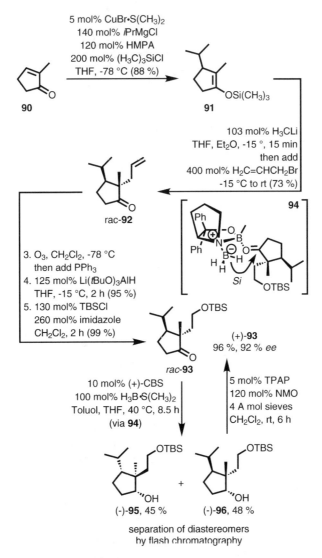

**Scheme 15** Synthesis of the non-racemic cyclopentanone (+)-93

quently in the total synthesis of terpenes (e.g. (±)-Crinipellin B [46], see Schemes 19, 30 and 35 for further examples). Racemic *rac*-93 was then converted into the enantiomerically enriched cyclopentanone (+)-93 utilizing the Corey-Bakshi-Shibata reduction [47–49]. In the event, the CBS-catalyst reduced the racemic cyclopentanone *rac*-93 in the presence of the borane dimethylsulfide complex to a nearly one to one mixture of the diastereomeric cyclopentanols (−)-95 and (−)-96. The reduction proceeded with a com-

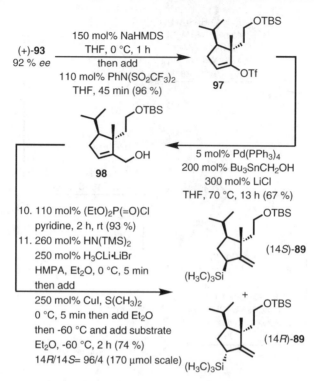

**Scheme 16** Synthesis of the allylic silane **89** by an $S_N2'$ displacement of an allylic phosphate

plete catalyst-induced diastereoselectivity (via **94**). Fortunately, the resulting diastereomers (**95, 96**) were separable by flash chromatography. Oxidation of the cyclopentanol (−)-**96** afforded the desired enantiomerically enriched (92% *ee*) cyclopentanone (+)-**93**.

The introduction of the allylic silane moiety required for the intermolecular Hosomi-Sakurai reaction is depicted in Scheme 16. Following the formation of the enol triflate **97**, a Stille coupling provided excess to the allylic alcohol **98** [51]. The allylic alcohol (**98**) was endowed with a phosphate leaving group for the subsequent allylic substitution. Utilizing a trimethylsilyl cuprate as nucleophile for the $S_N2'$ reaction, the allylic phosphate was converted into the allylic silane **89**. A useful substrate-induced diastereoselectivity in favour of (14*R*)-**89** was encountered at small scale but decreased significantly upon up-scaling.

The dihydropyran **88** served as the precursor for an oxocarbenium ion that was utilized as the acceptor for the intermolecular Hosomi-Sakurai reaction [53, 54]. Utilizing a second Hosomi-Sakurai reaction, pyran **88** was synthesized as outlined in Scheme 17 [53, 54]. Easy accessible MOM protect-

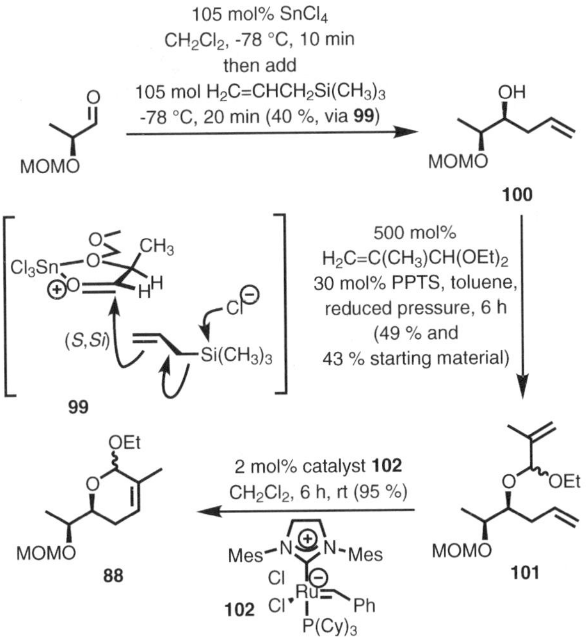

**Scheme 17** Synthesis of the dihydropyran **88** from MOM-protected lactic aldehyde (**91**)

ed lactic aldehyde was treated with an allylic silane in the presence of SnCl$_4$ to provide the homoallylic alcohol **100** as a single diastereomer in low yield. The substrate-induced diastereoselectivity can be explained by the reaction between a chelated aldehyde moiety and the allylic silane through an open antiperiplanar arrangement (**99**). One could assume that the unfortunate combination of a strong Lewis acid and a labile acetal protecting group was responsible for the low yield. Nevertheless, following the formation of the mixed acetal **101**, the ring closing olefin metathesis using Grubb's N-heterocyclic carbene catalyst (**102**) afforded the dihydropyran **88** as a 1/1 mixture of diastereomers [55, 56].

The intermolecular Hosomi-Sakurai reaction between the allylic silane (14R)-**89** and the oxocarbenium ion (see **103**) generated from the acetal **88** by treatment with a substoichiometric amount of BF$_3$ was employed to connect the A- and the B-ring (Scheme 18) [53, 54]. The substrate-induced diastereoselectivity was independent of the absolute configuration at C$^{14}$ of the allylic silane in so far the (14S)-configured allylic silane (14S)-**89** afforded the product **104** with same absolute configuration at C$^3$. The TBS ether and the MOM acetal were then cleaved under acidic conditions. Oxidation of the resulting diol afforded the keto aldehyde **87**. The decisive ring-closing pinacol coupling reaction was realized by treatment of the keto aldehyde **87** with low valent titanium generated by the reduction of Ti$^{III}$ with the zinc-

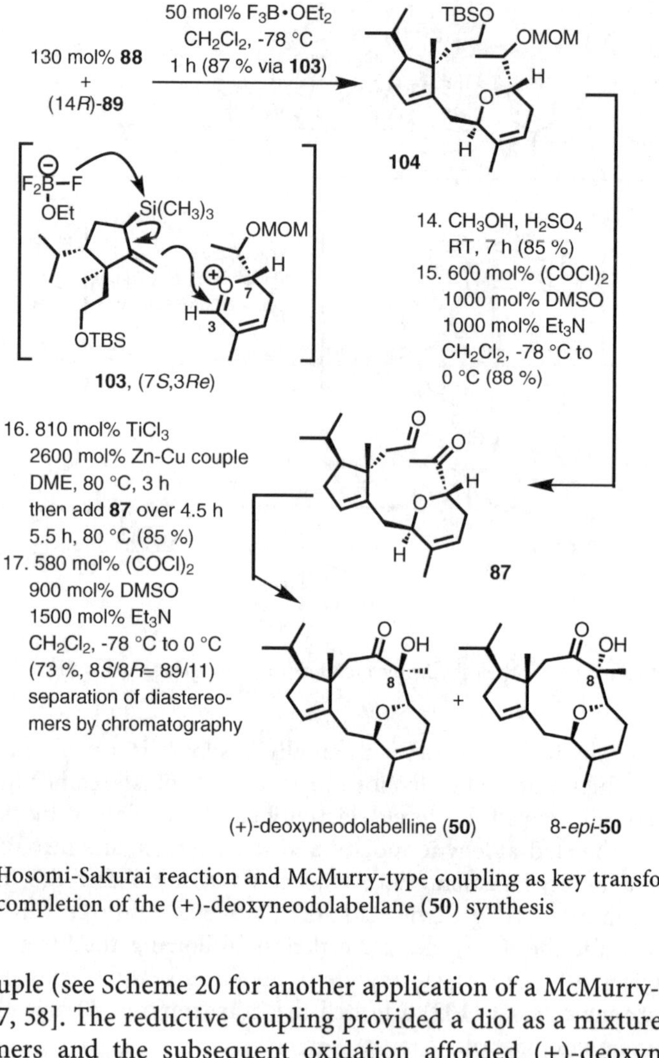

**Scheme 18** Hosomi-Sakurai reaction and McMurry-type coupling as key transformations toward the completion of the (+)-deoxyneodolabellane (**50**) synthesis

copper couple (see Scheme 20 for another application of a McMurry-type reaction) [57, 58]. The reductive coupling provided a diol as a mixture of four diastereomers and the subsequent oxidation afforded (+)-deoxyneodolabelline (**50**) and the $C^8$ epimer (8-*epi*-**50**) in a ratio of 89/11.

### 3.2.2.2
### (±)-Neodolabellenol

The total synthesis of racemic neodolabellenol *(rac-**45**)* was reported by the Williams group in 1995 as a short communication [59]. To gain a complete overview, it is advisable to consider a publication from 1993 in which Williams described the synthesis of an advanced precursor as well as the aforementioned full publication from 2003 [45, 60]. As summarized in Scheme 19, the synthesis of the bicyclic neodolabellenol (**45**) utilized an in-

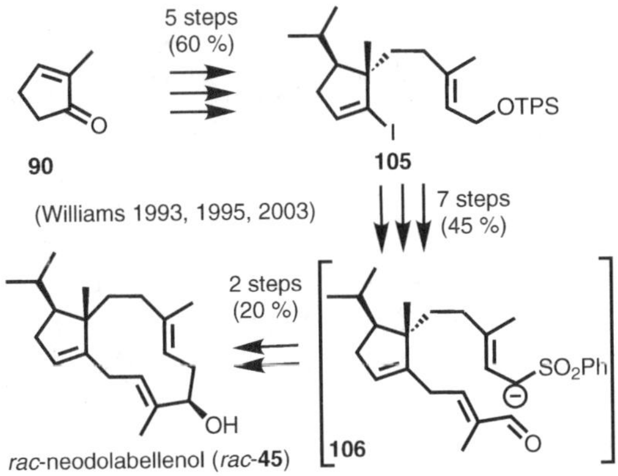

**Scheme 19** Summary of the synthesis of (±)-neodolabellenol according to Williams (1995)

tramolecular nucleophilic 1,2-addition between a sulfone-stabilized carbanion and an $\alpha,\beta$-unsaturated aldehyde (**106**). The macrocyclization proceeded with poor yield (25–35%) but acceptable diastereoselectivity (6/1, details not depicted). The details for the synthesis of the vinyl iodide **105** from 2-methyl-cyclopent-2-enone (**90**) can be found in Williams' full publication from 2003 [45].

# 4
# Dolastane Diterpenes

## 4.1
## Isolation, Structure and Biological Activities

About 25 diterpenes (**107–130**) of the dolastane-type featuring a hydrobenzo[f]azulene ring system have been isolated exclusively from marine sources (Fig. 14) [61–76].

The structural diversity of the dolastanes rests on the number and the position of double bonds and hydroxyl functions within the basic carbon skeleton. The dolasta-1(15),8-diene (**131**) represents the dominating carbon skeleton (Fig. 15).

The *trans* configuration of the two angular methyl groups at $C^5$ and $C^{12}$ is conserved. The BC-ring system is usually *trans* fused. The vast majority of the dolastanes have been isolated from brown soft algae of the genus *Dictyota* (family: *Dictyotaceae*). The relative configuration has been safely assigned as depicted by extensive use of X-ray crystal structure analysis and NMR studies. The absolute configuration was deduced for selected dolastanes by X-ray analysis (**119, 127**) and c.d. data (**110**) as well as enantioselec-

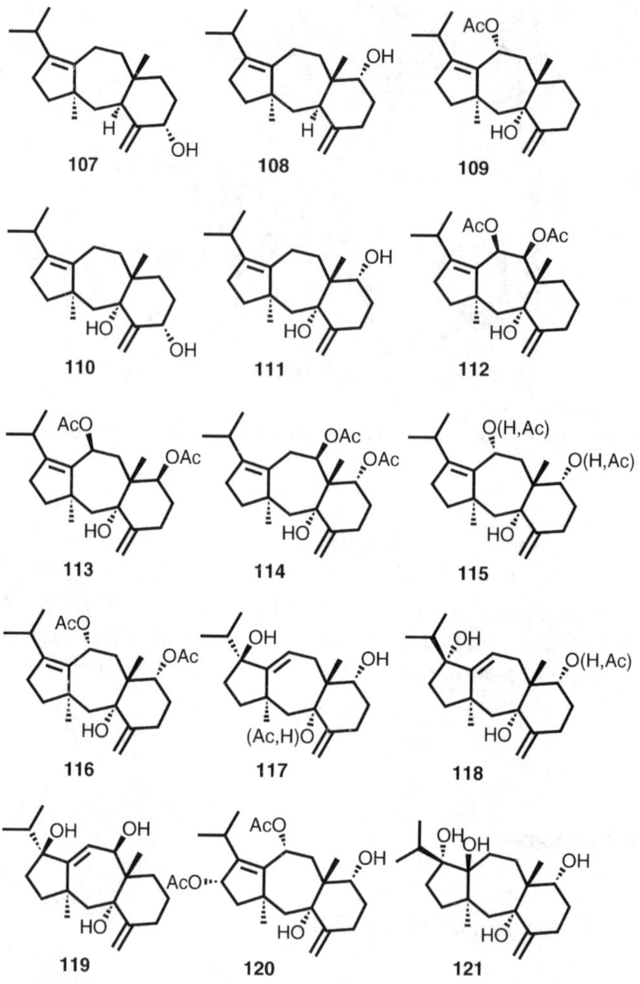

**Fig. 14** Naturally occurring dolastanes

tive total synthesis (**107, 129**). The first dolastane diterpene (**119** and 6-Ac-**119**) was isolated in 1976 from the sea hare *Dolabella auricularia* [61]. It was claimed that dolastanes isolated from the sea hare are of algal origin because *Dolabella sp.* is known to crop algae. Antimicrobial and cytotoxic properties of dolastanes have been described. Interestingly, diterpenes with the dolastane-1(15),17-diene framework (**132**) have been isolated from the soft coral *Clavularia inflata* (order: *Alcyonacea*, family: *Clavulariidae*) collected off the north coast of Papua-New Guinea [62, 65]. These dolastanes (**128–130**) were named clavularanes and it was shown by empirical methods and enantioselective total synthesis that they possess the opposite absolute

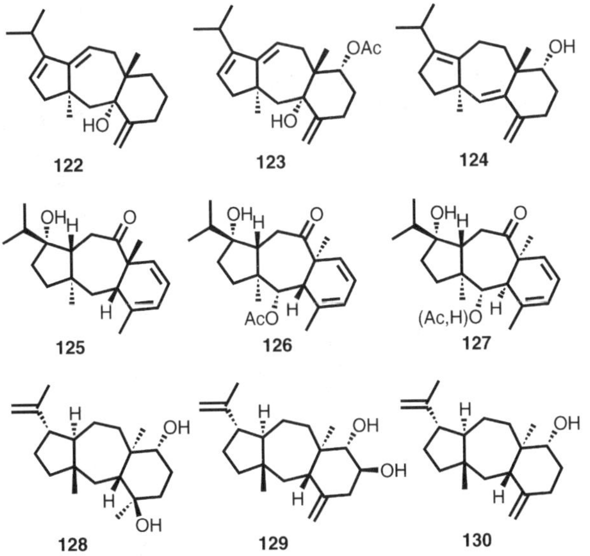

**Fig. 14** (continued)

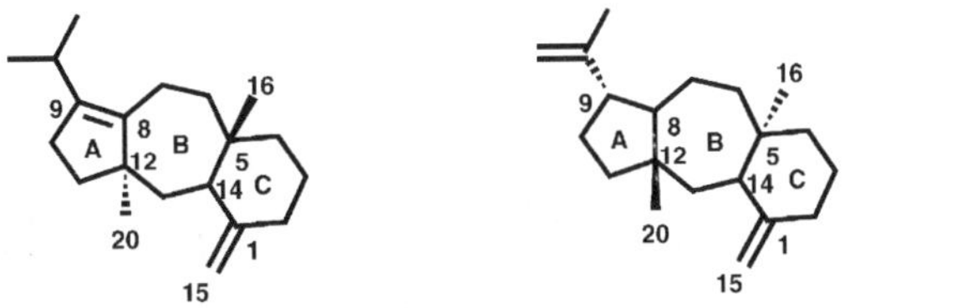

**Fig. 15** Basic carbon skeleton of dolastanes

configuration compared to the dolastanes isolated from *Dicyota* or *Dolabella*.

## 4.2
### Total Syntheses of Dolastanes

Between 1986 and 1993, the research group of Pattenden ((±)-**110**, 1986), Paquette ((±)-7,4-*epi*-**109**, 1986), Piers ((±)-**122** and (±)-**124**, 1986 and 1988), Mehta (*ent*-**110**, *ent*-**122**, 1987), Majetich ((±)-**107**, 1991) and Williams (*ent*-**129**, 1993) reported completed total syntheses of dolastanes. The total syntheses published earlier than 1993 will not be covered extensively in this review. Nevertheless, a brief summary of the "earlier" strategies toward the dolastane carbocyclic ring system is instructive and will be outlined in the following section.

The first total synthesis of a racemic dolastane (*rac*-110) was apparently published by Pattenden as a short communication 1986 followed in 1988 by a full paper [77, 78]. Pattenden's strategy is based on an intramolecular [2+2]-photocycloaddition/cyclobutane fragmentation to transform the diene 133 into the hydroazulene 136 (Scheme 20). In between the cycloaddition

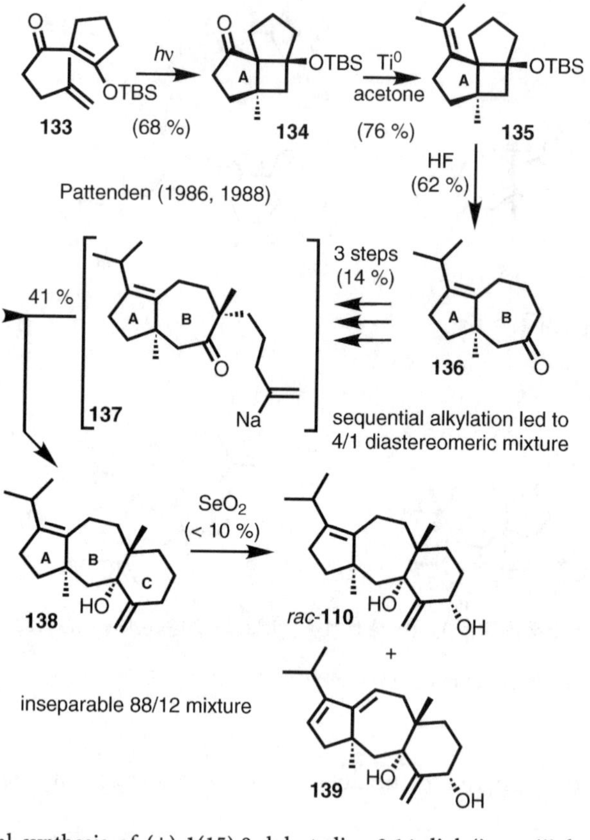

**Scheme 20** Total synthesis of (±)-1(15),8-dolastadien-2,14-diol (isoamijiol, 110) according to Pattenden (1986)

and the fragmentation, the required isopropyl group was introduced by a McMurry reaction between the ketone 134 and acetone in the presence of in situ generated titanium(0) (prepared from $TiCl_3$ and Li, see Scheme 18 for another application of the McMurry reaction) [57, 58]. Sequential regioselective alkylation of the hydroazulene 136 afforded a terminal alkyne that served as substrate for a reductive cyclization using the sodium naphthalene radical anion as reagent to generate the vinyl sodium intermediate **137**. The annulation afforded exclusively the *trans* fused ring system albeit in low chemical yield. Though not explicitly mentioned in the article, one can imagine that a competing acid/base reaction between the vinyl anion and the keto function may be responsible for the low yield. This reductive cyclization of an in situ generated vinyl anion has been frequently utilized in diterpene syntheses (see Schemes 21 and 31 for further applications). The non-natural (±)-1(15),8-dolastadien-14-ol (**138**) was converted into the natural (±)-1(15),8-dolastadien-2,14-diol (isoamijiol, **110**) by allylic oxidation with SeO$_2$ and *tert*-butyl hydroperoxide [79]. This procedure is obviously hampered by a very low yield and the formation of the non-natural (±)-1(15),7,9-dolastatrien-2,14-diol (**139**). Furthermore, the separation of the two dolastanes could not be achieved by Pattenden and coworkers (but compare to Scheme 24).

The total synthesis of the natural (±)-1(15),7,9-dolastatrien-14-ol ((±)-**122**) was reported in 1986 by Piers and Friesen (Scheme 21) [80]. The hydroazulene **143** served as key intermediate and the C-ring was annulated by an intramolecular Grignard reaction of the vinyl magnesium ketone **144** to provide the desired dolastane (±)-**122** in moderate yield as a single

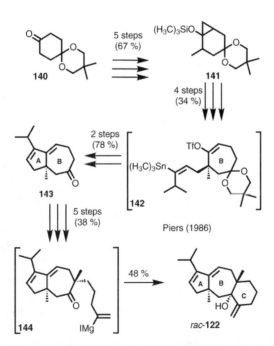

**Scheme 21** Total synthesis of (±)-1(15),7,9-dolastatrien-14-ol (*rac*-**122**) according to Piers (1986)

diastereomer. Sequential alkylation of the anion derived from the dimethyl hydrazone of the ketone **143** provided the precursor for the intramolecular Grignard addition. The hydroazulene **143** was synthesized by an intramolecular Stille coupling of the stannane **142** [50, 51]. A ring opening reaction of the bicyclic siloxy cyclopropane **141** served as key step for the synthesis of the vinyl triflate **142**. The siloxy cyclopropane **141** was synthesized from the cyclohexanone **140** using a cyclopropanation under Furukawa's conditions as the key transformation [81].

Starting with the intermediate (±)-**145** of the aforementioned total synthesis of (±)-**122**, Piers realized the synthetic access to the racemic dolastane (±)-**124** (Scheme 22) [82, 83]. An intramolecular Stille coupling served as key step, this time for the annulation of the C-ring.

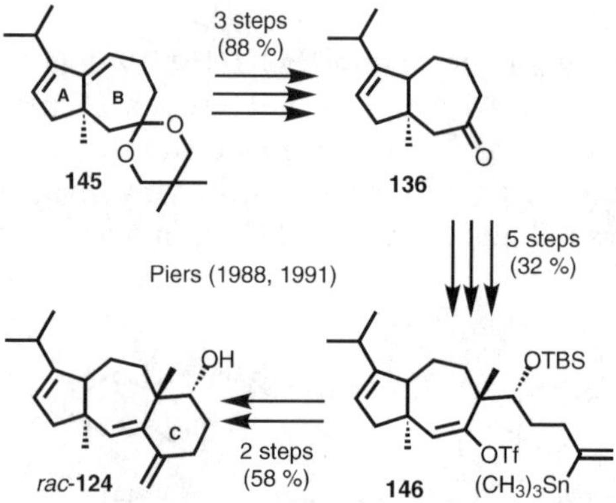

**Scheme 22** Total synthesis of (±)-1(15),9,13-dolastatrien-4-ol (*rac*-**124**) according to Piers (1988)

The synthesis of the non-natural (±)-7,14-*epi*-1(15),8-dolastadien-7,14-ol (*rac*-7,14-*epi*-**109**) was published by Paquette in 1986 and is highlighted by a photochemical rearrangement of the 6,6,6-tricyclic α,β-epoxy ketone **148** into the 5,7,6-tricyclic dolastane skeleton (**149**) (Scheme 23) [84]. The succeeding hydroxylation of carbon atom $C^{14}$ by photo oxygenation with singlet oxygen as well as a DIBAH reduction of a $C^7$ keto function proceeded with an undesired substrate-induced diastereoselectivity to provide the racemic 7,14-epimer of the natural dolastane **109**.

The enantioselective ex-chiral-pool synthesis of the enantiomer of the natural (+)-dolasta-1(15),7,9-trien-14-ol (*ent*-**122**) was achieved by Mehta and coworkers in 1987 (Scheme 24) [85, 86]. From the hydroazulene **136**, the synthesis proceeded analogous to the synthesis of Pattenden (Scheme 20) and provided the dolastane *ent*-**122** as a mixture with the non-natural dolastanes **153** and **154**. However, in contrast to Pattenden's work, Mehta and

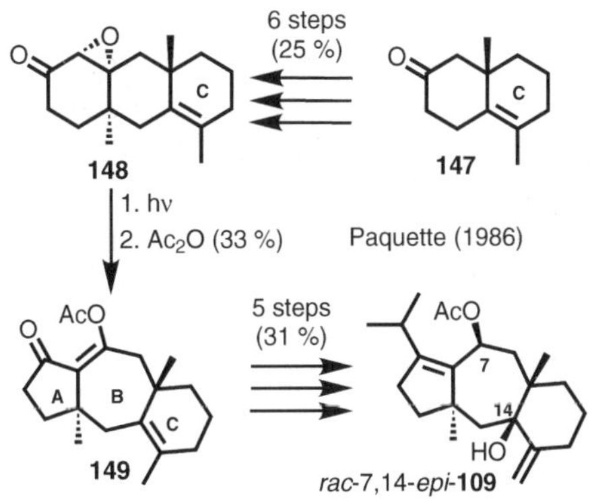

**Scheme 23** The synthesis of the non-natural (±)-7,14-*epi*-1(15),8-dolastadiene-7,14-ol (*rac*-7,14-*epi*-**109**) according to Paquette (1986)

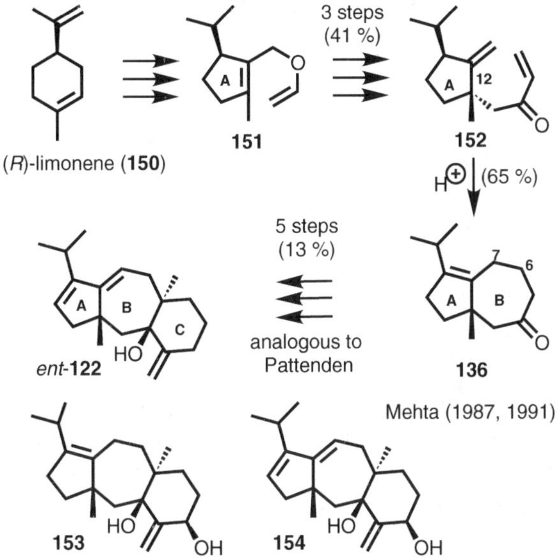

**Scheme 24** Synthesis of (+)-dolasta-1(15),7,9-trien-14-ol (*ent*-**122**) achieved by Mehta (1987)

coworkers were able to separate the three dolastanes (**122**, **153**, **154**) through "repeated column chromatography on $AgNO_3$-$SiO_2$". The ex-chiral-pool strategy of Mehta utilized (*R*)-limonene (**150**) and converted it into the allyl vinyl ether **151**. Thermal Claisen rearrangement of **151** proceeded with complete substrate-induced diastereoselectivity to afford the dienone **152** that was converted into the hydroazulene **136** by an acid-promoted ene type cyclization.

A B-ring-last-strategy was applied by Majetich and coworkers for their synthesis of the natural (±)-1(15),8-dolastadien-2-ol (±)-**107** (desoxyisoamijiol, 1991) [87, 88]. A Lewis acid mediated intramolecular allylsilane 1,6-addition of the dienone **157** afforded the 5,7,6-tricyclic carbon framework of the dolastanes (**158**) (Scheme 25). The missing $C^{15}$ carbon atom was intro-

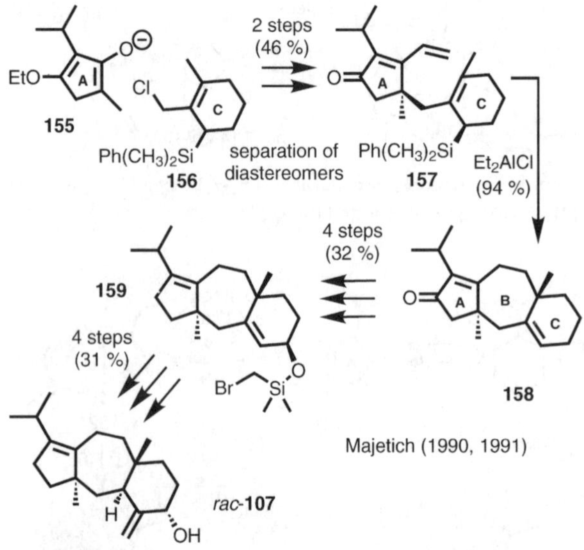

**Scheme 25** Total synthesis of (±)-1(15),8-dolastadien-2-ol (*rac*-**107**, desoxyisoamijiol) according to Majetich (1991)

duced by an intramolecular addition of a radical generated from the silicon-tethered bromomethylene moiety in **159** onto $C^1$ of the neighboured double bond [89]. The configuration of the hydroxyl function at $C^2$ was inverted by a sulfoxide-sulfenate rearrangement [90].

## 4.2.1
## Williams' Total Synthesis

The apparently latest total synthesis of a dolastane diterpene was published by Williams and coworkers in 1993 as a short communication (Fig. 16) [91]. (−)-Clavulara-1(15),17-dien-3,4-diol (**129**) was synthesized using a strategy that relied on the availability of the enantiomerically pure building block **162** from (+)-9,10-dibromocamphor (**163**) (Fig. 16). Cornerstones of the synthesis are a macrocyclization that afforded the 11-membered (A+B)C-ring (**160**) and a transannular cyclization that converted a bicyclic into a tricyclic ring system. Two of the seven chirality centres in the synthetic clavu-

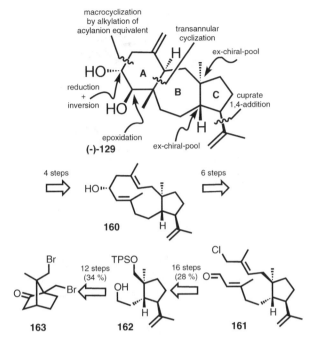

**Fig. 16** Synthetic strategy toward (−)-clavulara-1(15),17-dien-3,4-diol (**129**) by Williams (1993)

larane (−)-**129** were donated from the chiral pool and the relative configuration of the remaining five stereogenic carbon atoms was controlled by substrate-induced diastereoselectivity.

Following Money's procedure, (+)-9,10-dibromocamphor (**163**) was directly converted into the cyclopentane **165** featuring the correct relative and absolute configuration. (Scheme 26) [92]. The A-ring building block (**165**) contained the crucial quaternary chiral carbon atom $C^{12}$ and a $C_2$ unit at $C^8$. The exo-methylene group of **165** was utilized to introduce the required isopropenyl unit at $C^9$. However, the corresponding procedure required a lengthy reaction sequence and diminished the efficiency of the otherwise remarkable exploitation of an ex-chiral-pool strategy. The hydroxy acid **165** was transformed into the cyclopentanone **166** by a sequence consisting of protecting group and redox chemistry. Saegusa oxidation of **166** followed by a diastereoselective 1,4-addition of an isopropenyl cuprate afforded the cyclopentanone **167** [93]. Finally, an intramolecular iodo etherification protected the isopropenyl double bond and afforded the bicycle **169**.

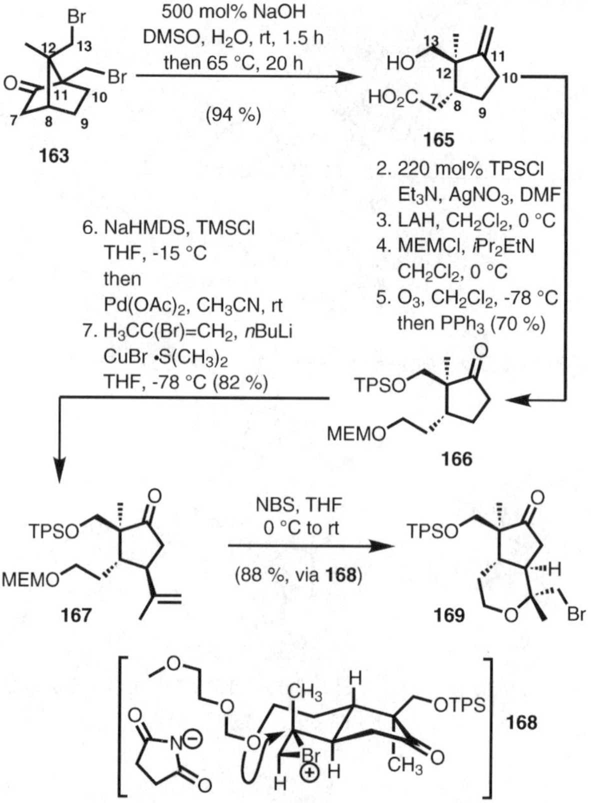

**Scheme 26** Exploitation of (+)-9,10-dibromocamphor (**163**) as ex-chiral-pool starting material

The superfluous $C^{11}$ carbonyl oxygen atom was removed by carbonyl reduction to provide the $C^{11}$ alcohol **171**, subsequent Chugaev elimination (via **172** to **173**) and double bond hydrogenation with in situ generated diimide (Scheme 27) [94]. The isopropenyl double bond was finally re-established by reductive cleavage of the α-bromo ether unit in **173** to afford the fully functionalized enantiomerically pure A-ring building block (**162**).

A multi-step reaction sequence was then realized to prepare the precursor (**178**) for the pivotal macrocyclization reaction. Alternate stepwise chain elongations were achieved according to Schemes 28 and 29. Reaction of the tosylate prepared from the alcohol **162** with lithium acetylide afforded the alkyne **174** (Scheme 28). Following the introduction of a tosylate at the upper branch, a one-carbon chain elongation of the terminal alkyne afforded the methyl alkynoate **175**. A methyl cuprate 1,4-addition was used to construct the tri-substituted $C^{5/4}$ double bond stereoselectively. For this purpose, the alkynoate **175** was initially transformed into the Z-configured α,β-unsat-

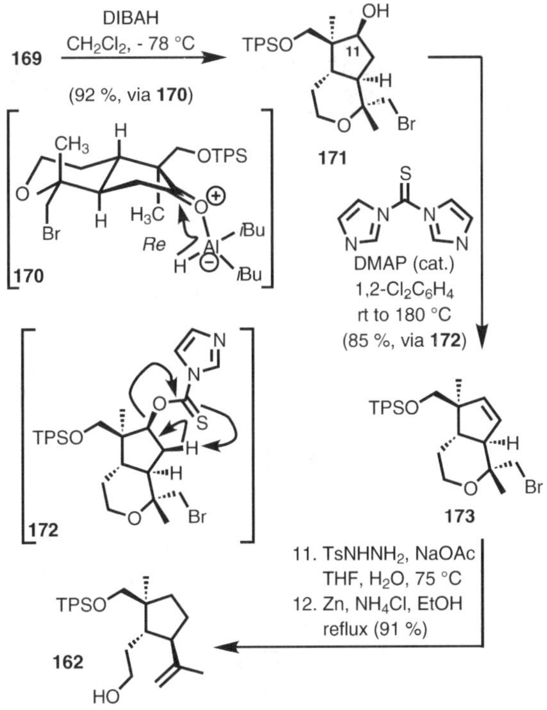

**Scheme 27** Removal of the superfluous $C^{11}$ carbonyl oxygen atom

urated β-phenylthio-substituted ester **176** by treatment with potassium thiophenolate under conditions that left the primary tosylate at the upper branch untouched [95, 96]. Methyl cuprate 1,4-addition/thiophenolate elimination proceeded under retention of the double configuration to afford the Z-configured enoate **177** which was reduced to an allylic alcohol [95, 96]. One carbon chain elongation at the upper branch was realized by tosylate displacement with cyanide followed by reduction to provide the hydroxy aldehyde **178**.

Olefination of the Aldehyde **178** using a stabilized Wittig reagent followed by protecting group chemistry at the lower branch and reduction of the α,β-unsaturated ester afforded the allylic alcohol **179** (Scheme 29). The allylic alcohol **179** was then converted into an allylic chloride and the hydroxyl function at the lower branch was deprotected and subsequently oxidized to provide the corresponding aldehyde **161** [42]. The aldehyde **161** was treated with trimethylsilyl cyanide to afford the cyanohydrin that was transformed into the cyano acetal **180**. The decisive intramolecular alkylation was realized by treatment of the cyano acetal **180** with sodium bis(trimethylsilyl)amide. Subsequent treatment of the alkylated cyano acetal **182** with acid (to **183**) and base afforded the bicyclo[9.3.0]tetradecane **184**.

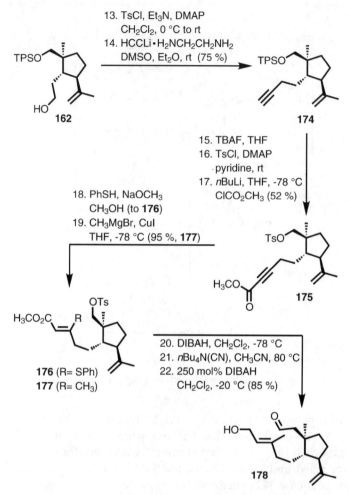

**Scheme 28** Stepwise chain elongation toward the macrocycle

Reduction of the $C^3$ ketone afforded the alcohol with the undesired absolute configuration. Nevertheless, Mitsunobu inversion with subsequent reduction delivered the desired (3R)-configuration of the allylic alcohol **160** (Scheme 30) [98, 99]. The regioselective epoxidation of the $C^5/C^4$ double was achieved by exploiting the directing effect of the allylic hydroxyl group [44]. Protonation of the epoxide (**185**) followed by ring opening to a tertiary carbocation triggered the transannular cyclization event. The positive outcome of the transannular reaction was made possible by the rigidity of the 11-membered ring and the proximity between the participating functional groups. Separation of the regioisomers and saponification finally afforded the two regioisomeric dolastanes *ent*-**129** and **186** in moderate yield. A com-

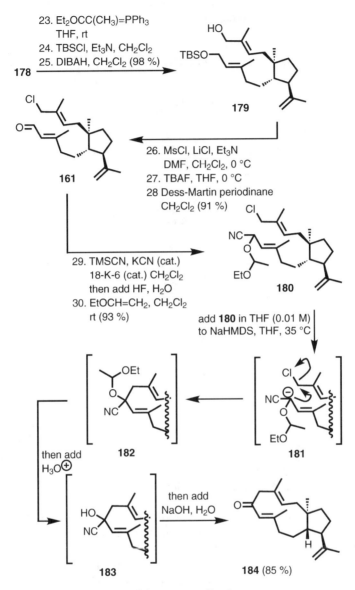

**Scheme 29** Set-up and realization of the macrocyclization

parison of the optical rotation of the synthetic dolastane (−)-**129** and the natural product (+)-**129** isolated from the soft coral *Clavularia inflata* proved the absolute configuration of the natural product as depicted in Fig. 14.

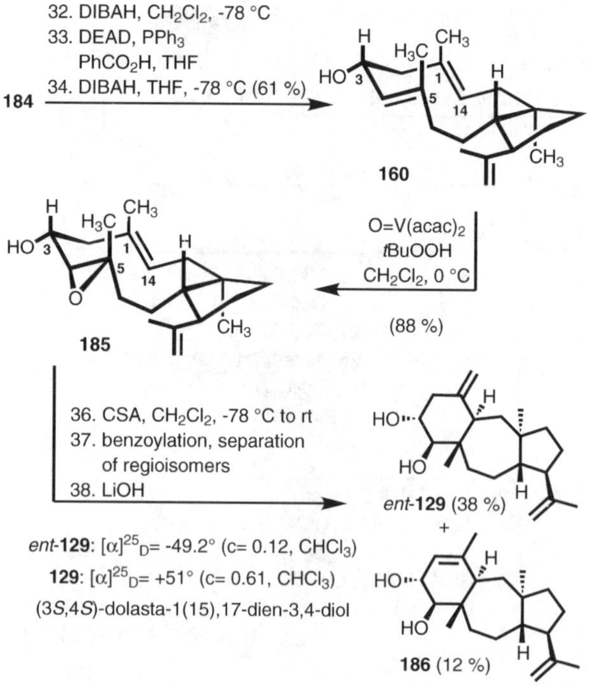

**Scheme 30** Transannular cyclization as key transformation toward *ent*-129 (Williams 1993)

# 5
# Neodolastane Diterpenes

## 5.1
## Isolation, Structure and Biological Activities

Tropical plants contain an unprecedented richness and abundance of fungal endophytes [100]. Consequently, the search for new secondary metabolites of endophytic fungi from tropical plants is very promising. In 2000, Clardy and coworkers isolated the secondary metabolite guanacastepene A ($G_A$, 187) from a culture of the endophytic fungus CR115 (Fig. 17) [101]. The fungus was isolated from a branch of a *Daphnopsis amercina* tree (*Thymeleaceae*) from the Area de Conservación Guanacaste in the extreme northwest of the province of Guanacaste (Costa Rica). Guanacastepene A (187) was the first characterized member of the family of neodolastane diterpenes. Phylogenetic studies suggest that CR 115 is a member of the class basidiomycete (kingdom fungi, phylum *basidiomycota*). Endophytic fungi intracellularly colonize plant host tissues without causing visible adverse effects. Depending on the participating organisms, the endophytic mycotoxins can protect

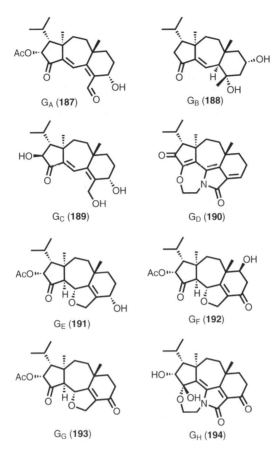

**Fig. 17** Guanacastepenes (187–201) produced by endophytic fungi from *Daphnopsis amercina*

the woody plant host against herbivores, nematodes and pathogens. The CR 115 fermentation extract showed moderate antibacterial activity against methicillin-resistant *Staphylococcus aureus* and vancomycin-resistant *Entereococcus faecium* as well as moderate antifungal activity against *Candida albicans*. However, the strong membrane damaging effects of guanacastepene A (**187**) on *Escherichia coli imp* and human red blood cells as well as the poor activity against Gram-negative bacteria clearly limit the pharmacological potential of guanacastepene A (**187**) as an antibiotic [102]. Nevertheless, guanacastepenes remain attractive synthetic targets due to the novelty of their structure and the possibility of yet unexplored biological activities. Significantly, the cultured fungus no longer produces guanacastepene A (**187**). The structure of guanacastepene A (**187**) was elucidated by NMR spectroscopy and X-ray crystallography. Further investigation of extracts from CR 115

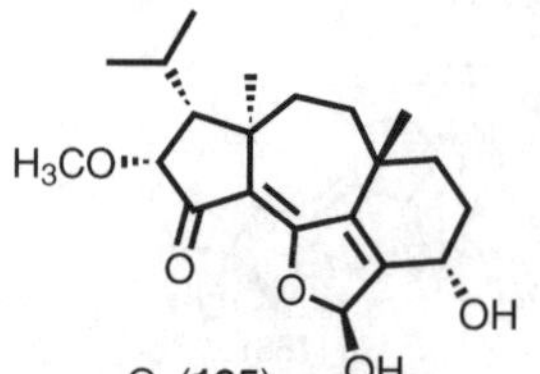

G$_I$ (**195**) 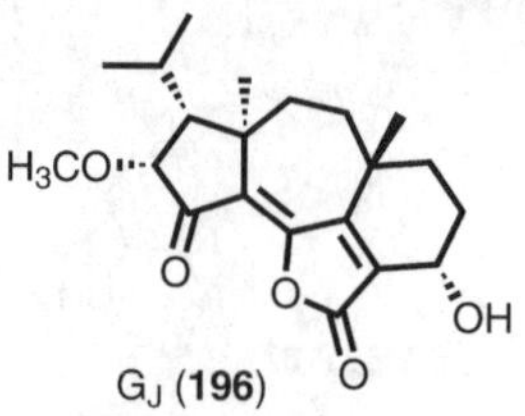　　G$_J$ (**196**)

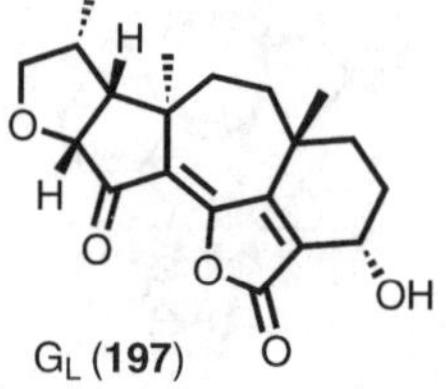

G$_L$ (**197**) 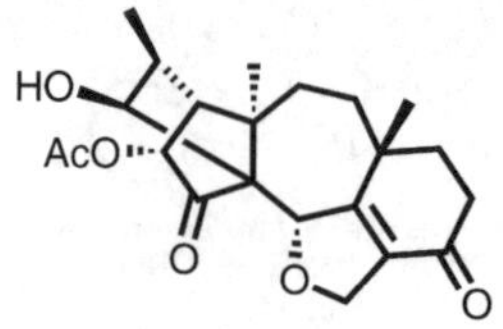　　G$_K$ (**198**)

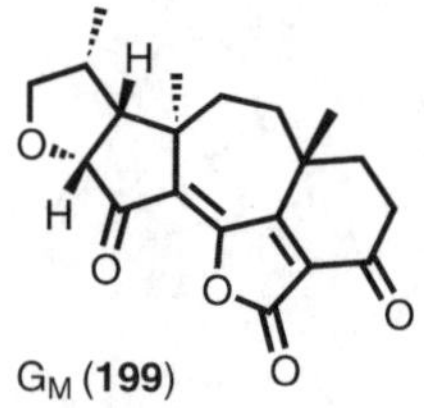

G$_M$ (**199**) 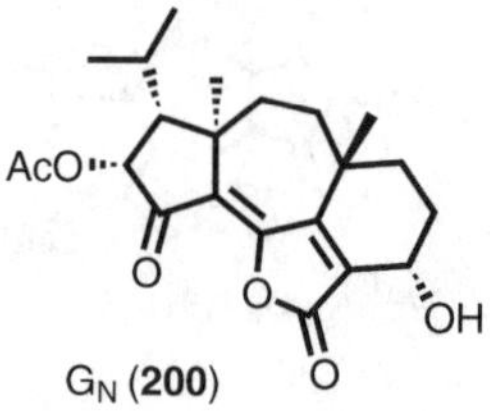　　G$_N$ (**200**)

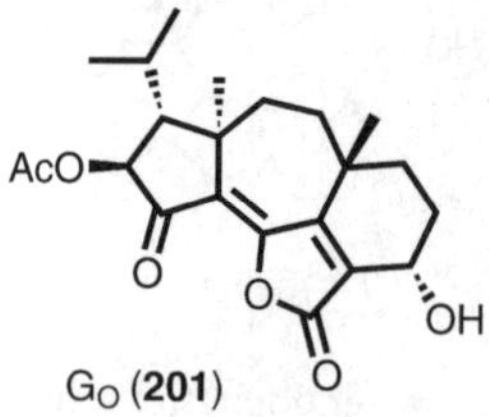

G$_O$ (**201**)

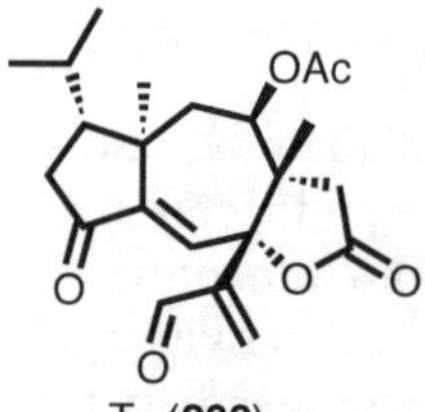

T$_A$ (**202**) 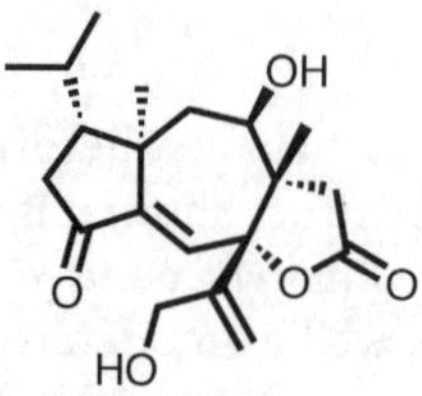　　T$_B$ (**203**)

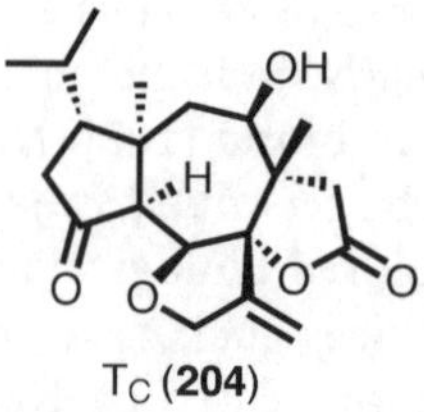

T$_C$ (**204**) 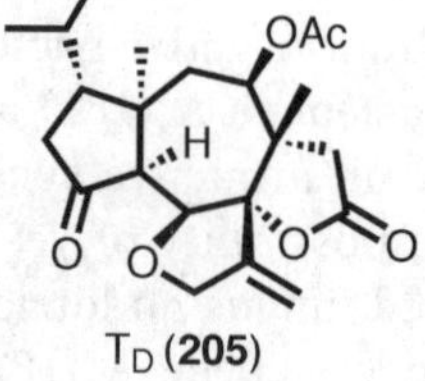　　T$_D$ (**205**)

**Fig. 17** (continued)

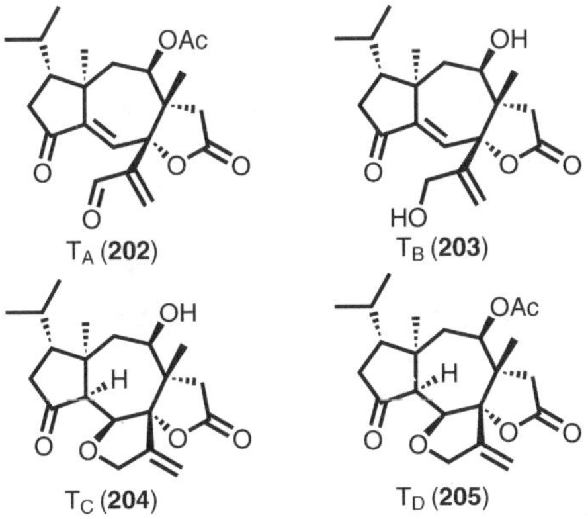

**Fig. 18** Trichoaurantianolide A-D (**202–205**) from the toadstools *Tricholoma aurantium* and *T. fracticum* (1995)

cultures afforded the 14 other guanacastepenes B-O ($G_{A-O}$, **187–201**, Fig. 17) [103].

However, and most notably, the guanacastepenes do not represent the first reported members of the neodolastane class of diterpenes. In 1995 Steglich and Vidari independently reported the isolation and structural elucidation of the oxidatively rearranged neodolastanes trichoaurantianolides A-D (**202–205**) (Steglich: "trichaurantin A") from the toadstools *Tricholoma aurantium* and *T. fracticum* (order: *Agaricales*) [104–106] (Fig. 18). The relative and absolute configuration was proved by an X-ray crystal structure analysis. Steglich and Vidari also coined the name neodolastane which is used throughout this review for this class of diterpenes. Trichoaurantianolide A (**202**) showed moderate antibiotic activity against *Bacillus subtilis* and *Staphylococcus aureus*.

## 5.2
## Completed and Formal Total Syntheses

### 5.2.1
### Danishefsky's Total Synthesis

The first diastereoselective total synthesis of (±)-guanacastepene A (*rac*-**187**) was published in a series of papers by Danishefsky and co-workers [107–110]. The tricyclic carbon framework was assembled by interplay of enolate alkylations and ring-closing carbon/carbon double bond forming reactions (Fig. 19). Commercially available cyclopentenone **90** was utilized as the A-ring precursor.

The assembling of the hydroazulene AB-ring system **210** set out with three consecutive C/C connecting transformations (Scheme 31). Copper catalysed

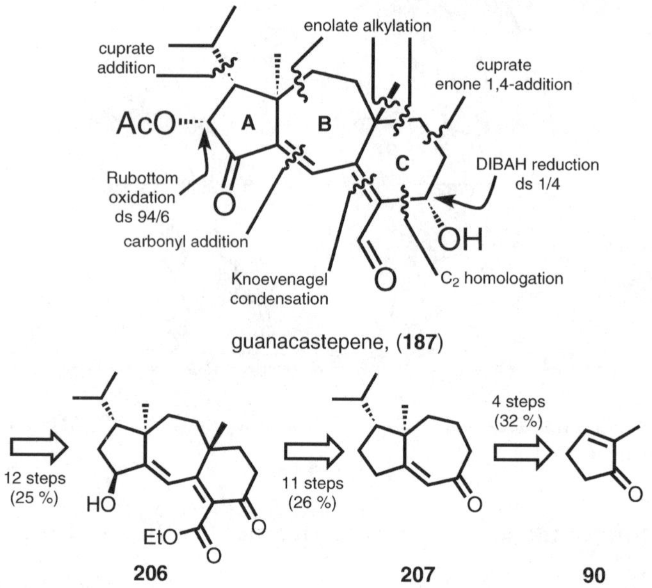

**Fig. 19** Retrosynthetic summary of the first (±)-guanacastepene A (*rac*-**187**) synthesis by Danishefsky (2001)


1,4-addition of a Grignard reagent was followed by in situ trapping and isolation of the silyl enol ether **91**. Not surprisingly, the alkylation of the enolate generated from silyl enol ether **91** by treatment with methyl lithium afforded the desired *trans* diastereomer **208** exclusively. This two step sequence was adapted from Piers' total synthesis of (±)-Crinipellin B, a tetraquinane diterpene and was also exploited by Williams for the synthesis of a neodolabellane diterpene (Scheme 15) [45, 46]. It was originally envisioned to install the seven-membered B-ring by an intramolecular olefination but the corresponding HWE reaction failed to deliver the desired product [107]. Alternatively, an intramolecular carbonyl 1,2-addition of a vinyl lithium species generated from **208** by iodine lithium exchange afforded the hydroazulene **210** (for related applications of this cyclization reaction see Schemes 20 and 21). The major by-product (**209**) was formed by an acid/base reaction between the vinyl lithium species and the ketone. The tertiary allylic alcohol **210** was then successfully converted into the enone **207** by an oxidative allylic transposition [39].

The diastereoselective generation of the quaternary carbon atom $C^8$ and the annulation of the C-ring on the hydroazulene AB-ring system were next on the agenda. Two sequential alkylations were originally envisioned for this task (Scheme 32). Unfortunately, though the first alkylation of the enolate of

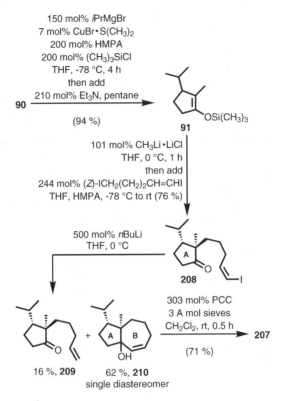

**Scheme 31** Toward the hydroazulene building block (207)

207 afforded the mono-alkylated product 214, the generation and alkylation of the enolate 213 could not be accomplished. Nevertheless, the desired alkylation product 215 was successfully synthesized by an alternative strategy. Alkylation of 207 with Eschenmoser's salt followed by a Cope elimination afforded dienone 211 [111]. 1,4-Addition of a vinyl cuprate on the terminal methylene moiety in the presence of trimethylsilylchloride afforded the silyl enol ether 212. Treatment of 212 with methyl lithium liberated the lithium enolate 213 that was methylated in a remarkable yield to afford the desired enone 215. The (8Si) topicity of the diastereoface-differentiating methylation can be explained by the axial position of the $C^{11}$ methyl group which efficiently blocks the Re face of the enolate.

A few functional group transformations were necessary to set the stage for the Koevenagel condensation of the β-keto ester 220 (Scheme 33). Ketone 215 was protected as a 1,3-dioxolane. Regioselective hydroboration of the terminal exocyclic double bond was then accomplished with 9-BBN and

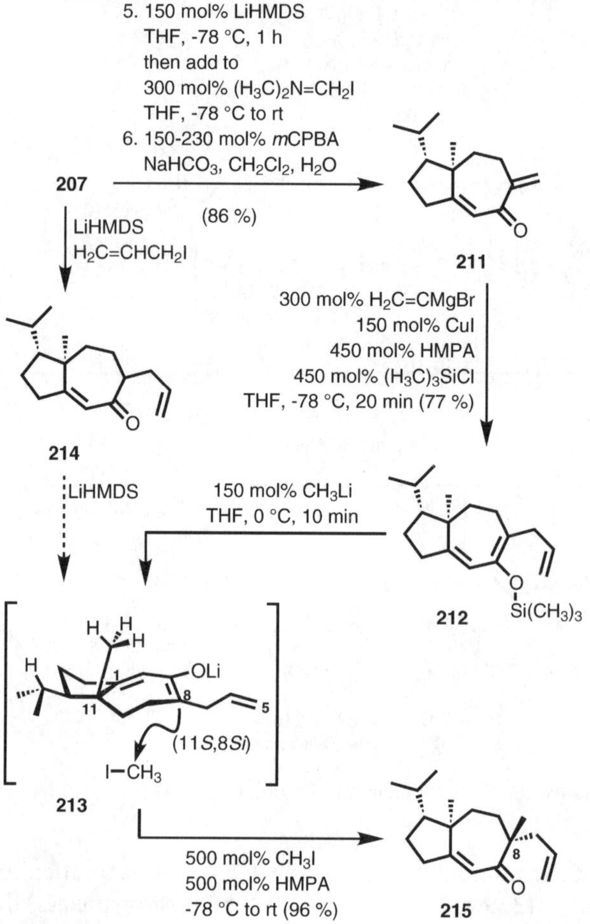

**Scheme 32** Generation of the quaternary $C^8$ by sequential cuprate addition/enolate alkylation

the resulting primary alcohol was oxidized to the aldehyde **216** using the Dess-Martin periodinane [42, 112]. The aldehyde **216** was directly converted into a β-keto ester using the method described by Roskamp [113]. Cleavage of the ketal protecting group under acidic conditions afforded the β,γ-unsaturated ketone **218**. The attempted Knoevenagel condensation of **218** failed due to the increased acidity of the α-methylene protons of the β,γ-unsaturated ketone moiety. This problem was elegantly circumvented by the epoxidation of the $C^1/C^{14}$ double bond with mCPBA [via **219**] to afford the epoxide **220** as single diastereomer. The (1Si,14Si) topicity of the epoxidation is enforced by the angular $C^{11}$ methyl group which effectively blocks the Re face of the $C^1/C^{14}$ double bond.

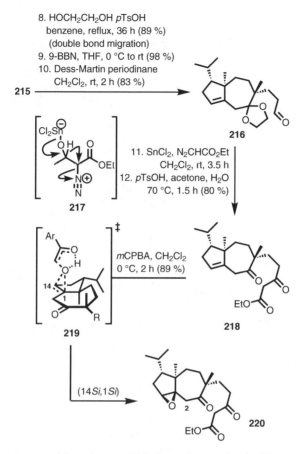

**Scheme 33** Preparation of the substrate (**220**) for an intramolecular Knoevenagel condensation

Treatment of the β-keto ester **220** with sodium ethoxide at elevated temperature triggered off an epoxide ring opening by β-elimination that was followed by the desired Knoevenagel condensation to afford the tricyclic product **206** (Scheme 34). The enone moiety in the intermediate **221** did not show a propensity for deprotonation and, therefore, the ketone carbonyl function of the enone moiety was available for a Knoevenagel condensation. The reduction of the β-keto ester (**206**) to the corresponding diol was the next objective. Treatment of the TES-protected β-keto ester (TES-**206**) with DIBAH afforded the diastereomeric diols **222** and **223** in a moderate diastereoselectivity in favour of the undesired diastereomer **222**. The diastereomers were separated and the undesired diastereomer **222** was epimerized to **223** by a sequence that consists of Mitsunobu inversion and benzoate ester reduction [98, 99].

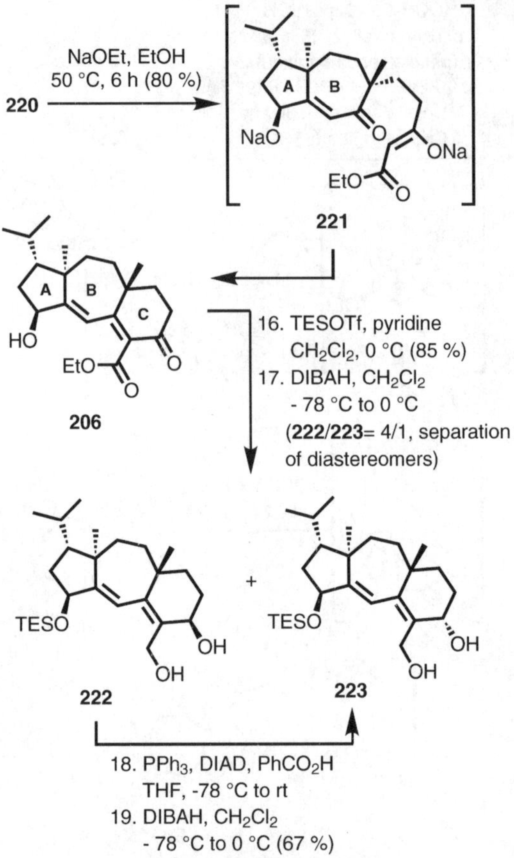

**Scheme 34** Annulation of the C-ring by a Knoevenagel condensation

The final eight steps of the synthesis were devoted to redox and protecting group chemistry in order to convert the tricyclic diol **223** into (±)-guanacastepene A (*rac*-**187**) (Scheme 35). The diol **223** was transformed into the dienone **224** by two protecting group transformations and a Dess-Martin oxidation [42]. The diastereoselective introduction of the $C^{13}$ hydroxyl group was realized by a Rubottom oxidation of the silyl enol ether **225** using dimethyldioxirane (DMDO) as the oxidant [114, 115]. The preferred (13*Re*,14*Re*) topicity of the epoxidation is remarkable because that required an approach of the DMDO *cis* to the $C^{12}$ isopropyl group and the $C^{11}$ methyl group. Acylation of the $C^{13}$ hydroxyl group followed by ketal cleavage and regioselective oxidation of the primary hydroxyl group concluded the first total synthesis of racemic guanacastepene A (**187**). The longest linear sequence consisted of 27 steps including seven C/C-connecting transformations. Substrate-induced diastereoselectivity served efficiently to control the

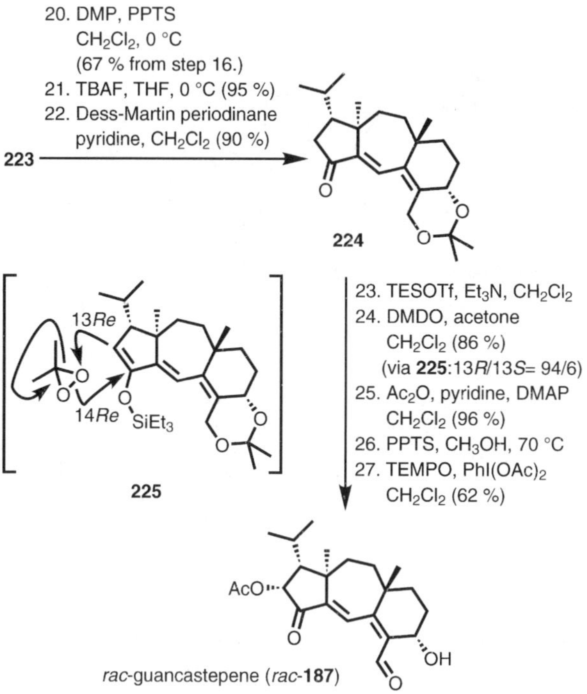

**Scheme 35** Final steps toward (±)-guanacastepene (*rac*-187) (Danishefsky 2002)

relative configuration of newly created chirality centres. The critical quaternary chiral carbon atoms $C^8$ and $C^{11}$ were build up by enolate methylation. Thirteen steps are connected with the introduction and differentiation of the four oxygen functionalities at $C^5$, $C^{13}$, $C^{14}$ and $C^{15}$. No transition metal catalysed or mediated reactions were used for the synthesis.

## 5.2.2
### Snider's Formal Total Synthesis

The formal total synthesis of racemic guanacastepene (*rac*-187) from Snider and co-workers (Fig. 20) was submitted six months later than the completed synthesis of Danishefsky's group [116–118]. The shortest sequence developed by the Snider group utilized the sequential cuprate addition/enolate alkylation of 2-methylcyclopent-2-enone **90** previously exploited by Piers, Williams and Danishefsky (Schemes 15 and 31). As outlined in Figs. 19 and 20, the strategies of Danishefsky and Snider are closely related. Both rely on stepwise annulations to build up the tricyclic ring system. They differ only in respect to the particular reactions that converted the monocyclic starting material (**90**) via bicyclic hydroazulenes (**207** vs **227**) into the desired tricyclic 5-7-6-system (**224**).

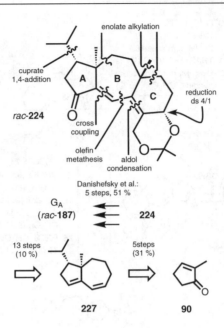

**Fig. 20** Summary of Snider's formal total synthesis of racemic guanacastepene (*rac*-187) (2002)

In contrast to Danishefsky's synthesis, Snider's approach to the B-ring of the hydroazulene AB-ring system is based on transition metal catalysed processes (Scheme 36). The enol triflate **230** prepared from the ketone **228** was subjected to the conditions of an intramolecular Mizoroki-Heck reaction [119, 120]. Unfortunately, the hydroindene **236** was formed rather then the desired hydroazulene **227**. The migratory insertion requires a coplanar arrangement between the C/C double bond and the C-Pd σ-bond in the four-membered transition state. Therefore, the transition state **234** could be favoured over **235** what would account for the preferred formation of **236** instead of the desired **227** [121]. Alternatively, a sequence of two transition metal catalysed C/C-connecting transformations was exploited. Murahashi-type coupling between **230** and vinyl magnesium bromide afforded the triene **233** [122]. However, the efficiency of this reaction is hampered by the formation of a significant amount of the reduced diene **232**. The triene **233** was then treated with Grubbs metathesis catalyst to afford the desired hydroazulene **227** [123–125].

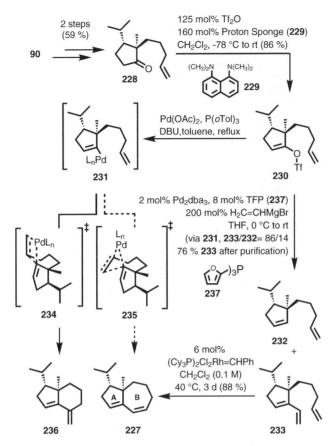

**Scheme 36** Hydroazulene synthesis by transition metal catalysed C/C-bond forming reactions

The 1,3-diene moiety in **227** which included the carbon atoms $C^{13}/C^{14}$ and $C^1/C^2$ was oxidized to the 1,4-dihydroxy-2-ene moiety in **238** that was further exploited to functionalise the A-ring as well as for the annulation of the C-ring (Scheme 37). The transformation of **227** into **238** was realized by a diastereoselective epoxidation of **227** to afford a vinyl epoxide (**241**) that was subjected to the conditions for a Palladium(0)-catalysed allylic substitution with the acetate ion [126]. The mechanism and the stereochemical course of the allylic substitution may be explained as depicted in Scheme 37. $S_N2'$ ring opening of the protonated vinyl epoxide **241** by an anionic Pd complex proceeded with a (3*Si*) topicity to the π-allyl Pd com-

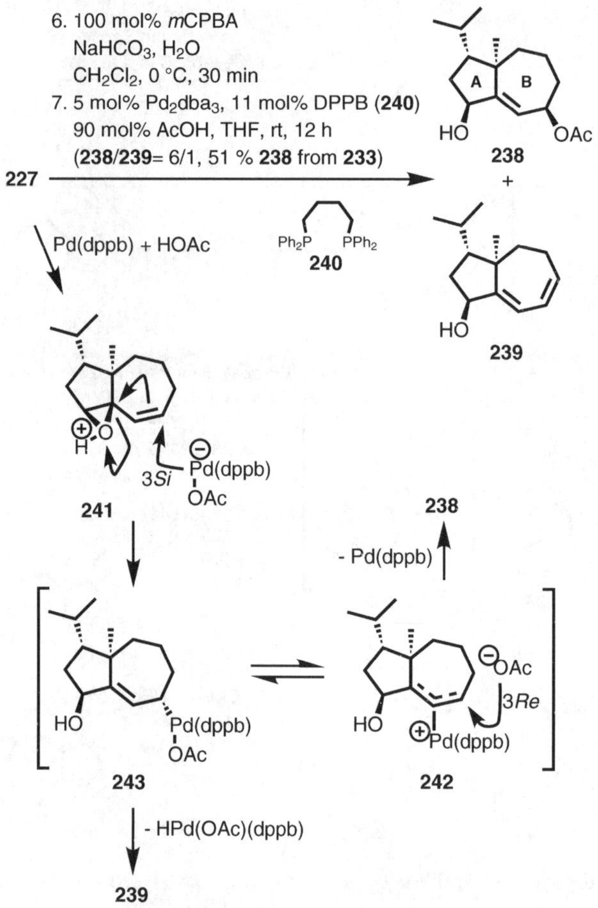

**Scheme 37** Pd⁰-catalysed oxidation of the 1,3-diene moiety of **227**

plex **242**. Nucleophilic attack of an acetate ion at the carbon atom $C^3$ proceeded from the *Re* face and afforded the desired olefin **238**. This mechanism accounts for the formation of the diene **239** as major by-product by a reductive elimination of HPd(OAc)(dppb) from the Pd-carbon σ-complex **243**.

Two consecutive enolate alkylations were utilized to generate the quaternary carbon atom $C^8$ (Scheme 38). Alcohol **238** was transformed into the protected hydroxy enone **244**. Regioselective deprotonation at the α-position of the ketone **244** led to a cross-conjugated enolate that was alkylated with the allylic iodide **245**. The vinyl silyl moiety in **245** represents a masked keto group [127]. The choice of the TBS protecting group for the hydroxyl group at $C^{13}$ of **244** was crucial in order to prevent the deprotonation at the γ-posi-

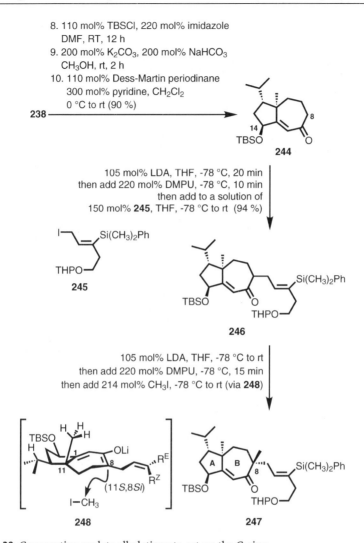

**Scheme 38** Consecutive enolate alkylations to set up the C-ring

tion to the keto function ($C^{14}$) what would lead to an extended conjugated ketone dienolate. The nature of the protecting group for the primary hydroxyl group of **245** was also of pivotal importance for the chemoselectivity of the alkylation reaction. The angular $C^{16}$ methyl group was finally introduced by the methylation of the enolate **246** to afford the vinyl silane **247**. The already present angular $C^{17}$ methyl group on $C^{11}$ blocked efficiently the *Re* face and prompted the methyl iodide to approach the enolate **248** from the *Si* face.

The vinyl silane 247 was unmasked to the ketone 250 by epoxidation, subsequent ring opening of the epoxide with HF in pyridine and concurrent cleavage of the THP and TBS protecting group (Scheme 39). The Stork-Jung annulation was completed through the treatment of the diketone 250 with sodium methoxide to mediate the intramolecular aldol con-

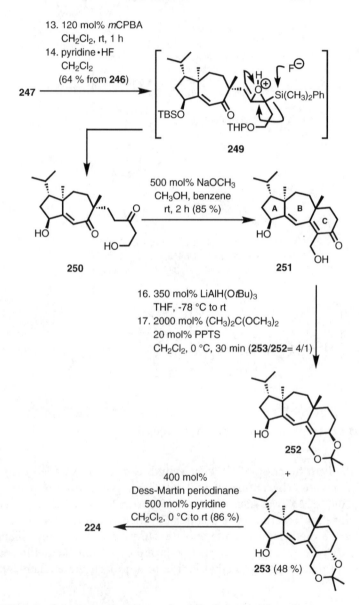

Scheme 39 A Stork-Jung annulation completed the formal total synthesis

densation affording tricycle **251** [127]. Reduction of the keto function at $C^5$ of **251** to the required hydroxyl function with LiAlH(O$t$-Bu)$_3$ proceeded with a moderate 4/1 diastereoselectivity in favour of the desired (5$S$)-configured diastereomer. The Danishefsky group had observed an apparently contradictory result: The structurally related $\beta$-keto ester **206** was reduced with a variety of different reducing agents with a 4/1 diastereoselectivity in favour of the *un*desired (5$R$)-configured alcohol **222** (Scheme 34). The 1,3-diol moiety was then protected and the diastereomeric 1,3-dioxanes **252** and **253** were separated. Finally, oxidation of the hydroxyl function at $C^{14}$ of **253** afforded the ketone **224** that can be transformed into guanacastepene A (**187**) via the sequence of five steps developed by Danishefsky (Scheme 35).

The linear sequence to the ketone **224** consisted of 18 steps with an overall yield of 3.8%. In comparison, Danishefsky's synthesis of the ketone **224** required 22 steps with an overall yield of 4.9%. Snider's formal synthesis rested on eight C/C-connecting transformations. In analogy to Danishefsky's synthesis, the two quaternary carbon atoms $C^8$ and $C^{11}$ were assembled by enolate alkylation chemistry. Both syntheses relied on a convergent approach by a stepwise annulation of the B- and C-ring on an appropriately functionalized A-ring. The syntheses can be distinguished by the method used for the annulation of the B-ring. Snider's group utilized "modern" methods (cross-coupling and olefin metathesis) whereas Danishefsky employed a rather "traditional" method (carbonyl addition).

## 5.3
## Approaches Toward Guanacastepene

### 5.3.1
### Strategies Based on Metathesis

The olefin metathesis has emerged as a powerful and accepted synthetic method for the construction of carbo- and heterocycles featuring various ring sizes [55, 123, 124]. The power of this C/C-connecting transformation rests on predictability, reliability and simplicity of the metathesis protocol. Given the fact that guanacastepene A (**187**) is a polycyclic compound containing two double bonds, it is not surprising that the metathesis has been exploited by more than one group. The formal total synthesis of guanacastepene A (**187**) from the Snider laboratory has already been dissected (see above) [116–118]. Snider was also the first who submitted a communication concerning the exploitation of an olefin metathesis for the construction of the B-ring of guanacastepene (**187**). In the event, diene **233** was treated with (Cy$_3$P)$_2$Cl$_2$Rh=CHPh to afford the metathesis product which was further transformed into the advanced intermediate **224** (Eq. 1):

$$\text{90} \Rrightarrow \text{233} \Rrightarrow \text{224} \tag{1}$$

Almost a year after Snider had submitted his publication, the Mehta group presented the first of three papers concerning their approach to the racemic guanacastepene-like ring system **256** (Eq. 2) [128–130]:

$$\text{254} \Rrightarrow \text{255} \Rrightarrow \text{256} \tag{2}$$

The B-ring was closed by a ring-closing metathesis of the diene **255** catalysed by $(Cy_3P)_2Cl_2Rh=CHPh$ [123]. The isopropyl group was introduced by a cuprate addition to the masked cyclopentadienone **254** and the quaternary carbon atom $C^8$ was assembled by a sequential enolate alkylation. The C-ring was closed by an intramolecular aldol condensation of a keto aldehyde. The quaternary carbon atom $C^5$ was also generated by sequential enolate alkylations.

The Trauner group utilized an olefin metathesis for the construction of the A-ring (Eq. 3) [131, 132]:

$$\text{257} \rightarrow \text{258} \quad \text{259} \leftarrow \text{260} \tag{3}$$

Triene **257** was transformed into the racemic cyclopentenone **258** by a ring closing metathesis using Grubbs' N-heterocyclic carbene catalyst (**102**) [55].

This approach sets the stage for an enantiotopos-differentiating olefin metathesis which would allow the enantioselective synthesis of **258**. However, the realization of such an approach has not yet been successful [132]. The second building block (**259**) containing the A ring was synthesized diastereoselectively by a diastereoface-differentiating intramolecular Heck-Mizoroki reaction of the enantiomerically enriched furan **260** [120].

The Tius group used Grubbs' N-heterocyclic carbene catalyst (**102**) for the ring closing metathesis of the triene **262** (Eq. 4) [55, 133]:

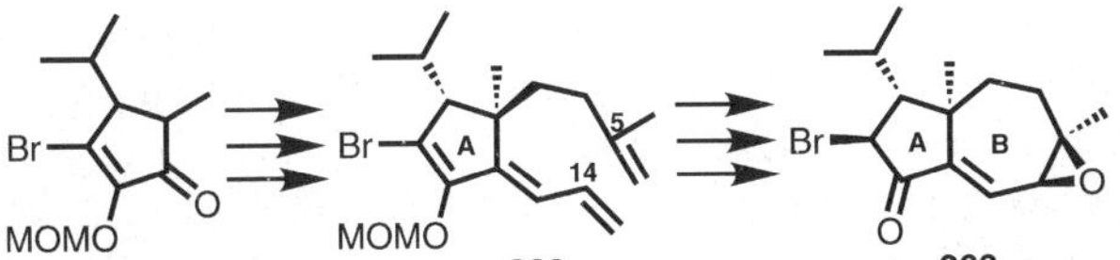

Recent Progress in the Total Synthesis of Dolabellane and Dolastane Diterpenes     129

(4)

This approach is related to Mehta's in so far as the carbon atoms $C^5$ and $C^{14}$ were connected to afford a desired hydroazulene (Eq. 2). Ensuing manipulation of functional groups led to the functionalized racemic hydroazulene **263**. The cyclopentenone **261** was synthesized from an acyclic precursor by a Nazarov cyclization [134].

A sophisticated exploitation of a ring closing metathesis strategy was delivered by Hanna (Eq. 5) [135]:

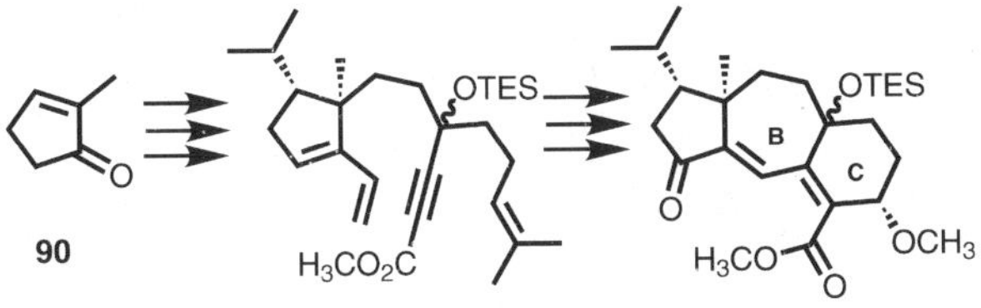

(5)

The sequential RCM of the dienyne **264** using Grubbs' *N*-heterocyclic carbene catalyst (**102**) afforded the racemic tricyclic ring system of guanacastepene A [55]. Further functionalization of the initial metathesis product provided the seemingly advanced intermediate **265**. The only drawback is that the synthetic strategy for the dienyne **265** did not consider the angular methyl group at $C^5$ in guanacastepene A (**187**). A triethylsiloxy group is instead connected to $C^5$. Nevertheless, this RCM strategy could be spectacular provided that Hanna's group will establish an efficient access to an appropriately functionalized (non-racemic) dienyne ($CH_3$ instead of OTES in **264**).

## 5.3.2
### Strategies Based on Cycloadditions

Magnus pursued a strategy in which an intramolecular [5+2]-cycloaddition of the pyrylium ylide **266** afforded the oxygen bridged hydroazulene **267** (Eq. 6) [136, 137]:

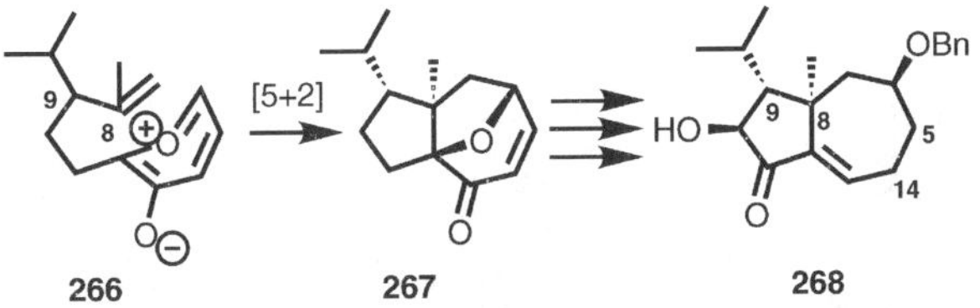

(6)

Ether cleavage and further functionalization afforded the intermediate **268**. The [5+2]-cycloaddition provided the hydroazulene **267** with the correct relative configuration at $C^8$ and $C^9$. Tracing back the synthesis of the pyrylium ylide **266** leads to the astonishing realization that 2-methyl cyclopent-2-enone (**90**) was the original cyclic starting material and that the $C^8$ methyl as well as the $C^9$ isopropyl group were introduced by a sequence of cuprate addition and enolate alkylation (see Schemes 15, 31 and 36 for comparison).

A highly convergent approach was published by Sorensen and coworker (Eq. 7) [138]:

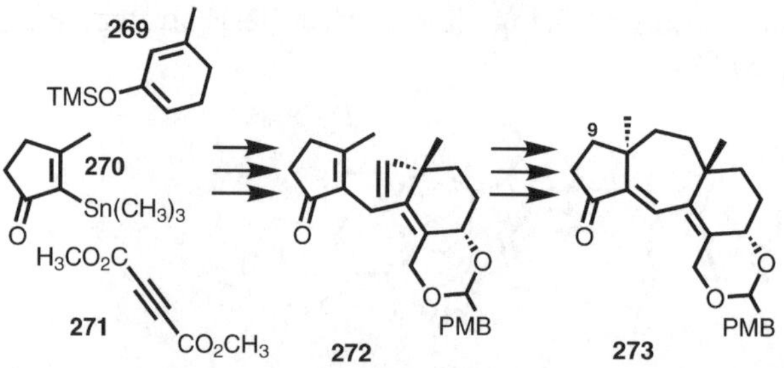

(7)

Sorensen's strategy utilized an intramolecular [2+2]-photocycloaddition of the enone **272** followed by a $SmI_2$-mediated reductive ring opening to afford a tricyclic carbon framework in moderate yield which was further functionalized to the dienone **273**. Compound **273** resembled the intermediate **224** of the syntheses from the Danishefsky and the Snider group (Figs. 19 and 20). However, the vinyl stannane **270** has to be modified in order to introduce the isopropyl group at $C^9$ of guanacastepene A (**187**). The substrate **272** for the crucial cycloaddition was synthesized from the three building blocks **269–271**. Key step was a Pd-catalysed Stille coupling to connect vinyl stannane **270** and a building block generated from the cycloaddition product between **269** and **271** [51].

## 5.3.3
**Miscellaneous**

A B-ring-last-strategy was developed by Lee and coworkers (Eq. 8) [139, 140]:

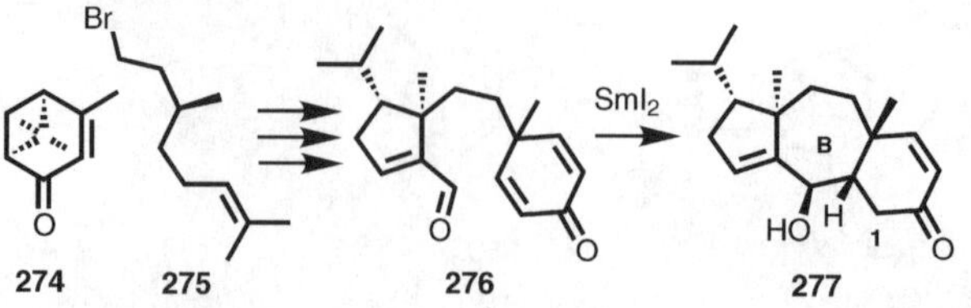

(8)

An SmI$_2$-mediated intramolecular ketyl-enone cyclization of the aldehyde **276** afforded the ABC-ring system **277** [141]. The carbon framework of **277** has the correct relative and absolute configuration. However, the C$^1$ formyl group of guanacastepene A (**187**) is missing and additional functional group manipulations will be necessary to complete an enantioselective guanacastepene A (**187**) total synthesis.

A recent approach toward (±)-guanacastepene (*rac*-**187**) published by Kwon in 2003 employed a sequential methyl cuprate 1,4-addition to the enone **281** followed by an intramolecular Mukaiyama aldol condensation to construct the B ring (Eq. 9) [142, 143]:

$$(9)$$

This strategy culminated in the synthesis of the racemic tricycle **282** as a pair of diastereomers that were separated by chromatography. The cyclopentenone **274** served as A-ring building block and the C-ring was synthesized utilizing a Diels-Alder reaction between **279** and **280** as the key step.

A linear strategy was recently published by Brummond (Eq. 10) [144]:

$$(10)$$

The key step employed a rhodium(I)-catalysed allenic Pauson-Khand reaction to generate the tricyclic ring system **276** from the allenyne **275**. However, all attempts to introduced the missing C$^8$ methyl group by a 1,4-addition protocol failed to provide the completed neodolastane framework. Allenyne **275** was synthesized from the enone **274** by a multistep procedure. The quaternary C$^5$ atom was constructed by sequential enolate alkylations.

# 6
# Conclusion

This review summarizes the synthetic efforts of about two dozen research groups toward four structurally related but nevertheless significantly distinguished classes of diterpenes over the last 17 years. Consequently, it is very difficult to draw a general conclusion. We have seen a variety of ingenuous C/C connecting transformations especially for the synthesis of highly substituted cyclopentanones and hydroazulenes. Driven by the complexity of the diterpene structure, fascinating (stepwise) annulations, macrocyclizations, transannular cyclizations and even the metal-catalysed one-pot generation of a tricycle from a monocyclic precursor have been designed and executed. The strengths and weaknesses of the ex-chiral-pool approach can be illustrated from syntheses presented in this chapter. The application of concepts from solid phase synthesis and of catalytic asymmetric C/C-bond forming reactions should further improve the efficiency of synthetic strategies toward dolastanes and dolabellanes. Further work in this area is undoubtedly worthwhile since it is foreseeable that new bioactive members of these classes of diterpenes will be isolated. Once again, synthetic organic chemistry is breaching the gap between natural product isolation, structural elucidation and the determination and understanding of biological activities on the molecular level.

**Acknowledgements** Financial support from the Deutsche Forschungsgemeinschaft (Heisenberg-Program) is gratefully acknowledged. MH thanks Prof. D. A. Evans for his hospitality during a stay at Harvard University as a visiting scholar in summer 2003.

**Note added in Proof** New work on the isolation of dolabellanes and neodolabellanes has been published [1001, 1002, 1003 and 1008]. Synthetic studies toward neodolabellanes are frequently published [1004, 1005, 1006, 1009].
    1001. Iguchi K, Fukaya T, Yasumoto A, Watanabe K (2004) J Nat Prod 67:577
    1002. Morikawa T, Fengming X, Yousuke K, Hisashi M, Kiyofumi N, Yoshikawa M (2004) Org Lett 6:869
    1003. Nagashimaa F, Takaoka S, Huneck S, Asakawa Y (1999) Phytochemistry 51:563
    1004. Srikrishna A, Dethe DH (2003) Org Lett 6:165
    1005. Sarpong R, Su JT, Stoltz BM (2003) J Am Chem Soc 125:13624
    1006. Chiu P, Li S (2004) Org Lett 6:613
    1007. Boyer F-D, Hanna I, Ricard L (2004) Org Lett 6:1817
    1008. Barbosa JP, Teixeira VL, Villaca R, Pereira RC, Abrantes JL, Frugulhetti ICPdP (2003) Biochem Syst Ecol 31:1451
    1009. Tsukamoto S, Macabalang AD, Nakatani K, Obara Y, Nakahata N, Ohta T (2003) J Nat Prod 66:1578

# References

1. Rodríguez AD, González E, Ramírez C (1998) Tetrahedron 54:11683
2. Davis EM, Croteau R (2000) Top Curr Chem 209:53
3. Bohlmann J, Meyer-Gauen G, Croteau R (1998) Proc Natl Acad Sci USA 95:4126

4. Dewick PM (2002) Nat Prod Rep 19:181
5. Hamano Y, Kuzuyama T, Itoh N, Furihata K, Seto H, Dairi T (2002) J Biol Chem 277:37098
6. Eisenreich W, Rieder C, Grammes C, Heßler G, Adam K-P, Becker H, Arigoni D, Bacher A (1999) J Biol Chem 274:36312
7. Peters RJ, Croteau RB (2002) Proc Natl Acad Sci USA 99:580
8. Coleman AC, Kerr RG (2000) Tetrahedron 56:9569
9. Eguchi T, Dekishima Y, Hamano Y, Dairi T, Seto H, Kakinuma K (2003) J Org Chem 68:5433
10. Fukuyama Y, Minami H, Kagawa M, Kodama M, Kawazu K (1999) J Nat Prod 62:337
11. Fukuyama K, Katsube Y, Kawazu K (1980) J Chem Soc Perkin Trans 2:1701
12. Fukuyama Y, Minami H, Takeuchi K, Kodama M, Kawazu K (1996) Tetrahedron Lett 37:6767
13. Hashimoto T, Toyota M, Koyama H, Kikkawa A, Yoshida M, Tanaka M, Takaoka S, Asakawa Y (1998) Tetrahedron Lett 39:579
14. Ramírez MDC, Toscano RA, Arnason J, Omar S, Cerda-García-Rojas CM, Mata R (2000) Tetrahedron 56:5085
15. Hinkley SF, Mazzola EP, Fettinger JC, Lam Y, Jarvis BB (2000) Phytochemistry 55:663
16. Duh C, Chia M, Wang S, Chen H, El-Gamal AAH, Dai C (2001) J Nat Prod 64:1028
17. Sonwa MM, König WA (2001) Phytochemistry 56:321
18. Iwashima M, Matsumoto Y, Takenaka Y, Iguchi K, Yamori T (2002) J Nat Prod 65:1441
19. Costantino V, Fattorusso E, Mangoni A, Rosa MD, Ianaro A, Aknin M, Gaydou EM (1999) Eur J Org Chem 227
20. Iguchi K, Sawai H, Nishimura H, Fujita M, Yamori T (2002) Bull Chem Soc Jpn 75:131
21. Chang K, Duh C, Chen I, Tsai I (2003) Planta Med 69:667
22. Ireland C, Faulkner DJ, Finer J, Clardy J (1976) J Am Chem Soc 98:4664
23. Mehta G, Karra SR, Krishnamurthy N (1994) Tetrahedron Lett 35:2761
24. Kato N, Higo A, Nakanishi K, Wu X, Takeshita H (1994) Chem Lett 1967
25. Jenny L, Borschberg H (1995) Helv Chim Acta 78:715
26. Corey EJ, Kania RS (1996) J Am Chem Soc 118:1229
27. Luker T, Whitby RJ (1996) Tetrahedron Lett 37:7661
28. Kato N, Higo A, Wu X, Takeshita H (1997) Heterocycles 46:123
29. Corey EJ, Kania RS (1998) Tetrahedron Lett 39:741
30. Zhu Q, Qiao L, Wu Y, Wu Y (1999) J Org Chem 64:2428
31. Zhu Q, Fan K, Ma H, Qiao L, Wu Y, Wu Y (1999) Org Lett 1:757
32. Zeng Z, Xu X (2000) Tetrahedron Lett 41:3459
33. Zhu Q, Qiao L, Wu Y, Wu Y (2001) J Org Chem 66:2692
34. Hu T, Corey EJ (2002) Org Lett 4:2441
35. Miyaoka H, Isaji Y, Kajiwara Y, Kunimune I, Yamada Y (1998) Tetrahedron Lett 39:6503
36. Miyaoka H, Isaji Y, Mitome H, Yamada Y (2003) Tetrahedron 59:61
37. Katsuki T, Sharpless KB (1980) J Am Chem Soc 102:5974
38. Katsuki T, Martin V (1996) Org React 48:1
39. Sundararaman P, Herz W (1977) J Org Chem 42:813
40. Miyaoka H, Baba T, Mitome H, Yamada Y (2001) Tetrahedron Lett 42:9233
41. Abushanab E, Vemishetti P, Leiby RW, Singh HK, Mikkilineni AB, Wu DCJ, Saibaba R, Panzica RP (1988) J Org Chem 53:2598
42. Dess DB, Martin JC (1983) J Org Chem 48:4155

43. Blanchette MA, Choy W, Davis T, Essenfeld AP, Masamune S, Roush WR, Sakai T (1984) Tetrahedron Lett 25:2183
44. Sharpless KB, Michaelson RC (1973) J Am Chem Soc 95:6136
45. Williams DR, Heidebrecht RW (2003) J Am Chem Soc 125:1843
46. Piers E, Renaud J, Rettig SJ (1998) Synthesis 590
47. Corey EJ, Bakshi RK, Shibata S (1987) J Am Chem Soc 109:5551
48. Corey EJ, Helal CJ (1998) Angew Chem Int Ed 37:1986–2012
49. Itsuno S (1998) Org React 52:395
50. Fugami K, Kosugi M (2002) Top Curr Chem 219:87
51. Farina V, Krishnamurthy V, Scott WJ (1997) Org React 50:1
53. Hosomi A, Sakurai H (1976) Tetrahedron Lett 17:1295
54. Fleming I, Barbero A, Walter D (1997) Chem Rev 97:2063
55. Scholl M, Ding S, Lee CW, Grubbs RH (1999) Org Lett 1:953
56. Enev VS, Kaehlig H, Mulzer J (2001) J Am Chem Soc 123:10764
57. McMurry JE, Lectka T, Rico JG (1989) J Org Chem 54:3748
58. McMurry JE (1989) Chem Rev 89:1513
59. Williams DR, Coleman PJ (1995) Tetrahedron Lett 36:35
60. Williams DR, Coleman PJ, Nevill CR, Robinson LA (1993) Tetrahedron Lett 34:7895
61. Pettit GR, Ode RH, Herald CL, Dreele RBV, Michel C (1976) J Am Chem Soc 98:4677
62. Braekman JC, Daloze D, Schubert R, Albericci M, Tursch B, Karlsson R (1978) Tetrahedron 34:1551
63. Ochi M, Watanabe M, Miura I, Taniguchi M, Tokoroyama T (1980) Chem Lett 1229
64. Ochi M, Watanabe M, Kido M, Ichikawa Y, Miura I, Tokoroyama T (1980) Chem Lett 1233
65. Bowden BF, Braekman J, Coll JC, Mitchell SJ (1980) Aust J Chem 33:927
66. Ochi M, Miura I, Tokoroyama T (1981) J Chem Soc Chem Commun 100
67. Sun HH, McConnell OJ, Fenical W, Hirotsu K, Clardy J (1981) Tetrahedron 37:1237
68. Crews P, Klein TE, Hogue ER, Myers BL (1982) J Org Chem 47:811
69. González AG, Norte JDMM, Rivera P, Perales A, Fayos J (1983) Tetrahedron 39:3355
70. Harada N, Yokota Y, Iwabuchi J, Uda H, Ochi M (1984) J Chem Soc Chem Commun 1220
71. Ochi M, Asao K, Kotsuki H, Miura I, Shibata K (1986) Bull Chem Soc Jpn 59:661
72. Teixeira VL, Tomassini T, Fleury BG, Kelecom A (1986) J Nat Prod 49:570
73. Rao CB, Pullaiah KC, Surapaneni RK, Sullivan BW, Albizati KF, Faulkner DJ, He C, Clardy J (1986) J Org Chem 51:2736
74. Kelecom A, Teixeira VL (1988) Phytochemistry 27:2907
75. Dunlop RW, Ghisalberti EL, Jefferies PR, Skelton BW, White AH (1989) Aust J Chem 42:315
76. König GM, Wright AD (1994) J Nat Prod 57:1529
77. Pattenden G, Robertson GM (1986) Tetrahedron Lett 27:399
78. Begley MJ, Pattenden G, Robertson GM (1988) J Chem Soc Perkin Trans 1 1085
79. Rabjohn N (1976) Org React 24:261
80. Piers E, Friesen RW (1986) J Org Chem 51:3405
81. Denmark SE, Edwards JP (1991) J Org Chem 56:6974
82. Piers E, Friesen RW (1988) J Chem Soc Chem Commun 125
83. Piers E, Friesen RW (1992) Can J Chem 70:1204
84. Paquette LA, Lin HS, Belmont DT, Springer JP (1986) J Org Chem 51:4807
85. Mehta G, Krishnamurthy N (1987) Tetrahedron Lett 28:5945
86. Mehta G, Krishnamurthy N, Karra SR (1991) J Am Chem Soc 113:5765
87. Majetich G, Song JS, Ringold C, Nemeth GA, Newton MG (1991) J Org Chem 56:3973

88. Majetich G, Song JS, Ringold C, Nemeth GA (1990) Tetrahedron Lett 31:2239
89. Nishiyama H, Kitajima T, Matsumoto M, Itoh K (1984) J Org Chem 49:2298
90. Evans DA, Andrews GC (1974) Acc Chem Res 7:147
91. Williams DR, Coleman PJ, Henry SS (1993) J Am Chem Soc 115:11654
92. Hutchinson JH, Money T, Piper SE (1986) Can J Chem 64:854
93. Ito Y, Hirao T, Saegusa T (1978) J Org Chem 43:1011
94. Pasto D, Taylor R (1991) Org React 40:92
95. Kobayashi S, Mukaiyama T (1974) Chem Lett 705
96. Dieter RK, Silks LA (1986) J Org Chem 51:4687
98. Hughes D (1992) Org React 42:335
99. Mitsunobu O (1981) Synthesis 1
100. Tan RX, Zou WX (2001) Nat Prod Rep 18:448
101. Brady SF, Singh MP, Janso JE, Clardy J (2000) J Am Chem Soc 122:2116
102. Singh MP, Janso JE, Luckman SW, Brady SF, Clardy J, Greenstein M, Maiese WM (2000) J Antibiotics 53:256
103. Brady SF, Bondi SM, Clardy J (2001) J Am Chem Soc 123:9900
104. Knops L, Nieger M, Steffan B, Steglich W (1995) Liebigs Ann 77
105. Invernizzi AG, Vidari G, Vita-Finzi P (1995) Tetrahedron Lett 36:1905
106. Benevelli F, Carugo O, Invernizzi AG, Vidari G (1995) Tetrahedron Lett 36:3035
107. Dudley GB, Danishefsky SJ (2001) Org Lett 3:2399
108. Dudley GB, Tan DS, Kim G, Tanski JM, Danishefsky SJ (2001) Tetrahedron Lett 42:6789
109. Lin S, Dudley GB, Tan DS, Danishefsky SJ (2002) Angew Chem Int Ed 41:2188
110. Tan DS, Dudley GB, Danishefsky SJ (2002) Angew Chem Int Ed 41:2185
111. Danishefsky S, Kitahara T, McKee R, Schuda PF (1976) J Am Chem Soc 98:6715
112. Brown HC, Knights EF, Scouten CG (1974) J Am Chem Soc 96:7765
113. Holmquist CR, Roskamp EJ (1989) J Org Chem 54:3258
114. Rubottom GM, Vazquez MA, Pelegrina DR (1974) Tetrahedron Lett 15:4319
115. Adam W, Saha-Möller CR, Zhao C-G (2003) Org React 61:219
116. Snider BB, Shi B (2001) Tetrahedron Lett 42:9123
117. Snider BB, Hawryluk NA (2001) Org Lett 3:569
118. Shi B, Hawryluk NA, Snider BB (2003) J Org Chem 68:1030
119. Heck RF (1979) Acc Chem Res 12:146
120. Link JT (2002) Org React 60:157
121. Beletskaya IP, Cheprakov AV (2000) Chem Rev 100:3009
122. Murahashi S, Yamamura M, Yanagisawa K, Mita N, Kondo K (1979) J Org Chem 44:2408
123. Nguyen ST, Grubbs RH, Ziller JW (1993) J Am Chem Soc 115:9858
124. Grubbs RH, Chang S (1998) Tetrahedron 54:4413
125. Fürstner A (2000) Angew Chem Int Ed 39:3012–3043
126. Trost BM, Angle SR (1985) J Am Chem Soc 107:6123
127. Stork G, Jung ME (1974) J Am Chem Soc 96:3682
128. Mehta G, Umarye JD, Gagliardini V (2002) Tetrahedron Lett 43:6975
129. Mehta G, Umarye JD (2002) Org Lett 4:1063
130. Mehta G, Umarye JD, Srinivas K (2003) Tetrahedron Lett 44:4233
131. Gradl SN, Kennedy-Smith JJ, Kim J, Trauner D (2002) Synlett 411
132. Hughes CC, Kennedy-Smith JJ, Trauner D (2003) Org Lett 5:4113
133. Nakazaki A, Sharma U, Tius MA (2002) Org Lett 4:3363
134. Tius MA (2003) Acc Chem Res 36:284
135. Boyera F, Hanna I (2002) Tetrahedron Lett 43:7469

136. Magnus P, Waring MJ, Ollivier C, Lynch V (2001) Tetrahedron Lett 42:4947
137. Magnus P, Ollivier C (2002) Tetrahedron Lett 43:9605
138. Shipe WD, Sorensen EJ (2002) Org Lett 4:2063
139. Nguyen TM, Lee D (2002) Tetrahedron Lett 43:4033
140. Nguyen TM, Seifert RJ, Mowrey DR, Lee D (2002) Org Lett 4:3959
141. Molander GA, Harris CR (1996) Chem Rev 96:307
142. Du X, Chu HV, Kwon O (2003) Org Lett 5:1923
143. Mukaiyama T, Banno K, Narasaka K (1974) J Am Chem Soc 96:7503
144. Brummond KM, Gao D (2003) Org Lett 5:3491

# Strategies for Total and Diversity-Oriented Synthesis of Natural Product(-Like) Macrocycles

Ludger A. Wessjohann (✉) · Eelco Ruijter

Leibniz-Institute of Plant Biochemistry, Weinberg 3, 06120 Halle (Saale), Germany
*wessjohann@ipb-halle.de*

| 1 | Introduction | 139 |
|---|---|---|
| 1.1 | Natural Product Macrocycles | 139 |
| 1.2 | Macrocyclization Strategies | 144 |
| 1.2.1 | Macrolactonization | 144 |
| 1.2.2 | Macrolactamization | 145 |
| 1.2.3 | Ring-Closing Metathesis | 145 |
| 1.2.4 | Macroaldolization | 147 |
| 1.2.5 | Palladium Coupling | 147 |
| 1.2.6 | Other Methods | 148 |
| 1.3 | Diversity-Oriented Synthesis | 150 |
| 1.3.1 | Solid Phase Diversity-Oriented Synthesis | 152 |
| 1.3.2 | Solution Phase Diversity-Oriented Synthesis | 154 |
| 2 | Classical Synthesis: Epothilones and Derivatives | 156 |
| 2.1 | Introduction | 156 |
| 2.2 | Retrosynthesis | 157 |
| 2.3 | Synthesis | 158 |
| 2.3.1 | Northern Half Synthesis | 158 |
| 2.3.2 | Southern Half Synthesis | 160 |
| 2.3.3 | Macrocyclization and Completion of the Synthesis | 160 |
| 2.3.4 | Synthesis of Epothilone $D_5$ Analogues | 161 |
| 2.4 | Conclusion | 163 |
| 3 | Rapid Synthesis of Peptoid Macrocycles by Multi Component Reactions | 163 |
| 3.1 | Synthesis of Cyclopeptide Alkaloids | 165 |
| 3.1.1 | Introduction | 165 |
| 3.1.2 | Synthetic Concept | 166 |
| 3.1.3 | Synthesis | 167 |
| 3.1.4 | Conclusion | 168 |
| 3.2 | Biaryl Ether Macrocycles | 169 |
| 3.2.1 | Introduction | 169 |
| 3.2.2 | Synthesis | 169 |
| 3.2.3 | Conclusion | 171 |
| 3.3 | Synthesis of Steroid-Derived Macrocycles | 171 |
| 3.3.1 | Introduction | 171 |
| 3.3.2 | Synthetic Concept | 172 |
| 3.3.3 | Synthesis | 172 |
| 3.3.4 | Conclusion | 178 |
| 4 | Postmodification Studies of Macrocycles | 178 |
| References | | 181 |

**Abstract** Numerous biologically active macrocycles, including antibiotic, antifungal, and antitumor compounds, have been isolated from natural sources. In recent years, the number of such structures has steadily increased, predominantly by polyketide- and peptide-derived compounds from various microorganisms. Macrocycles can combine the right amount of rigidity and flexibility and often exhibit unrivalled activity, thereby deviating from the current paradigm that medicinally active compounds should be small, nitrogen-rich heterocycles. Their challenging structures and intriguing activities have motivated organic chemists to find synthetic access to these compounds. Total synthesis plays a crucial role in the medicinal chemistry efforts towards macrocycles of already defined activity, as well as in the development of new and selective macrocyclization reactions. For lead discovery purposes, however, isolation or classical total synthesis may lack structural variability or prove to be too time consuming and impractical. A more rapid solution may be provided by diversity-oriented synthesis (DOS) of natural product-like molecules. A compromise between total synthesis and combinatorial chemistry, DOS concerns molecules displaying sufficient molecular complexity to resemble natural products, but features a more straightforward synthesis, thus allowing introduction of significant structural diversity. A brief review of flexible macrocyclization strategies and applications of DOS is given, as well as an overview of contributions to total and diversity-oriented synthesis of macrocycles from our laboratory.

**Keywords** Macrocycles · Natural products · Diversity-oriented synthesis · Multi component reactions

**List of Abbreviations**

| | |
|---|---|
| BCL | *Burkholderia cepacia* lipase (Amano PS) |
| BINAP | 2,2′-Bis(diphenylphosphino)-1,1′-binaphthyl |
| CAL-B | *Candida antarctica* lipase B |
| DOS | Diversity-oriented synthesis |
| EDCI | *N*-Ethyl-*N*′-(dimethylaminopropyl)carbodiimide·HCl |
| FDPP | Pentafluorophenyldiphenylphosphinate |
| HATU | *O*-(7-Azabenzotriazolyl)-*N*,*N*,*N*′,*N*′-tetramethyluronium hexafluorophosphate |
| MCR | Multi component reaction |
| MiB | Multiple Multicomponent Macrocyclization including Bifunctional Building Blocks |
| NaHMDS | Sodium hexamethyldisilazide |
| PyBOP | (Benzotriazol-1-yloxy)-tripyrrolidinophosphonium hexafluorophosphate |
| RCM | Ring-closing metathesis |
| rPLE | Recombinant pig liver esterase |
| TAS-F | Tris(dimethylamino)sulfonium difluorotrimethylsilicate |
| TBS | (= TBDMS) *tert*-Butyldimethylsilyl |
| U4CR | Ugi four component reaction |

# 1
# Introduction

## 1.1
## Natural Product Macrocycles

Natural products continue to play an extremely important role in drug development and are overproportionally successful. Together with natural product derivatives, mimics, and other compounds derived or inspired by natural products they comprise some 35–40% of all current drugs [1], but at the same time their proportion in screening efforts constitutes less then 1%. It was recently estimated that known natural products fit Lipinski's rule of five [2] for bioavailability equally well as current trade drugs, which is not surprising, since they are designed by nature to have a certain biological effect and are useless if they do not arrive where they are required to go [3]. However, natural products appear to have less nitrogen present and are less predominant in five- and six-membered rings than synthetic drugs [3]. Thus, natural products commonly contain medium to large ring systems [4]. Macrocycles with varying ring sizes and chemical constitution have been isolated from a wide variety of natural sources. A possible rationalization for the relatively high occurrence of macrocycles in nature is the fact that they constitute an equilibrium between conformational preorganization and flexibility to achieve optimal binding properties to their biological target. Indeed, many of them show remarkable biological activity ranging from antibiotic to antitumor activity and from insecticidal to antifungal properties. The structural diversity of macrocycles even exceeds the diversity in the currently known biological activity they display. Common substructural units may be derived from polyketide and isoprenoid metabolism as well as peptide and carbohydrate biosynthesis. Also, many macrocycles contain 5- or 6-membered heterocycles derived from internal ring closure, e.g., tetrahydrofurans or -pyrans from polyketide backbone cyclization or, e.g., oxazoles or thiazoles from cyclizations involving amino acids originating, e.g., from the starter units of polyketide biosynthesis. These heterocycles are either embedded in the macrocycle (e.g., pateamine, madumycin) or positioned in a side chain (e.g., epothilone B, rhizoxin, see Fig. 1) [4].

Naturally occurring macrocycles are attractive targets for the synthetic organic community, first of all because of their challenging structures. They have aided progress in the systematic approach to analyze and synthesize stereogenic triads by aldol or homoallyl alcohol methods [5, 6].

Second, many macrocycles possess biological activities, that are unrivaled by more traditional small ring medicinal compounds. This combination of high activity and especially selectivity gives macrocycles an enormous potential in drug development, which is illustrated by the great importance of the macrolide antibiotics such as erythromycin or vancomycin for the treat-

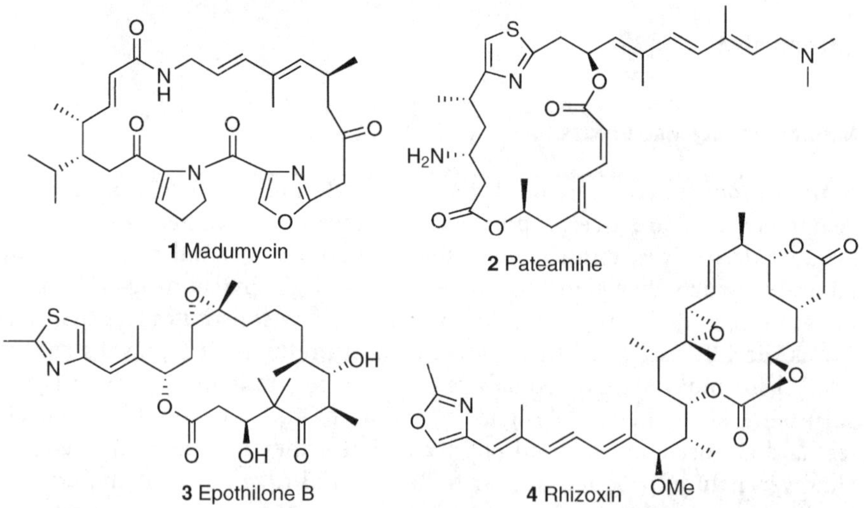

**Fig. 1** Naturally occurring oxazole- and thiazole-containing macrocycles

ment of bacterial infections, nystatin as antifungal compound [7], FK506 as immunosuppressant [8], and recently the epothilones, which are currently in clinical trials for cancer therapy [9–13]. The reliable and versatile synthetic access to these functionalized macrocycles enables the synthesis of derivatives and thereby lead optimization. Finally, syntheses of these molecules also help us to gain an understanding of their actions on a molecular and supramolecular level. Concerning the former, insight can be gathered on how substitution pattern and stereochemistry affect the overall conformation of the molecule. In the latter case, information can be obtained about the way these molecules bind to their cellular targets (usually proteins) and eventually even the characterization of (new) cellular processes will be possible.

However, classical total synthesis typically involves long reaction sequences and the use of exotic reagents or catalysts. Particularly from the viewpoint of drug development, these methods are not always practical for financial and technical reasons, and especially because of the long development times. There is a continuous demand for structurally novel lead compounds. In the development of antibiotics in particular, and to a lesser extent also for antitumor and antifungal agents, resistance is a growing concern. Furthermore, many cellular processes remain to be uncovered that could be attractive targets in drug development. In this context, Schreiber introduced the concept of "chemical genetics" [14, 15]: using chemical diversity arising from both nature and modern synthetic chemistry to identify thus far unknown cellular processes and the proteins and genes involved in them. In the past few years a lot of progress has been made in this field [16] and new cellular processes and biochemical tools to control them continue to be uncovered [17, 18].

In recent years, developments in high-throughput screening inspired many pharmaceutical companies to focus and rely on combinatorial chemistry, especially massive parallel synthesis, to find new lead structures. The employed chemistry is often simple and the concept depends on sheer numbers for success. The main research areas were heterocyclic and peptide chemistry, and the resulting structures often lacked complexity and diversity, and most importantly the chance to utilize the evolutionary advantage of natural products with their privileged structures.

A reasonable compromise between total synthesis of complex natural products and the simple but massive synthesis of combinatorial chemistry is the synthesis of natural product-like molecules [19, 20], giving up a degree of complexity to gain accessibility and at the same time a high level of diversity by making full use of modern synthetic methods. Doing so is also in accordance with the lessons to be learned from nature. After all, nature has been doing combinatorial chemistry for millions of years to achieve the ultimate ligand for a whole variety of receptors [4].

Some basic, repetitive processes, most of an iterative nature, cover a large part of what we consider the ultimate chemical diversity known from natural products [4, 21]. Basic biosynthetic pathways of nature include (1) polyketide paths, attaching mostly acetate, malonate, and propionate units in aldol (Claisen) type reactions, related to the very similar fatty acid and sugar biosynthesis; (2) isoprenoid paths from isopentenyl and dimethylallyl diphosphate building blocks condensed mostly by Friedel-Crafts-like reactions; (3) peptide paths using amino acids condensed to form amide bonds; and (4) sugar condensation to give oligo- and polysaccharides by acetal formation. The structural diversity achieved by these primarily linear processes is incredibly enhanced by subsequent cyclization reactions to give, e.g., (1) aromatic or macrocyclic polyketides; (2) terpenoids and steroids; or (3) alkaloids or cyclopeptides, respectively (Fig. 2).

A third step in diversity enhancement involves post-modification, mostly by oxidation or reduction, elimination, or side chain attachment.

Another method of nature to boost structural diversity is the mixing of natural product building blocks derived from any of the major pathways to form conjugates. Especially sugars are often involved in such processes, e.g., polyketide glycosides like erythromycin, steroid glycosides like digitoxin, and glycopeptides like vancomycin. These four steps of natural diversity enhancement are schematically represented in Fig. 2.

The natural antibiotic novobiocin (Fig. 3) is an example of diversity enhancement by the combination of building blocks from different pathways, such as aromatic rings, isoprenoid, amino acid, and sugar components, finished by post-modification [22].

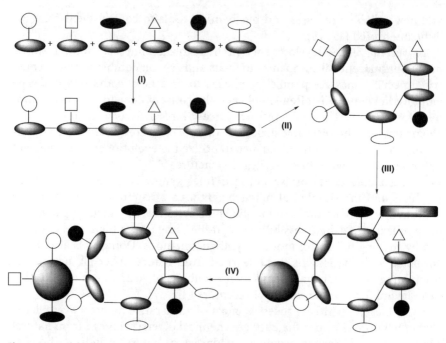

**Fig. 2** Typical biosynthesis of a complex natural product: (I) iterative oligomerization, (II) cyclization, (III) combination/conjugation, and/or (IV) (post-)modification

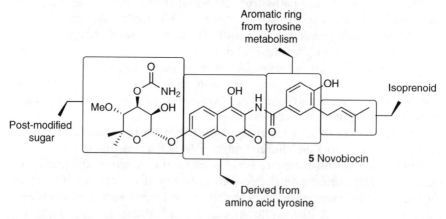

**Fig. 3** Novobiocin

There are, however, some limitations to this, involving, e.g., limitations of the metabolism of the organism in question. Where nature can sometimes do what is very difficult for the organic chemist, the reverse is also true in some cases, and synthetic organic chemistry may add a level of diversity that nature will not easily achieve. In addition, the clinical target of a biologically active natural product is rarely identical with the natural target. While targets may coincide, e.g., for *Streptomyces*-produced antibiotics, an antitumor compound such as epothilone B (see later) from the myxobacterium *Sorangium cellulosum* is hardly produced to fight human cancer. Often, though, targets of such compounds are related to important basic (and highly conserved) cellular processes, such as those involved in cell division, and they may display activity over a wide variety of organisms. It can therefore be expected that a slightly modified derivative of the natural product is a much better ligand for a slightly different receptor than an artificial one might be. The underlying reason on a molecular level is found in the evolution and construction of proteins based on domains (see below).

Waldmann recently suggested that natural products are ideal scaffolds for diversity-oriented synthesis, since they are biologically validated structures, i.e., they are already the product of nature's combinatorial chemistry and have proven their ability to bind to proteins and protein domains [23, 24]. Thus, in the infinite space of chemical diversity, natural products are a good starting point. The majority of current drugs exert their effect through binding to a receptor protein. Proteins are built up in a modular fashion and have evolved through variation and recombination of modules (domains), which fold in a certain way, more or less independent of the rest of the protein. Although the total number of proteins in humans is estimated to be extremely high (100,000–450,000), the number of folding types is estimated to be 600–8000. The number of sequence families in protein domains is estimated to be 4000–60,000. It is therefore possible that proteins with similar folding types (though not necessarily functionally related) are inhibited by the same class of small molecule compounds. Indeed, it has been demonstrated that derivatives of natural products that bind to a certain receptor can inhibit a structurally but not functionally related protein selectively [25, 26]. This observation validates the use of natural products as starting point for diversity-oriented or combinatorial chemistry, even if their native biological activity is not desired.

Combinatorial chemistry as it has been performed in the 1990s, relying solely on high numbers, rarely led to the expected success, probably because the majority of the synthesized compounds were not biologically relevant [23, 24]. There is a need for sensible starting points, and natural products in general and macrocycles in particular provide such starting points. In short, the objective is to achieve a synthesis fast, flexible, and economically feasible enough to keep it attractive for drug development, while keeping in mind the lessons learned from nature and making full use of modern methods in organic chemistry.

## 1.2
## Macrocyclization Strategies

In the case of macrocycles, even in total synthesis, the most demanding task may be the selective formation of the large ring. The difficulty resides in the preferential formation of an intramolecular bond, rather than giving the competing intermolecular reaction which will result in cyclic or linear oligomers. Unless templates or prefolding aid the intramolecular reaction, statistics and other factors like ring strain are in favor of the undesired oligomerization. Also, equilibrium conditions are normally unsuitable to form disfavored macrocycles, but in special cases they also can be useful (see below, dynamic library formation). In a combinatorial approach, it becomes increasingly difficult to find a macrocyclization method that is practical and general enough to work reliably with a wide variety of substrates, functional groups, and folding preferences. At the same time, other constraints amount, like demands to minimize steps, avoid protective groups, or problematic reagents (e.g., Sn-, Hg-compounds etc.).

### 1.2.1
### Macrolactonization

Since lactones are common in naturally occurring macrocycles [4], macrolactonization (Fig. 4) is an attractive way of ring closure. Many examples have

**Fig. 4** General mechanism macrolactonizaton

been described in the literature, using e.g., Yamaguchi (2,4,6-trichlorobenzoyl chloride, $i$-Pr$_2$NEt, DMAP) [27], Keck (DCC, DMAP, DMAP·HCl) [28], or Mitsunobu (Ph$_3$P, DEAD) [29] protocols. All three are common and reliable methods, and can be applied to a wide variety of substrates. An obvious drawback of this method is the fact that (in principal) all alcohol functionalities other than the one involved in lactone formation should be protected. An example of the use of a Yamaguchi macrolactonization was presented in a total synthesis of the microtubule-stabilizing agent laulimalide (Scheme 1) [27].

In rare cases, lactone formation can also be achieved by nucleophilic displacement with a cesium carboxylate onto an activated leaving group, thus circumventing the polyol-problem [30]. An alternative to the macrocyclic

**Scheme 1** *a* 2,4,6-trichlorobenzoyl chloride, *i*-Pr$_2$NEt, THF, then DMAP, benzene, 68%

adaptations of classical esterifications are biocatalytic methods utilizing hydrolases in inverse mode. Enzymes have been successfully applied for macrolactonization [31, 32], and macrolactamizations [33] and acetalizations (cf. cyclodextrin formation) [34] are also known. A wider applicability in complex macrocyclizations for the nucleophilic and enzymatic methods remains to be proven.

### 1.2.2
### Macrolactamization

Although in medicinal chemistry macrolactams are perhaps not as important as macrolactones, they still constitute an important class of biologically active macrocycles, even without counting the significant occurrence of cyclopeptides and glycopeptides [4, 33]. Since amines are even better nucleophiles than alcohols, macrolactamization is an obvious means of macrocyclization for these compounds. The development of solid phase peptide synthesis has led to the introduction of a variety of highly efficient peptide coupling agents, including HATU, PyBOP, and FDPP. Such reagents have largely replaced the classical Steglich conditions (DCC/DMAP) and act as activating and dehydrating agent simultaneously. They have also proved effective in the synthesis of macrolactams, e.g., in the total synthesis of the cyclopeptide ceratospongamide [35].

Because of the strong nucleophilicity of the amino group, weak activation of the carboxylic acid is often sufficient to effect macrocyclization. For example, the pentacyclic core of the manzamines was constructed by connection of a secondary amine and a pentafluorophenyl ester [36].

Oxygen- and sulfur-assisted methods have been described to synthesize strained macrolactams by ring contraction through attack of the amine on the intermediate macro(thio)lactone [37].

### 1.2.3
### Ring-Closing Metathesis

The recent enormous development in metathesis catalyst activity and stability has made ring-closing metathesis (RCM, Fig. 5) one of the most popular macrocyclization methods [38]. The enormous potential of this method lies in

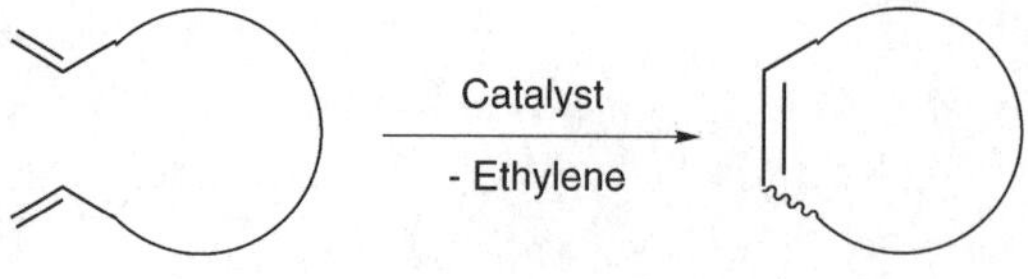

**Fig. 5** General mechanism ring-closing metathesis (RCM)

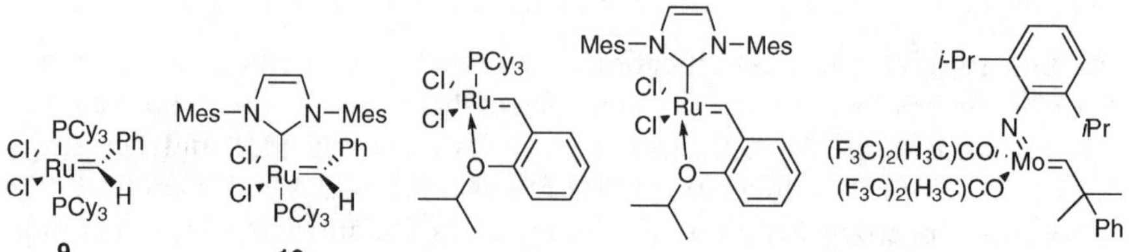

**Fig. 6** Commonly used metathesis catalysts

the fact that the RCM-reacting (preferably monosubstituted) carbon-carbon double bonds are not affected by most reaction conditions and can therefore be introduced at any desired stage in the synthesis, whereas under RCM conditions most other functionalities are unaffected and macrocycles can be obtained in generally good to excellent yields. Also, the resulting double bond can be further modified by hydrogenation, dihydroxylation, or other reactions.

The most widely used catalysts for RCM are Grubbs' ruthenium catalyst 9 and its second generation analogue 10, as well as first and second generation Hoveyda-Grubbs catalysts 11 and 12 (Fig. 6) [38]. The latter have superior stability and reactivity, expanding the applicability of the method considerably. Schrock molybdenum catalyst 13 has also been described for macrocyclization [38].

RCM can be used not only for five- or six-membered rings, but also for medium sized rings and macrocycles. After optimization of reaction conditions and catalyst, the product is often obtained in excellent yield and, as shown in the example below (Smith's synthesis of salicylhalamide A, 16, Scheme 2 [39]), even in high $E$-selectivity.

One disadvantage in some cases (especially with rings larger than ten-membered) is the lack of control over the stereochemistry of the double

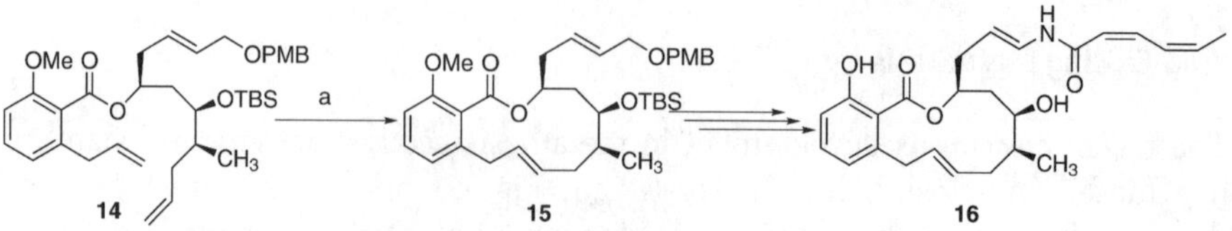

**Scheme 2** *a* Grubbs' catalyst, $CH_2Cl_2$, 85%, $E/Z$=10:1

bond. Fürstner et al. have reported the use of alkyne metathesis in the synthesis of epothilone A and C to circumvent the lack of E/Z selectivity [40]. The resulting cycloalkyne can subsequently be converted to either the E or the Z double bond. However, alkyne metathesis is currently not straightforward.

Another drawback of RCM applies in case of certain substitution patterns that lead to preferential complexation of the catalyst by substrate functional groups, thereby rendering it inactive. This may be overcome either by a more active or different catalyst or by precomplexation of the donor groups, e.g. with titanium salts [41].

An especially attractive strategy is simultaneous cyclization/cleavage by RCM [42, 43]. This concept was first introduced by van Maarseveen in the synthesis of seven-membered lactams [44] and subsequently employed by many groups. Recently, this strategy was employed in the solid phase synthesis of epothilone A [45, 46] and dysidiolide analogues [47, 48].

## 1.2.4
## Macroaldolization

Aldol reactions have also been used as a means of macrocyclization in total synthesis and were quite successful in some cases. However, over a broader spectrum of substrates, the results are unpredictable at best and yields and stereochemical outcome vary greatly. The predominant reasons are difficulties in selective enolate formation in multi-carbonyl compounds, competing and equilibrating retro-aldolizations—especially with polyketides, which often possess several aldol moieties—and intermolecular instead of intramolecular reaction preference. Whereas most of these drawbacks may be overcome, substrate-independent stereocontrol plays a crucial role. At least one new stereocenter is formed during a macroaldolization, and because of the folding constraints involved, its configuration cannot be adequately predicted. Therefore, this can be useful in special cases but with the current possibilities is not the method of choice for a general diversity-oriented synthesis.

One example of the successful use of a macroaldolization was described by Danishefsky et al. in their early synthesis of epothilone B [49]. Another one by us is reported below [50].

## 1.2.5
## Palladium Coupling

The selectivity, functional group tolerance, and mild reaction conditions of the various $Pd^0$ cross-coupling reactions have made them invaluable tools for carbon-carbon bond formation. Intramolecular variations to construct macrocycles have also been described.

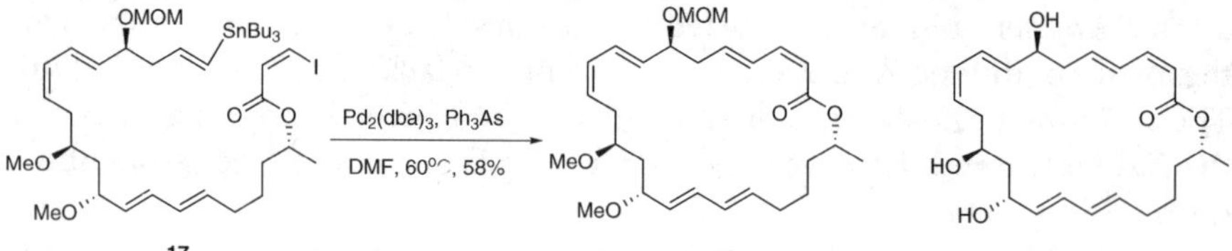

**Scheme 3** Macrocylization by intramolecular Stille reaction to protected macrolactin A by Pattenden

An example is the intramolecular Stille reaction described by Pattenden in his synthetic studies towards macrolactin A (**19**, Scheme 3) [51].

Other $Pd^0$ cross-coupling reactions such as Heck [52] and Suzuki [53] reactions have also been used for macrocyclizations. The main drawback for $Pd^0$ catalyzed macrocyclization is the yield, that is often somewhat disappointing if compared with other established methods. Also, the introduction of the required coupling components (e.g., trialkyltin group, vinylic iodide) can be difficult in some compounds. In other cases, Pd-catalyzed side reactions such as double bond migration or allylic activation can occur.

## 1.2.6
## Other Methods

Many other reactions have been used to construct macrocycles, and many are related to specific bond formations and therefore not very widely applicable. Since bisaryl ethers and biphenyl units are relatively common in naturally occurring macrocycles [4] (both units can be found in the important antibiotic vancomycin, **20**, Fig. 7), some approaches to their synthesis will

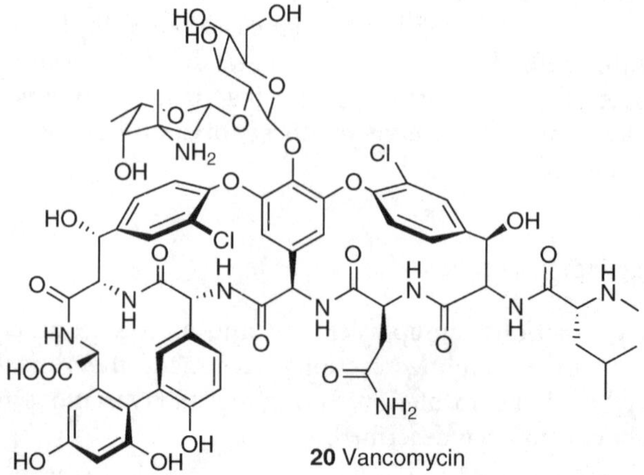

**Fig. 7** Structure of the clinically used antibiotic vancomycin

be briefly discussed. Most of these methods are based on $S_NAr$ reactions, often in a biomimetic fashion.

Schreiber proposed the Lipschutz method (oxidative coupling of two aryl-Cu(I) moieties) to generate a library of biaryl-containing macrocycles [54]. Biaryl macrocycles were obtained in good to excellent yield where other methods ($Pd^0$ coupling etc.) failed or gave inferior results (Scheme 4).

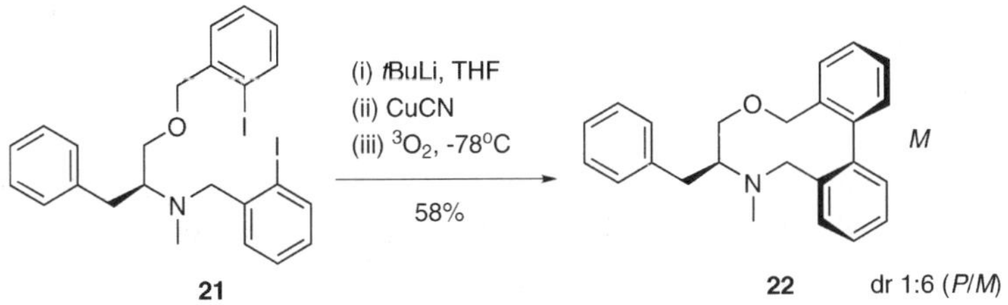

**Scheme 4** Formation of macrocyclic biaryl bond through oxidation of arylcopper intermediate

Zhu and co-workers used $S_NAr$ reactions to create the biaryl ether bond in their synthesis of cyclopeptide alkaloids [55–57] (Scheme 5). The electrophile in this case is an aryl fluoride, often additionally activated by an electron-withdrawing group such as a nitro group. The necessity of such an activation auxiliary obviously has a negative impact on general applicability of this method.

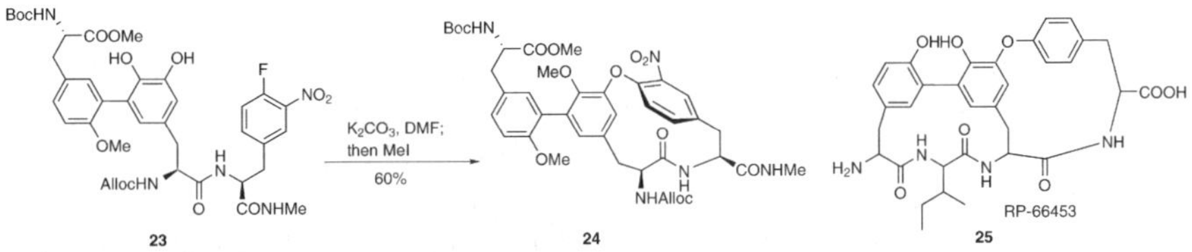

**Scheme 5** Macrocyclic biaryl ether formation by $S_NAr$ reaction in synthetic studies towards RP-66453 (25)

Venkatraman et al. recently reported intramolecular coupling of a phenol to an aryl chloride complexed to ruthenium [58] (see Scheme 6), a method employed also in approaches to vancomycin and the similar antibiotic ristotecin A [59–61]. The advantage of this method is that no fluorine substituent is required and no additional activating group had to be introduced and removed.

Another recent concept is the template-directed synthesis from dynamic combinatorial libraries. Here, one or several building blocks that are prone

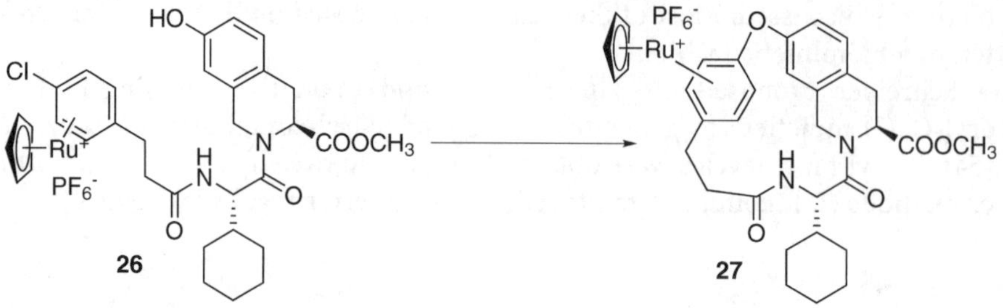

**Scheme 6** Conditions: 5.0 equiv. CsCO$_3$, DMF, Δ

to reversible (cyclo)oligo- or polymerization are combined in the presence of a template, which may be a host or guest molecule or ion, so that one of the possible structures is favored [62, 63]. If the reversible reaction is then terminated (and the dynamic equilibrium frozen) and the template removed, the template-induced structure ideally should prevail and can be isolated. A good example is the synthesis of pseudocyclopeptide lithium receptor. A linear monomer forms a whole range of cyclic oligomers when exposed to dilute acid, but in the presence of lithium ion as template, the cyclotrimer is strongly favored [64].

## 1.3
## Diversity-Oriented Synthesis

The term diversity-oriented synthesis (DOS) is relatively new and, as mentioned above, is usually defined as the synthesis of complex, natural product-like molecules using a combinatorial approach and employing the full palette of modern organic reactions. It may be a subject of discussion what exactly qualifies a molecule as being "natural product-like" [4], and in most cases the similarity to an actual natural product seems reciprocal to the number of synthesized compounds. However, even in less complex cases, the products may be highly substituted polycyclic structures with defined stereochemistry, reminiscent of natural products [19, 20]. In these cases, a moderately complex backbone structure is subsequently modified with a well-established set of selective reactions to introduce diversity.

In general, DOS demands well-defined, selective reactions that are still able to introduce a significant degree of complexity and diversity. Multicomponent and tandem processes fit this profile very well [65, 66], and several groups have used this approach [41, 67–71].

Sharpless et al. recently introduced the term "click chemistry" for the use of simple (but selective), high-yielding reactions forming mostly carbon-heteroatom bonds, preferably in aqueous medium, using only easily available compounds [72]. The authors urge the synthetic community to concentrate on the development of new properties rather than new compounds, and believe that synthetic chemistry as we know it in modern-day total synthesis is too inefficient for that purpose. This idea follows nature's principle that only function or properties are important, not what, e.g., chemists consider an exciting structure. Using a modular approach, diversity can be introduced easier and cheaper, and thus in greater numbers and/or scale.

The basis for the concept is addition of a heteroatom species to olefins to give, e.g., epoxides or aziridines, which can subsequently be ring-opened. Other "click chemistry" transformations include nucleophilic substitution of activated species, addition to (thio)cyanates, 1,3-dipolar cycloadditions, etc. followed by modification of the scaffold, enabling the creation of libraries without cumbersome purifications and resynthesis of a lead compound in large scale for further development. This is certainly an interesting concept, especially in the context of DOS, although it might be imprudent to ignore all the merits of modern synthetic chemistry, such as (stereo)selective C-C coupling reactions. Also, most click chemistry procedures, i.e., simple bulk-suitable reactions, are derived, developed, and optimized out of our two centuries old coal and petrochemical heritage. The available compounds and processes only in some cases (e.g., epoxidation) overlap with the much more oxygen rich chemistry nature uses preferentially. Thus, this approach may lead into the same trap the number crunching combinatorial chemistry encountered: that of low biological significance.

As a consequence, it must be asked if we can learn more from nature than important structural features for lead compounds. Can we also learn about the types of chemistries which deliver these compounds [4]? Can we design a next generation "click chemistry" or iterative or diversity-generating processes based on the concepts or reactions of nature rather than only on its compound leads? In case of the reactions, our petrochemistry-optimized knowledge often is still limiting such an approach, but the necessary improvements to follow nature's path, especially in catalysis, are encouraging. At this time, however, it appears easier to learn from nature's principles, thereby still applying the established classical, reliable reactions preferentially. Such principles are given in Fig. 2 (see above), and include the synthesis at spatially defined locations, e.g., cell compartments or membranes. The synthetic chemists analogue to a cell compartmentalized synthesis is solid phase-bound chemistry. A second important feature of the first step is the iterative oligomerization, which allows the production of defined homomers (e.g., linear oligoprenyl diphosphates like farnesyl diphosphate as sesquiterpene precursor) or non-repetitive heteromers (e.g., peptides, polyketides). The first non-natural example of such a process is Merrifield's peptide synthesis [73]. However, in contrast to nature, bulk production by iterative processes is still problematic [4, 5].

In recent years, the synthesis of all kinds of compounds on solid support has been the method of choice for DOS, both in industrial and academic laboratories. It is especially valuable when medium to large numbers of compounds are to be synthesized, the required amount is small, and the need for synthesis automation is high. In other cases, however, solution phase chemistry was highly successful, and parallel automated synthesis gains popularity. Some illustrative examples of both approaches are given below.

## 1.3.1
### Solid Phase Diversity-Oriented Synthesis

Solid phase synthesis has many advantages where library creation is concerned, the most important of which is the possibility of synthesis automation, providing strictly defined standardized reaction conditions. In the case of macrocycle synthesis, another advantage is the inherent creation of (pseudo-)dilution conditions required for the formation of macrocycles rather than di- or oligomers. In rigid or low loading resins, functional groups are generally so far away from each other that the system resembles a very dilute solution. In more flexible resins, site-site interactions may play a significant role and can actually be exploited [74]. If the cyclization reaction is connected to the resin-release reaction, an inherent purification is achieved (selective release strategy, Fig. 8): In case of a successful macrocyclization, the product is released, whereas in case of an intermolecular (site-site) reaction or any other reaction of the free end of the linear precursor, the byproduct still resides on the polymer. Thus, linear oligomers are not eluted. Cyclic oligomers, however, may be formed to a minor extent.

Solid phase DOS of natural product-like molecules in many cases concerns polycyclic compounds, but macrocycles have also been reported [4]. In some cases, it is difficult to draw a line between DOS and simple derivati-

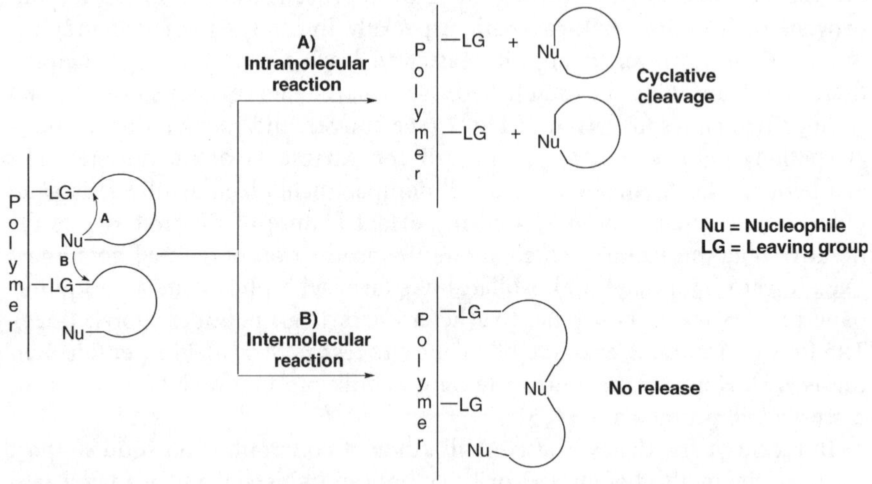

**Fig. 8** Example of a cyclative cleavage strategy

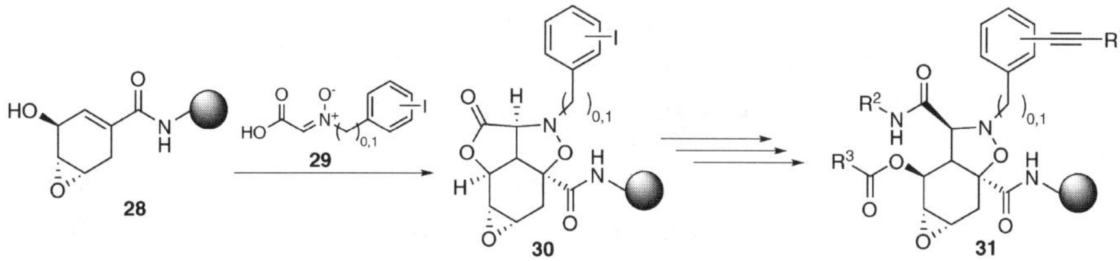

**Fig. 9** Solid phase DOS of partially "natural product-like" compounds by modification of a tricyclic scaffold

zation, since many reports feature the binding of a scaffold molecule to solid support and subsequent modification to create diversity, not unlike the post-modification approach described below. Usually, however, the modifications introduce sufficient new properties to fit the DOS definition.

Tan et al. reported the stereoselective synthesis of over two million compounds derived from solid phase-bound tricycle **30** (synthesized from shikimic acid derivative **28** and nitrone **29**, Fig. 9) [75, 76]. The aromatic iodide was modified by Sonogashira reaction with alkynes $R^1C\equiv CH$, followed by lactone ring opening by amines $R^2NH_2$ and acylation of the resulting alcohols with acids $R^3COOH$ to give compounds **31**. Further possibilities of modification are epoxide opening and acylation of the resulting alcohols. The term "natural product-like" stressed by the authors, however, can only relate to the starting material, shikimic acid, and the high functionalization

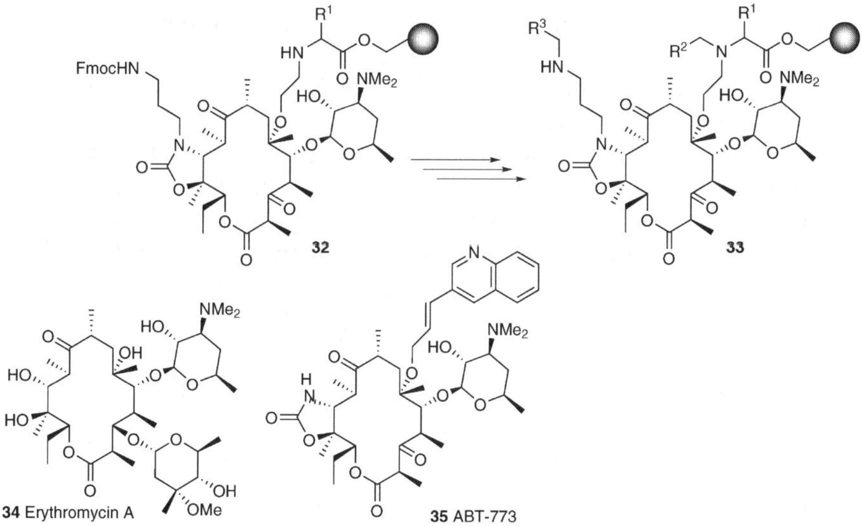

**Fig. 10** Solid phase synthesis of a library of derivatives of third-generation macrolide antibiotics

density of the product, but neither the products themselves nor the reactions are closely related to natural ones. Thus, this result can be considered an excellent example of a combined natural product-like and chemistry-driven DOS.

Another example of solid phase DOS involves post-modification of the natural product macrolide antibiotic erythromycin (**34**) [77]. Erythromycin was first converted to analogue **32** which resembles a third generation macrolide antibiotic with high activity against resistant strains (ABT-773, **35**), but is attached to solid phase-bound amino acids by reductive amination. Two further reductive amination steps and cleavage from solid support form a library of compounds of type **33** (Fig. 10).

## 1.3.2
### Solution Phase Diversity-Oriented Synthesis

An important difference between combinatorial or parallel DOS and, for example, total synthesis is the high number of synthesized compounds, which renders conventional purification procedures such as flash chromatography impractical, at least if they have to be performed after every step. Synthesis on solid support avoids these problems to some extent if high conversions can be achieved with excess reagent or selective release strategies (see above). To do solution phase DOS, selective, high yield reactions or selective purification processes are required.

Also reactions that rapidly create more diversity are sought after. In recent years, more and more multicomponent reactions (MCRs) have been developed and combined with modern synthetic methods such as RCM or Pd$^0$ couplings to provide rapid access to complex, natural product-like structures in solution.

Perhaps the most useful and most widely employed multicomponent reaction in this context is the Ugi four-component reaction (U4CR and variations, Fig. 11), where an isonitrile, an aldehyde, an amine, and a carboxylic acid react to form a single product [65, 66]. The principal product is a dipeptide or dipeptide derivative and a high degree of diversity can be introduced by substituents (each of the four components can be varied), subsequent reactions, or by other reaction paths, which can lead to a vast varia-

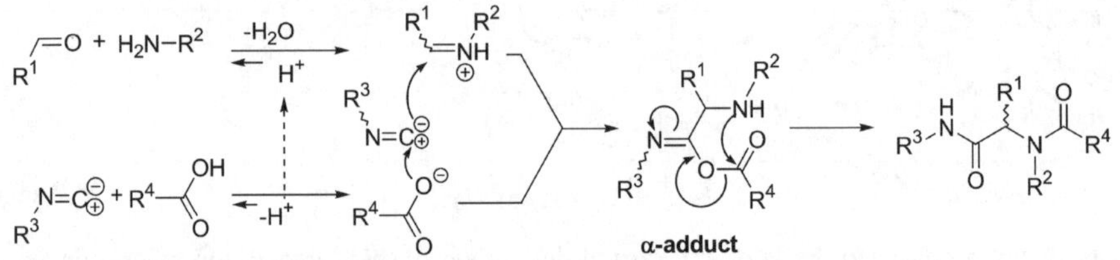

**Fig. 11** Putative mechanism of the Ugi four-component reaction

tion also in the backbone [78]. It provides an interesting basis for DOS of peptide-based natural product-like compounds. An additional advantage of this method is the high functional group tolerance, and therefore the possibility of directly modifying these functional groups after the Ugi reaction.

The possibilities of the Ugi reaction increase when it is combined with other reactions to increase molecular complexity. One example was described by Lee et al. and involved an Ugi reaction followed by a spontaneous intramolecular Diels-Alder reaction between the furan diene and the fumarate double bond [79]. After double amide allylation (KHMDS, allyl bromide), the resulting compound was treated with second generation Grubbs' catalyst 10, leading to a ring-opening/ring closure reaction and the formation of 7-5-5-7 polycyclic system 40, which was then desilylated to give 41 (Scheme 7).

**Scheme 7** *a* MeOH/THF, 48 h, 67%; *b* KHMDS, allyl bromide, rt, 89%; *c* metathesis catalyst, CH$_2$Cl$_2$, Δ, 69%; *d* HF·py, 95%

Other examples of Ugi reactions combined with RCM have been described in the literature. Hebach and Kazmaier reported the synthesis of conformationally fixed cyclic peptides [70] and Beck and Dömling synthesized biaryl-containing natural product-like macrocycles using this method [41]. The same group also reported combination of Passerini and Horner-Wadsworth-Emmons reactions to obtain butenolides [67] and another variation for the combinatorial synthesis of thiazoles [69].

Gámez-Montaño et al. described an Ugi variation where the carboxylic acid was replaced with an amide [68]. The amide oxygen is nucleophilic enough to effect ring closure to oxazole intermediates 44, which then undergo aza-Diels-Alder reaction with the double bond of the allylic amine component to form oxa-bridged heterocycles 45, which can either be isolated as a separate class of compounds or converted to pyrrolopyridines 46 by treatment with TFA (Scheme 8).

These examples illustrate the power and flexibility of the Ugi reaction in combination with additional complexity-generating reactions. These possi-

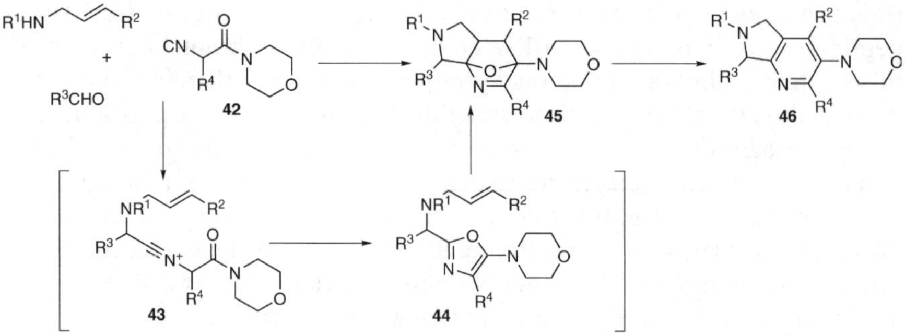

**Scheme 8** Ugi-type multicomponent synthesis of pyrrolopyridines

bilities were also exploited in efforts in the synthesis of peptoid macrocycles (see the third main section of this chapter [78]).

DOS seems mainly important for lead identification. Based on the lead structure, additional structural complexity can be introduced in a direct manner by generating focused libraries. For the synthesis of complex molecules, total synthesis as it exists today and the development of selective reactions that accompany, it remain of major importance. Even in case of the successful identification of advanced leads or drug candidates from DOS, an efficient total synthesis is always required in order to provide sufficient amounts of a limited set of derivatives for detailed medicinal chemistry studies up to clinical trials.

Total synthesis of complex (macrocyclic) natural products using fast and flexible strategies and diversity-oriented synthesis of natural product-like macrocycles are important research topics in our laboratory. The following sections describe the total synthesis of epothilone D and epothilone $D_5$ analogues, DOS of cyclopeptide alkaloid analogues, of biaryl ether macrocycles, and of steroid/peptide hybrid macrocycles, respectively.

# 2
# Classical Synthesis: Epothilones and Derivatives

## 2.1
## Introduction

Epothilones A–E (**3, 17, 48–50**, Fig. 12) were isolated by Höfle and Reichenbach and co-workers from the myxobacterium *Sorangium cellulosum* [80, 81]. After initial selection for their antifungal activity, the epothilones were soon shown to possess significant cytotoxicity against mammalian cells, epothilone B being the most active one [82]. Their mode of action was shown to be microtubule stabilization in a fashion very similar to that of

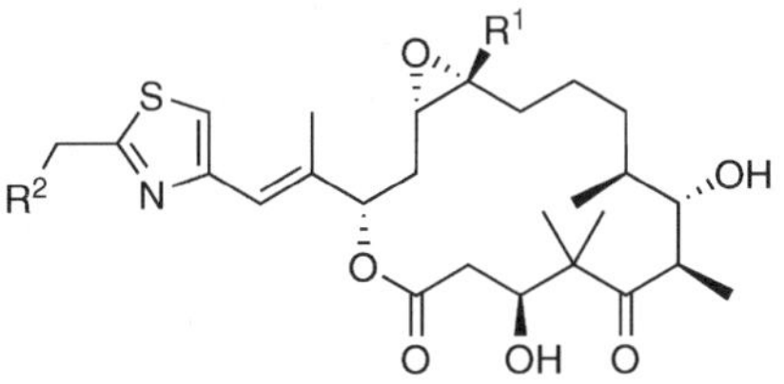

**47** Epothilone A: $R^1=R^2=H$
**3** Epothilone B: $R^1=Me$, $R^2=H$
**50** Epothilone E: $R^1=H$, $R^2=OH$

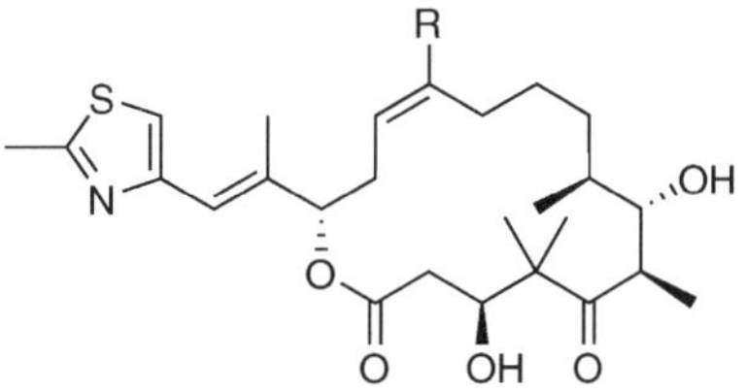

**48** Epothilone C: R=H
**49** Epothilone D: R=Me

**Fig. 12** Epothilones A–E

paclitaxel (Taxol, **51**). However, the epothilones have greater activity than paclitaxel, and maintain activity against paclitaxel-resistant cell lines. Evidence has been presented that epothilone and paclitaxel binding sites in tubulin at least partly overlap [82].

The extraordinary biological activity of epothilones has spurred interest of scientists around the world. Indeed, several epothilones and many derivatives are currently in different phases of clinical trials for the treatment of various forms of cancer. Also the synthetic community has given a great deal of attention to these remarkable compounds, probably more than to any other compound in the last ten years. This is not very surprising, because in comparison to paclitaxel (which until recently was one of the main success stories of natural products research), the epothilones have a relatively simple structure, which allows easier modification, and they display higher in vitro activity as well as better pharmacokinetic properties.

The first total syntheses of epothilones A and B were published by the groups of Danishefsky [49] and Nicolaou [83] only shortly after publication of the absolute stereochemistry of the natural products. Various other (formal) total syntheses followed, in which many different approaches and reaction types were used. Also, a variety of analogues was synthesized and even libraries were created [9, 84]. Several reviews on the properties and synthesis of epothilones and derivatives have been published [9–13].

## 2.2
## Retrosynthesis

Figure 13 outlines our retrosynthetic analysis of epothilone D (**49**). Since epoxidation of epothilone D to give epothilone B (**3**) is the last step in all total syntheses of the latter, this is considered to be trivial. Furthermore, epothilone D shows a lower non-specific toxicity relative to its antimitotic properties.

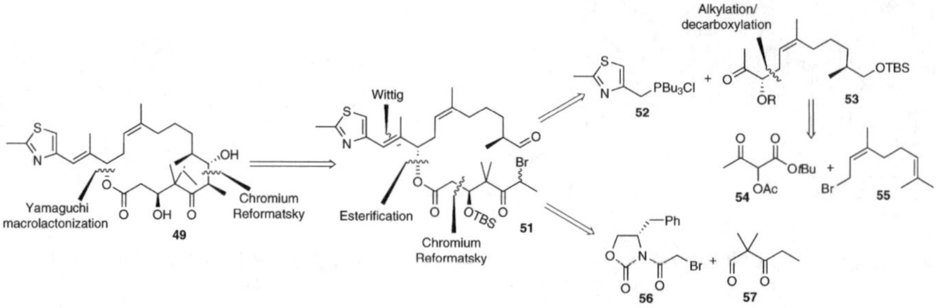

**Fig. 13** Retrosynthetic analysis of epothilone D by Wessjohann et al. [13]

The first three retrosynthetic cleavages are formation of the C7-C8 aldol by intramolecular chromium-Reformatsky reaction of linear precursor **51**, esterification between northern and southern half building blocks and Wittig reaction of phosphonium salt **52** known from Mulzer's work [85] and northern half precursor **53**. The final disconnections were placed at the C2-C3 aldol in **51** (again to be formed by chromium-Reformatsky reaction, here between bromoacetimide **56** and aldehyde **57**) and the C14-C15 bond by alkylation of acetoacetate **54** with neryl bromide **55**.

## 2.3
## Synthesis

### 2.3.1
### Northern Half Synthesis

Synthesis of northern half building block **53** (Scheme 9) commenced with the acetoxylation of commercially available *tert*-butyl acetoacetate to **54** by bromination and in situ displacement of the bromide by acetate. Deprotonation with NaH and alkylation with neryl bromide (**55**) afforded **58** in excellent yield. Catalytic $SeO_2$ oxidation led to selective oxidation of the terminal *E*-methyl group and gave **59** in 45% yield. Since the remainder is mainly unreacted starting material, this can be re-used. Reduction of the C8-C9 double

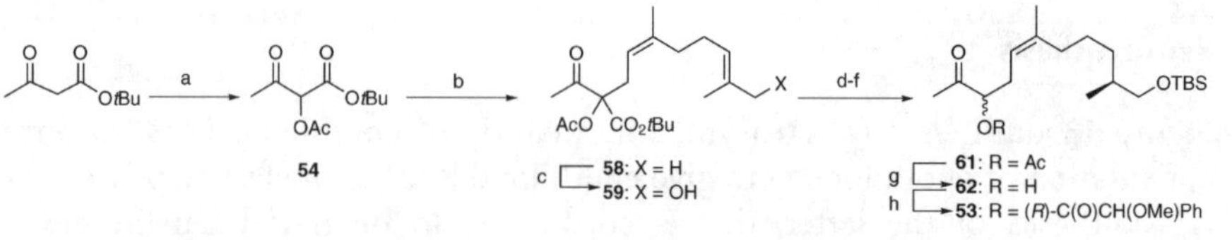

**Scheme 9** *a* $Br_2$, NaOAc, DMF, 64%; *b* NaH, neryl bromide, DMF, 96%; *c* cat. $SeO_2$, *t*-BuOOH, $CH_2Cl_2$, 45%; *d* 100 bar $H_2$, cat. Ru[(*S*)-BINAP](OAc)$_2$, MeOH/$H_2O$, 89% (8*S*:8*R*=81:19); *e* TFA, $CH_2Cl_2$, 1 h, 75%; *f* TBSCl, Et$_3$N, cat. DMAP, $CH_2Cl_2$, 80%; *g* $K_2CO_3$, MeOH, 97%; *h* (*R*)-α-methoxyphenylacetic acid, EDCI, cat. DMAP, $CH_2Cl_2$, 80%

bond by enantioselective catalytic hydrogenation with Ru[(S)-BINAP](OAc)$_2$ furnished primary alcohol **60** (not shown, 89% yield, 8S:8R=81:19). Decarboxylation by TFA treatment and subsequent TBS protection of the primary alcohol led to **61** with racemic C15 stereocenter. After acetate cleavage, the C15 alcohol was re-esterified with (R)-α-methoxyphenylacetic acid, allowing separation of diastereomers of **53** [87].

With the correct C15 stereochemistry established, Wittig reaction of **53** with known phosphonium salt **52** followed by saponification completed the synthesis of northern half building block **64** (Scheme 10) [50].

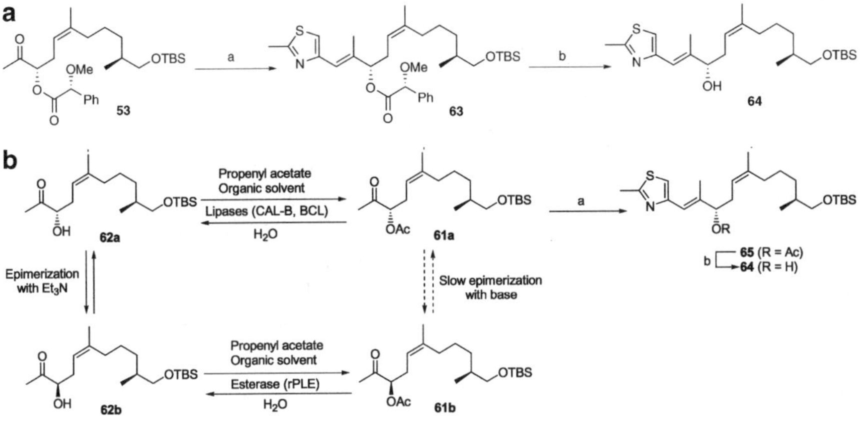

**Scheme 10 a** Northern half precursor **64** by chemical resolution: *a* **52**; NaHMDS, THF, −63 °C, 96%, chromatogr. sepn.; *b* K$_2$CO$_3$, MeOH, 72%; **b** Northern half precursor **64** by enzymatic resolution: *a* **52**; NaHMDS, THF, −63 °C, 97%; *b* K$_2$CO$_3$, MeOH

Alternatively, enzymatic resolution of **61** by hydrolysis or of **62** by enzymatic esterification could be achieved with >99% ee and enantioselectivities of E≥200, e.g. hydrolysis with common lipases like CAL-B or BCL (Amano PS) [86–88]. Wittig reaction and deprotection led to **64**. Enzymatic resolution is also possible at the stage of C15-racemic **65** [86–88].

Thus, racemic **62** can be resolved enzymatically to give either C15 diastereomer of **61**, simply by using the appropriate enzyme. Because it was observed that the C15 stereocenter of **62** epimerizes faster in the presence of base [87, 88] (e.g., Et$_3$N) than that of **61**, even an enantioconvergent process is possible, where **62a** and **62b** are in constant equilibrium, but only **62a** is converted to the corresponding acetate. Enantiomerically pure **61a** is then readily converted to **64**.

## 2.3.2
### Southern Half Synthesis

The next task was to form the C2-C3 aldol bond stereoselectively. However, asymmetric coupling of acetate derivatives to aldehydes is often accompanied by poor β-induction [89]. Moreover, the C3-C4 bond is particularly sensitive to retro-aldol reaction, especially under basic conditions. In the natural products, this was observed to be the main decomposition reaction. The first total syntheses of epothilones circumvented this problem by constructing this part of the molecule in an indirect manner, e.g., by using reduced forms at C1 or C5. We decided to employ our chromium-Reformatsky methodology, which avoids these problems and allows the direct use of reagents in the correct oxidation state. The non-basic reaction conditions, the intermediacy of a chromium(III) aldolate that is resistant to retro-aldol reaction, and the potential of a direct asymmetric carboxymethyl ("acetate") transfer favor the use of this method [90].

Thus, aldol **66** was obtained from Evans' amide **56** and aldehyde **57** in 76% yield and 84% de [91, 92]. After TBS protection of the free alcohol and cleavage of the chiral auxiliary (LiOH, $H_2O_2$, $THF/H_2O$), the resulting free acid was to be brominated at C6 to provide the α-bromoketone required for chromium-Reformatsky coupling to northern half building block **64**. After evaluation of many different brominating agents, phenyltrimethylammonium tribromide proved to be the reagent of choice for the selective bromination at C6, affording **68** almost quantitatively (Scheme 11).

**Scheme 11** *a* 2.2 equiv. $CrCl_2$, 0.1 equiv. LiI, THF, 63% (84% de); *b* TBSOTf, 2,6-lutidine, $CH_2Cl_2$, 93%; *c* LiOH, $H_2O_2$, $THF/H_2O$, 90%; *d* $PhNMe_3Br_3$, THF, 98%

## 2.3.3
### Macrocyclization and Completion of the Synthesis

Several macrocyclization methods were pursued, including macrolactonization and the construction of the C2-C3 bond, but it proved most efficient to form the epothilone macrocycle through formation of the C6-C7 aldol

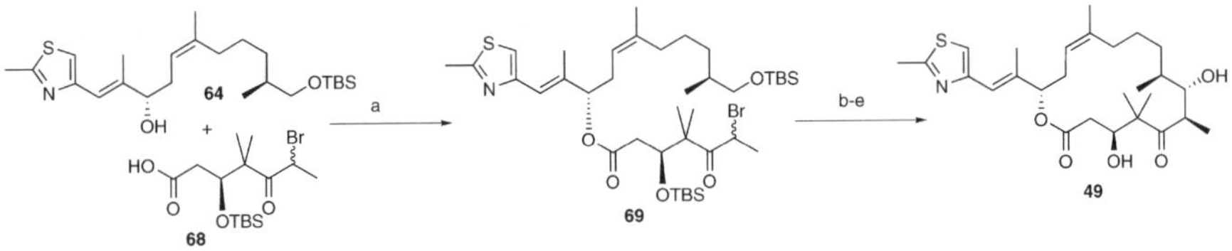

**Scheme 12** *a* EDCI, DMAP, CH$_2$Cl$_2$, 80%; *b* CSA, CH$_2$Cl$_2$/MeOH, 91%; *c* Py·SO$_3$, Et$_3$N, DMSO, CH$_2$Cl$_2$, 88%; *d* 2.2 equiv. CrCl$_2$, 0.1 equiv. LiI, THF, 29%; *e* TFA, CH$_2$Cl$_2$, 79%

(Scheme 12). For this purpose, acid **68** was esterified with alcohol **64** to give the linear epothilone precursor **69**. After selective deprotection of the primary TBS ether (CSA, CH$_2$Cl$_2$/MeOH) and oxidation, precursor **51** (Fig. 13), suitable for ring-closing chromium-Reformatsky reaction, was obtained. After treatment of **51** with CrCl$_2$ and LiI in THF, the desired 6R,7S isomer was obtained exclusively, whereas the diastereomeric precursor with the incorrect stereochemistry at C8 did not cyclize with the same efficiency, rate and selectivity.

Final deprotection of the C3 alcohol then afforded epothilone D (**49**) (Scheme 12). The synthetic product was shown to be identical to the natural material in all respects.

Alternatively, a macrolactonization route can be followed, where formation of the C6-C7 aldol by chromium-Reformatsky and esterification are interchanged. However, this route is longer and less selective in formation of the C6 and C7 stereocenters, giving both *syn* products [50, 86–88].

### 2.3.4
### Synthesis of Epothilone D$_5$ Analogues

Recently, Höfle and Reichenbach also described the isolation of some further components of the epothilone family [93]. One of these was epothilone D$_5$ (Fig. 14), which differs from epothilone D by the presence of an additional unsaturation between C8 and C9. Since this double bond is inherent to our synthesis of epothilone D, we also decided to pursue the total synthesis of epothilone D$_5$ and its diastereomers.

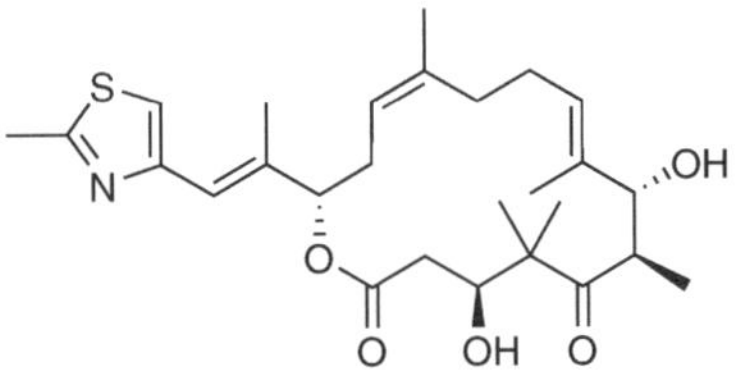

Epothilone D$_5$ (**70**)

**Fig. 14** Structure of epothilone D$_5$

Synthesis of the northern half building block **64** was the same up to compound **59**. Its asymmetric hydrogenation was omitted, and the subsequent steps were performed in analogy to those of the synthesis of the northern half of epothilone D (Scheme 13).

**Scheme 13** *a* TFA; *b* TBSCl, Et$_3$N, cat. DMAP; *c* K$_2$CO$_3$, MeOH; *d* (R)-α-methoxyphenylacetic acid, EDCI, DMAP; *e* **52**, NaHMDS, –78 °C; *f* K$_2$CO$_3$, MeOH

The southern half of epothilone D$_5$ is identical to that of epothilone D, and was coupled to northern half alcohol **77** in moderate yield, presumably because of a competing intramolecular cyclization of **68**. Selective cleavage of the primary TBS ether, followed by Doering oxidation, smoothly provided the cyclization precursor. Intramolecular chromium-Reformatsky reaction led to the formation of a single macrocyclic product in a similar yield as for the saturated counterpart. The final TBS deprotection proved troublesome. Many fluoride-based methods (TBAF, AcOH/TBAF, Py·HF, TAS-F) failed or led to decomposition, whereas acidic cleavage with TFA in CH$_2$Cl$_2$ led to rearrangement of the allylic C7 alcohol and formation of **80**. Finally, treatment of **78** with HF in acetonitrile smoothly cleaved the C3 TBS ether (Scheme 14). However, the isolated product **79** was not identical to natural epothilone D$_5$. The sharp contrast with the epothilone D case is currently the subject of investigation.

**Scheme 14** *a* EDCI, cat. DMAP; *b* CSA; *c* Py·SO$_3$, Et$_3$N, DMSO; *d* 2.2 equiv. CrCl$_2$, 0.1 equiv. LiI; *e* 40% aq. HF; *f* TFA

In order to overcome the problem, the inverse strategy was followed. Chromium-Reformatsky reaction between **76**-derived C8 aldehyde and a **68**-derived ester afforded all four possible C6,C7-diastereomers, which can be independently processed to epothilone $D_5$ diastereomers (including the natural one) by the macrolactonization route.

## 2.4
## Conclusion

We have reported one of the shortest total syntheses of epothilone D so far. A large moiety of the 16-membered macrocycle is derived from the easily available isoprenoid nerol. Only four C-C bond formations and three protective groups are required. Only one step requires the use of a (recoverable) chiral auxiliary. The other stereocenters are constructed by catalytic hydrogenation (C8), resolution (C15), and induction (C6, C7). In particular the northern half synthesis is well suited for large-scale preparation because of the combination of cheap reagents and catalytic processes. Using essentially the same method, a number of epothilone $D_5$ analogues were synthesized.

## 3
## Rapid Synthesis of Peptoid Macrocycles by Multi Component Reactions

In rapid, diversity-oriented synthesis of natural product-like compounds, multicomponent reactions have enormous potential. The Ugi four-component reaction (U4CR) belongs to the most widely used ones because of mild reaction conditions, a high variability of the four components, and the fact that the dipeptide(-like) product is a common structural motif in many natural products. Several groups have demonstrated the use of the U4CR in the synthesis of macrocyclic or polycyclic molecules, mostly combining the multicomponent reaction with a subsequent cyclization reaction, comparable to nature's modular strategy followed by cyclization. However, multicomponent reactions themselves also offer ample possibilities for (macro) cyclizations. This approach requires at least one bifunctionalized building block. However, in most cases, more bifunctional building blocks will be employed, and such reactions will be referred to as Multiple Multicomponent Macrocyclizations including Bifunctional Building Blocks (MMMiBBBs, $M^3B^3$s, or MiBs) [78]. There are many possibilities to achieve such reactions, depending not only on the type and number of MCRs and the number of components involved, but also on the directionality of the formed bonds. For simplicity reasons, the U4CR will be the only MCR discussed in the following. Macrocyclization through a single Ugi reaction requires only three components, one of which is an asymmetrically bifunctionalized building block (see Fig. 15).

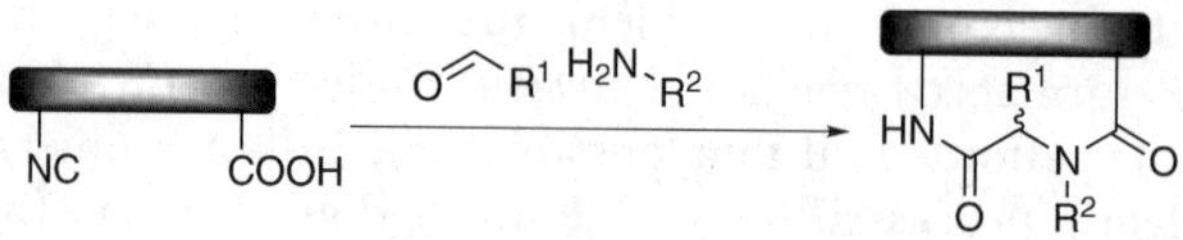

**Fig. 15** An example of a mono-Ugi macrocyclization

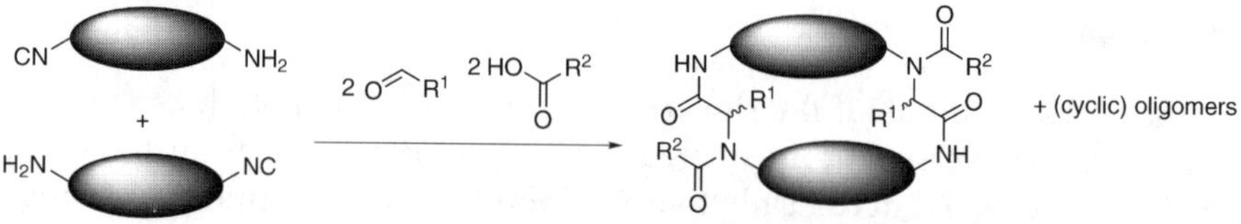

**Fig. 16** An example of a unidirectional MiB

In case of macrocyclization through two or more MCRs, the resulting macrocycle may be uni- or bidirectional, i.e., the resulting dipeptide moieties (*N*- to *C*-terminal direction) run in the same (parallel) or in opposite (counter) direction, respectively. Figure 16 shows an example of a unidirectional MiB (the isonitrile and amine function are too far apart for intramolecular reaction, and thus, a double U4CR results).

An alternative strategy employs bifunctionalized building blocks with the same functionality on either side, leading to a bidirectional macrocycle. The main advantage of this method with respect to the unidirectional reaction is the more straightforward bifunctional building block synthesis. Since only one type of functional group is required on each building block, problems with functional group incompatibility are avoided.

The major problem of these strategies is the competing (cyclizing) oligomerization and polymerization that may occur in competition with the desired double Ugi macrocyclization. Such side reactions can be hard to control. Possible solutions are extreme dilution or pseudo-dilution conditions by slow addition of a component, solid phase pseudo-dilution (see above) or the use of a template that pre-organizes the dimeric structure, thus making it the lower energy product. A third possibility is the use of relatively rigid bifunctional building blocks that favor a structural pre-organization that leads preferentially to the desired macrocycle, such as steroid-derived building blocks (see below).

As shown in Fig. 17, multiple Ugi macrocyclizations using bifunctional isonitriles (MUMBIs) are an interesting form of MiBs, offering many possibilities for macrocyclization and structural variation.

A number of different products may be formed, depending on the symmetry of the bifunctionalized building blocks. When at least one of the bifunctionalized building blocks is symmetrical, only one structural isomer of the macrocycle is formed. When both are unsymmetrical, head-to-head and head-

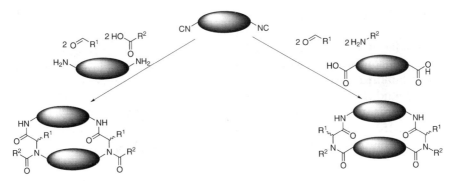

**Fig. 17** Bidirectional double Ugi macrocyclizations (*left*: diisonitrile/diamine; *right*: diisonitrile/diacid)

to-tail isomers are possible. The number of possible products increases further when one (or both) of the bifunctionalized building blocks contain two different functionalities (see below). A detailed analysis is given in ref. [78].

## 3.1
## Synthesis of Cyclopeptide Alkaloids

### 3.1.1
### Introduction

The cyclopeptide alkaloids are a family of 13- to 15-membered macrocycles isolated from different plant sources. They display a variety of interesting biological activities, including antibacterial, insecticidal and sedative properties. The most striking structural features are the macrocyclic aryl ether and, in many derivatives an enamide moiety. Most of the naturally occurring cyclopeptide alkaloids (including the sanjoinines, condaline A, frangulanine, and integerrine) are of type **81** (Fig. 18). The paracyclophane ring structure is probably derived from an oxidatively decarboxylated tyrosine residue. $R^1$ is an amino acid side chain, $R^2$ is usually an *N*-alkylated mono- or dipeptide, and $R^3$ is the rest of the side chain of the $\beta$-oxidized amino acid that forms the aryl ether. In some cases, $R^2$ and $R^3$ together form an oxidized proline

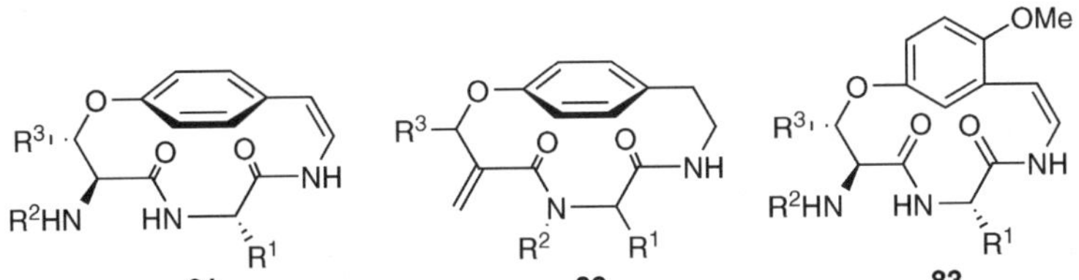

**Fig. 18** Cyclopeptide alkaloids

residue. Some cyclopeptide alkaloids such as the paliurines share basic structure 83, having a 13-membered metacyclophane, where $R^1$, $R^2$, and $R^3$ are similar to those of the 14-membered cyclopeptide alkaloids [4].

A multicomponent reaction/cyclization strategy was employed to synthesize simplified cyclopeptide alkaloid analogues 82. The enamide double bond found in many natural derivatives is missing, but biologically active cyclopeptide alkaloids with a hydrated enamide double bond (like sanjoinine G1 [55, 56]) are known. The synthesis is considerably simplified by omitting this unsaturation, obviously not required for biological activity.

### 3.1.2
### Synthetic Concept

The obvious disconnection in cyclopeptide alkaloids and indeed the strategy employed in most total syntheses of this type of compound is the formation of the aryl ether bond. Many groups chose to form the macrocyclic aryl ether by $S_NAr$ reaction. We decided to follow the inverse strategy, i.e., displacement of an allylic leaving group by a phenolate. In case of the natural cyclopeptide alkaloids, this would involve activation of $\beta$-hydroxy-$\alpha$-amino acids, which is likely to be accompanied by extensive elimination. Elimination is not possible when $\alpha$-methylene-$\beta$-hydroxy acids are used. Additionally, the double bond activates the leaving group and provides a handle for a possible later side chain attachment.

The linear macrocycle precursor 84 is formed in one step by Ugi reaction between tyrosamine-derived isonitrile 85, $\alpha$-methylene acids 86, and simple aldehydes and amines (route A). This synthesis plan combines the generation of significant molecular complexity with a short synthetic sequence and several possibilities for variation.

The alternative route (B, Fig. 19) features isocyano acid 87 and subsequent ring-closing Ugi reaction. However, early studies showed that the 14-

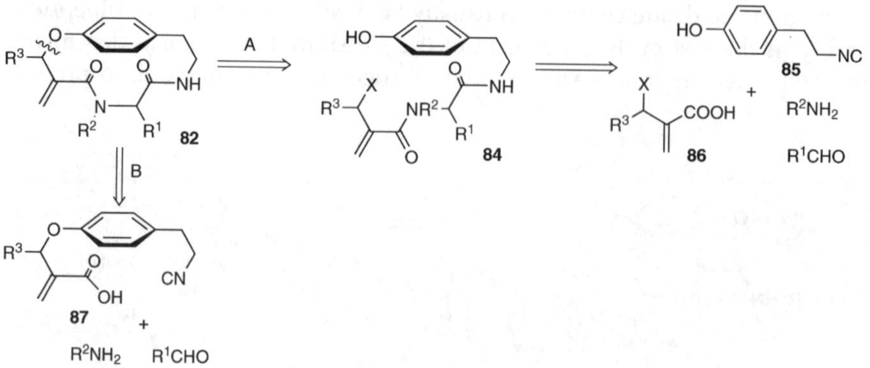

**Fig. 19** Retrosynthesis of cyclopeptide alkaloid analogues

membered structures **82** were not formed (probably because of severe ring strain in the product and intermediate 13-membered paracyclophane), and the 28-membered double Ugi macrocycle seemed the main product. Also, the required isocyano acids **87** are difficult to synthesize and proved to be instable, which renders them useless for an automated approach.

### 3.1.3
### Synthesis

The four components required for the key Ugi reaction are simple aldehydes and amines such as isopropylamine and isobutyraldehyde, isonitrile **85** and $\alpha$-(bromomethyl)acrylic acid and derivatives (**86**). Isonitrile **85** was synthesized from tyrosamine in three straightforward steps [94].

$\alpha$-Methylene-$\beta$-hydroxy acids can be synthesized by the Baylis-Hillman reaction (Fig. 20) from acrylates and aldehydes. The Baylis-Hillman reaction affords $\alpha$-methylene-$\beta$-hydroxy acids that are not readily synthesized by other methods in a single step under mild conditions. The main drawback of this reaction are the generally low reaction rates and therefore long reaction times. However, faster and stereoselective methods are under development [95, 96]. The hydroxy group is converted to a leaving group, e.g., by tosylation. Alternatively, $\alpha$-(bromomethyl)acrylic acid can be employed.

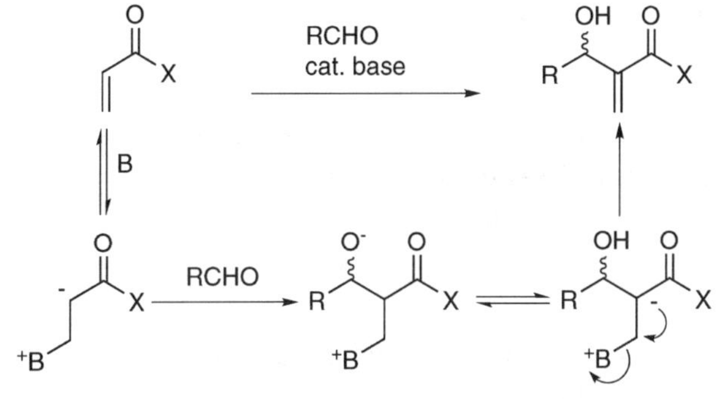

**Fig. 20** Baylis-Hillman reaction

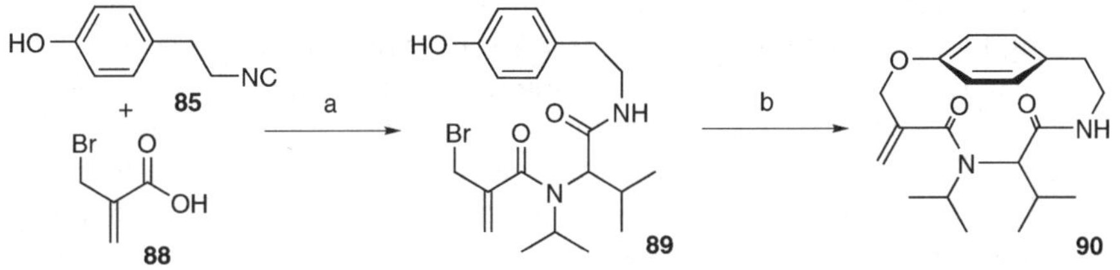

**Scheme 15** *a* i-PrNH$_2$, i-PrCHO, MeOH, 67%; *b* 0.6 mmol/l, K$_2$CO$_3$, cat. 18-crown-6, acetone, 96%

**Table 1** Cyclopeptide alkaloid library using 85 and 88

| Compound | R$^1$ | R$^2$ |
|---|---|---|
| 90 | $i$-Pr | $i$-Pr |
| 91 | $i$-Pr | CHCH$_3$C(CH$_3$)$_2$ |
| 92 | Ph | H |
| 93 | Ph | $i$-Pr |
| 94 | Ph | CH$_2$Ph |
| 95 | Ph | $i$-Bu |
| 96 | CH$_2$Ph | $i$-Pr |

When the R group in the Baylis-Hillman products is not H, a problem might arise in the allylic ring-closing etherification (direct vs allylic attack of the phenolate), leading to mixtures of regio-isomers. Therefore, in an initial study we used parent acid **88** to avoid these problems (Scheme 15).

Ugi reaction of acid **88** with isonitrile **85**, isobutyraldehyde and isopropylamine furnished dipeptide **89** in 67% yield. Similar Ugi reactions with other components afforded linear cyclization precursors in yields up to 98%. The final macrocyclization was not straightforward (no similar reactions were described in literature), but after optimization of the reaction conditions (varying base, solvent, concentration and reaction time) cyclopeptide alkaloid analogue **90** was obtained in 96% yield after treatment with K$_2$CO$_3$ and catalytic 18-crown-6 in acetone.

Interestingly, even ammonia could be successfully employed in the Ugi reaction, thus giving the parent cyclopeptides without additional protection/deprotection.

Cyclopeptide alkaloids **91–96** were synthesized in analogy to **90** (Table 1). Isonitrile derivatives of **85** could be used in the same way.

### 3.1.4
### Conclusion

Synthesis of cyclopeptide alkaloid analogues of type **82** was achieved in a short synthetic route. Based on the building blocks, only two to three steps (Ugi reaction, macrocyclization and sometimes activation or deprotection or post-modification) can deliver libraries of strained *ansa*-macrocycles using a multicomponent/macroetherification strategy. Even if the synthesis of building blocks is included, the longest linear sequence comprises only six or less steps. The method allows variation in each of the four components of the key Ugi reaction as well as post-macrocyclization modification, e.g., by Michael addition to the $\alpha,\beta$-unsaturated amide functionality. In summary, high diversity can be introduced in complex molecules strongly resembling natural products in few synthetic steps.

## 3.2
## Biaryl Ether Macrocycles

### 3.2.1
### Introduction

Several biologically active naturally occurring macrocycles contain biaryl ether moieties, the most prominent representative of this class being the (former) "last line" antibiotic vancomycin (**20**, see also above) [4]. The pharmacological function of these domains is poorly understood, but it is clear that they have a pronounced influence on the overall conformation of the macrocycle. Because of our interest in both macrocycles and multicomponent reactions and recent success in employing MCRs in the synthesis of cyclopeptide alkaloid analogues, we decided to use an MCR strategy in constructing small libraries of biaryl ether macrocycles.

### 3.2.2
### Synthesis

Key components for the synthesis of biaryl ether macrocycles were diisonitriles **98** and **101**. The former was synthesized in two steps (61% yield) from commercially available diamine **97** by bisformamide formation and subsequent dehydration with phosphorus oxychloride (Scheme 16). Starting ma-

**Scheme 16** *a* Ethyl formate, Δ; *b* POCl$_3$, Et$_3$N, THF, 62% over two steps for **98**, 78% for **101**; *c* p-fluoronitrobenzene, K$_2$CO$_3$, DMF, 61% over two steps; *d* H$_2$, Raney nickel, MeOH, 97%

terial for the synthesis of **101** was tyramine (**99**), which was converted to the corresponding formamide and subsequently coupled with *p*-fluoronitrobenzene to give **100**. Reduction of the nitro group (H$_2$, Raney nickel) followed by the formamide formation/dehydration sequence then afforded diisonitrile **101** in good yield.

Diisonitriles **98** and **101** were subsequently employed in a double Ugi macrocyclization reaction under pseudo-dilution conditions, where the di-

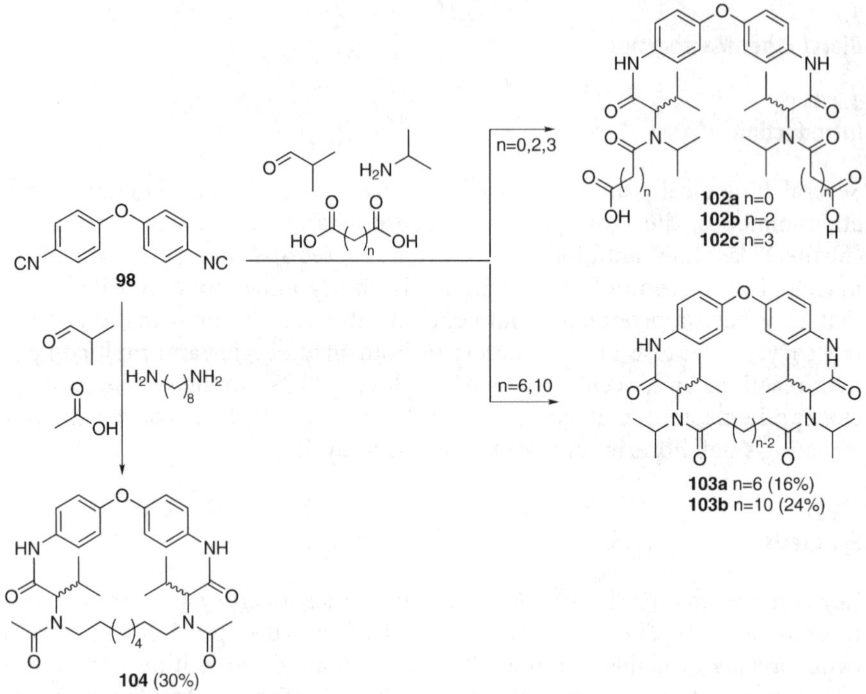

**Scheme 17**

isonitrile is added to the reaction mixture via syringe pump over several days. Thus, diisonitrile **98** was reacted with diacids (oxalic acid, succinic acid, and glutaric acid) and isobutyraldehyde and isopropylamine (Scheme 17). In these cases, however, none of the expected macrocycles were isolated and the linear Ugi products **102a–c** were the main products. A possible cause for this observation is the rigidity of the aromate-conjugated amide bonds. When longer-chain diacids were used (octanedioic acid, dodecanedioic acid), the expected macrocycles **103a** and **103b** were isolated in moderate yield. Similarly, when **98** was reacted with 1,8-diaminooctane, isobutyraldehyde and acetic acid, macrocycle **104** was isolated in 30% yield (Scheme 17).

For unsymmetrical diisonitrile **101**, reaction with both glutaric acid and octanedioic acid, and isobutyraldehyde and isopropylamine afforded macrocycle **105a** and **105b**, respectively, in reasonable yield. Reaction of **101** with 1,6-diaminohexane, acetic acid and isobutyraldehyde yielded **106** in a very acceptable 51% yield (Scheme 18).

It was then demonstrated that the mixed aromatic and aliphatic isonitrile in **101** displays a differential reactivity at both ends. Ugi reaction of **101** with isopropylamine, isobutyraldehyde, and acetic acid yields the mono-Ugi

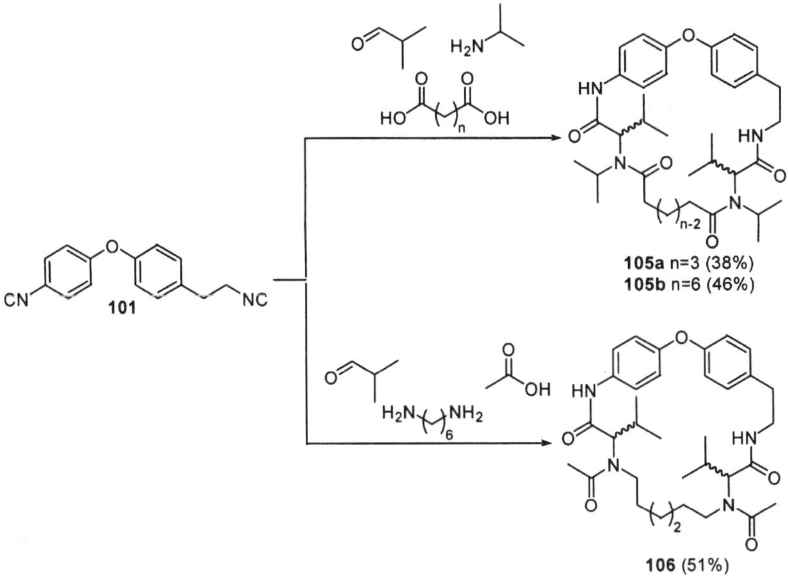

Strategies for Total and Diversity-Oriented Synthesis 171

**Scheme 18**

product (with unreacted aromatic isonitrile or Passerini-product at the other end) unless more stringent conditions force a twofold Ugi-4CR to occur.

### 3.2.3
### Conclusion

The described synthesis illustrates that MCRs are not only a means of introducing structural diversity, but can also effect macrocyclization [78]. Such macrocyclizations, using a multiple MCR-DOS had not been previously reported in the literature. This strategy allows the fast and efficient construction of macrocycles of various sizes, simultaneously introducing a high degree of diversity by variation of the components.

### 3.3
### Synthesis of Steroid-Derived Macrocycles

### 3.3.1
### Introduction

Encouraged by the success of MUMBIs to synthesize biaryl ether macrocycles, we endeavored to combine this methodology with experience in steroid synthesis to obtain very large macrocycles with both peptide and steroid moieties (Fig. 21). These molecules bear no direct resemblance to any known natural products, but they do display a high degree of structural

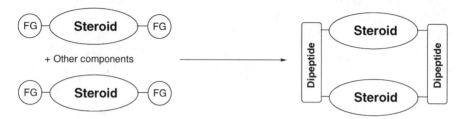

**Fig. 21** Concept towards steroid-derived macrocycles via Mibs.

complexity. Moreover, the fact that, to the best of our knowledge, no peptide/steroid hybrid macrocycles are known from nature does not evidence that they will lack biologically relevant properties, but rather that a combination of metabolic pathways leading to these compounds has not been developed during evolution, for whatever reason.

The size of these macrocycles and their cage-like structure renders them suitable to serve as host molecules or artificial receptors for small molecules, in addition to having possible biological activity.

### 3.3.2
### Synthetic Concept

The ring-closing reaction in this case is a double or fourfold Ugi reaction featuring at least two bifunctionalized building blocks (e.g., steroid-derived) [78]. Because the steroid skeleton is large and often umbrella-shaped, a single dipeptide formed in one Ugi reaction is not long enough to span the distance between a steroids A- and D-ring functional groups, in cases where two different functional groups are present on one steroid building block.

The steroid may be bifunctionalized with any of the functional groups participating in the Ugi reaction, i.e., isonitrile, amine, aldehyde (ketone), and carboxylic acid, and having two different functionalities on one steroid building block even increases the possibilities.

### 3.3.3
### Synthesis

A variety of bifunctionalized steroid building blocks were synthesized from the cheap, commercially available lithocholic acid (**107**, see Fig. 22). Diacid **108**, diamine **109**, and diisonitrile **110** were all successfully used in double or fourfold Ugi macrocyclization reactions [78, 97].

The first successful attempt involved diisonitrile **110**. Ugi reaction of **110** with succinic acid, isopropylamine, and isobutyraldehyde led to the formation of macrocycle **111** (Scheme 19) and its head-to-head coupled isomer **112** (not shown). All of the obtained disteroid macrocycles were mixtures of

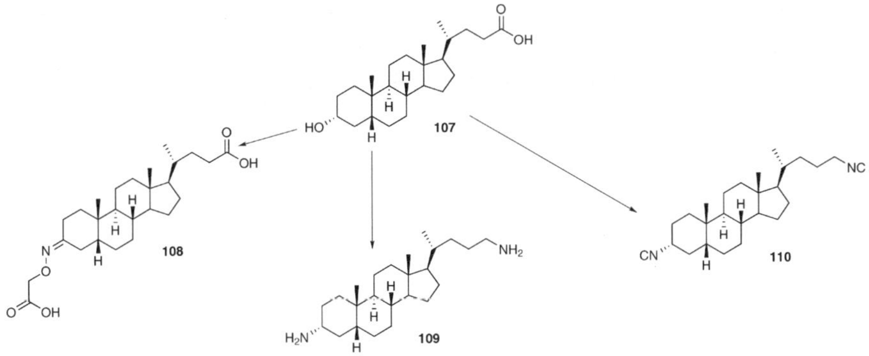

**Fig. 22** Lithocholic acid-derived homo-bifunctional building blocks

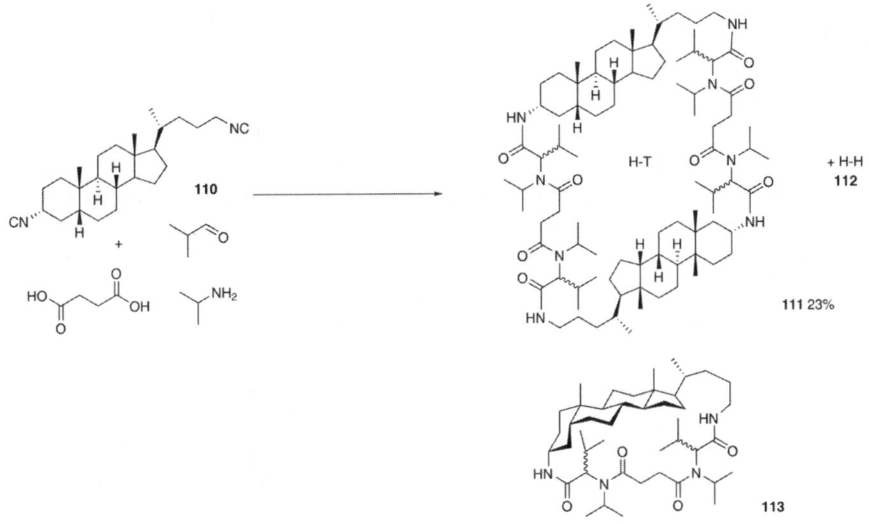

**Scheme 19**

head-head (H-H) and head-tail (H-T) coupling. For simplicity reasons, only the head-tail isomers will be shown.

Interestingly, the smaller macrocycle **113** was not detected. Even though the succinate-bridged double dipeptide should be long enough to span the distance between the two functionalities on the steroid, a macrocycle like **113** would be quite strained. The bulky isopropyl groups might cause significant steric hindrance too. Additionally, formation of the smaller macrocycle would require at least one energetically disfavored *cis* amide bond.

A similar experiment using a diamine (ethylenediamine) as the bridging agent instead of a diacid, in combination with isobutyraldehyde and acetic acid, led to the isolation of macrocycle **114** (Scheme 20).

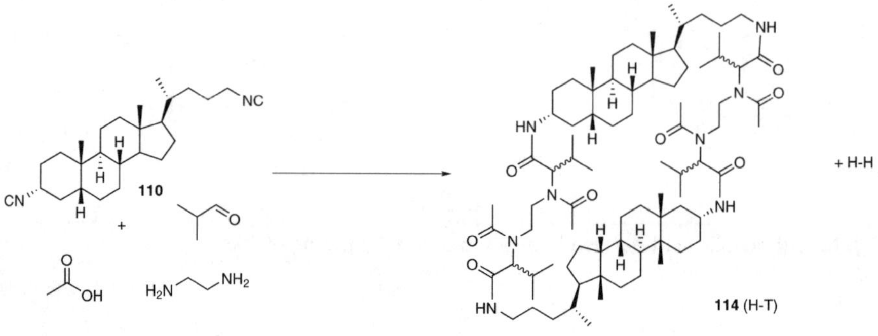

**Scheme 20**

In addition to this fourfold Ugi macrocyclization using two equivalents of one bifunctionalized steroid building block and another (simpler) bifunctionalized building block, two different steroid building blocks can be combined to form a macrocycle in a combined Ugi 5-CR/Ugi 4-CR. For example, diacid **108** and diisonitrile **110**, generated by condensation of a steroidal keto-acid with o-(carboxymethyl)-hydroxylamine, react with isobutyraldehyde and isopropylamine to form macrocycle **115** (Scheme 21).

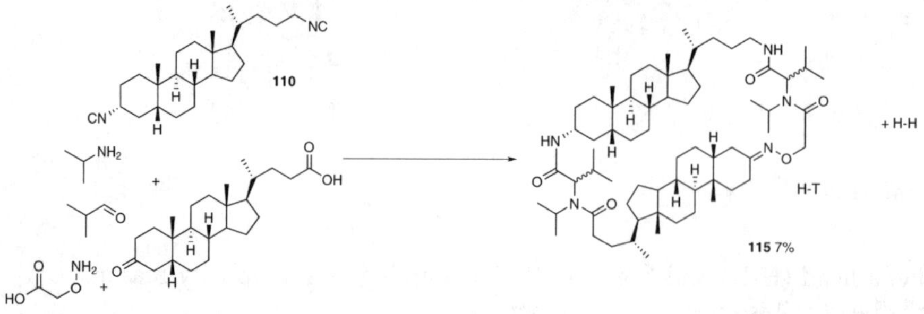

**Scheme 21**

Similarly, reaction of diisonitrile **110** and diamine **109** with isobutyraldehyde and acetic acid affords **116** in a double U4CR (Scheme 22).

In all of the cases described above, characterization of the product macrocycles was complicated by the presence of up to two diastereomers per U4CR. Induction of stereochemistry is typically low in Ugi reactions, and the formation of four new stereocenters in **114**, for example, probably

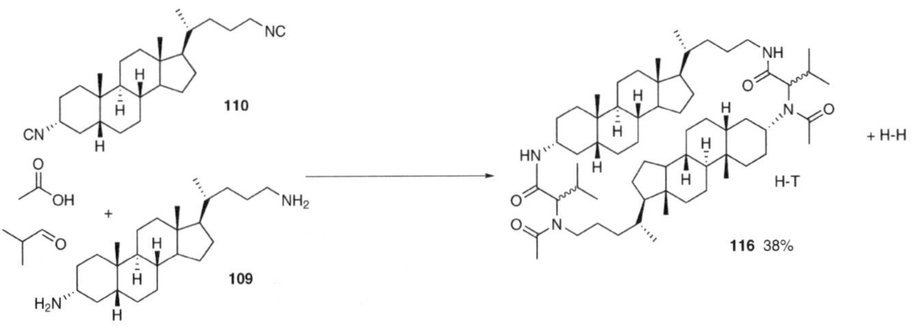

**Scheme 22**

leads to generation of all 16 possible stereoisomers (in addition to the possible structural isomers!). The use of (para)formaldehyde as aldehyde component in the Ugi reaction does avoid the formation of new stereocenters. Macrocycles 117–121 (Scheme 23) were prepared in analogy to the reactions

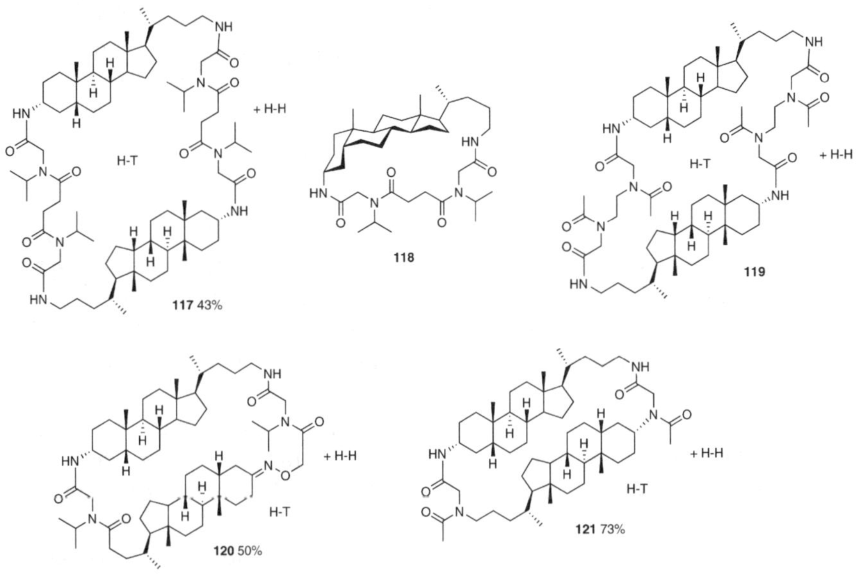

**Scheme 23**

above, using paraformaldehyde instead of isobutyraldehyde as the aldehyde component. The products were formed in yields up to 73%, which is excellent for a one-step synthesis of such complex macrocycles. Because of reduction in steric hindrance, yields of macrocyclization using paraformaldehyde are typically doubled with respect to isobutyraldehyde reactions.

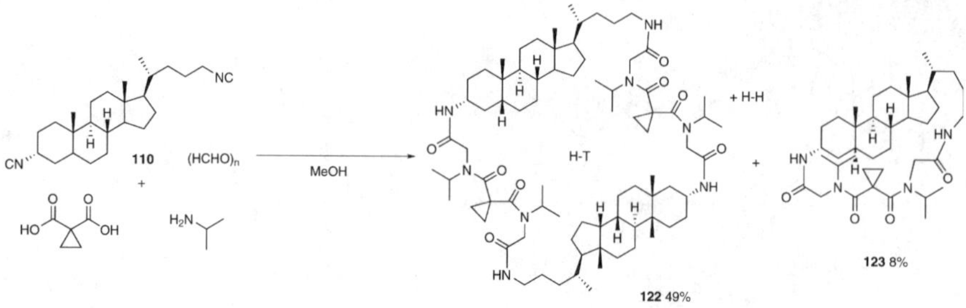

**Scheme 24**

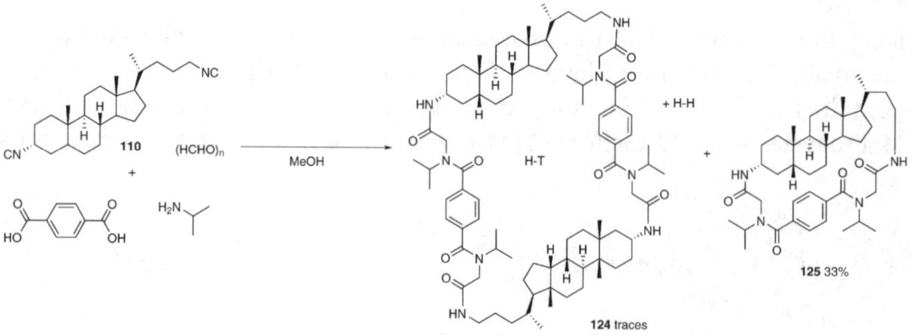

**Scheme 25**

It is noteworthy that in the synthesis of **117**, the smaller Ugi macrocycle **118** was also formed. The lack of sterically demanding isopropyl groups in comparison to **113** allows the formation of this somewhat strained macrocycle. Because of this intriguing result, the diacid component in this reaction was varied to cyclopropane-1,1-dicarboxylic acid (Scheme 24) and terephthalic acid (Scheme 25).

Interestingly, in the former case the fourfold Ugi "large" macrocycles were the main products (49% yield), while the smaller macrocycle was also isolated in smaller amounts (8%). In the latter case, however, the "smaller" macrocycle **125** was the main product (33% yield).

In order to increase structural diversity and to create a polar domain in the cage-like interior of these steroid/peptide macrocycles, dihydroxylated diacid **127** was synthesized in two steps from cholic acid (**126**) and reacted with diisonitrile **110**, paraformaldehyde, and isopropylamine to form macrocycle **128** and its head-head isomer (Scheme 26).

The same dihydroxylated steroid diacid was reacted with biarylether diisonitrile building block **98** (see above) and paraformaldehyde and isopropylamine (Scheme 27). Double Ugi macrocycle **129** was the main product, but extremely large fourfold Ugi macrocycles (**130** and head-head isomer) were also isolated.

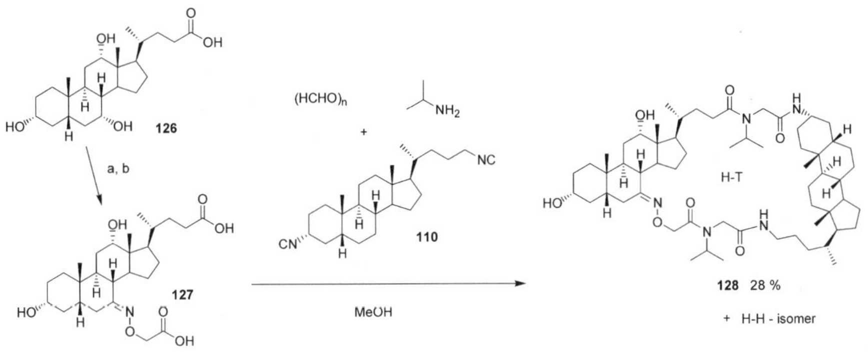

**Scheme 26**  *a* CrO$_3$, py; *b* H$_2$N-OCH$_2$CO$_2$H, 25% over two steps

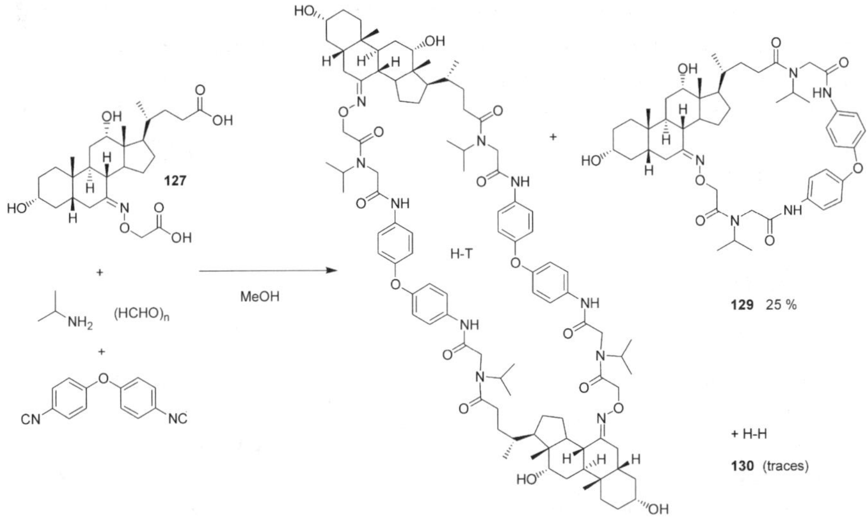

**Scheme 27**

Reaction of diisonitrile **110** and oximcarboxylic acid **131** (Scheme 28) illustrates the versatility of this method by extending the number of different functional groups on the steroid building blocks. The oxime reacts as the imine-component that is usually formed from the aldehyde and the amine in the Ugi reaction.

In the previous cases, the two or four Ugi reactions taking place were either identical or similar, but in this case, the reaction taking place at the oxime side involves other components than the reaction on the carboxylic acid side. In effect, two four-component reactions are combined to give a five-component macrocyclization reaction.

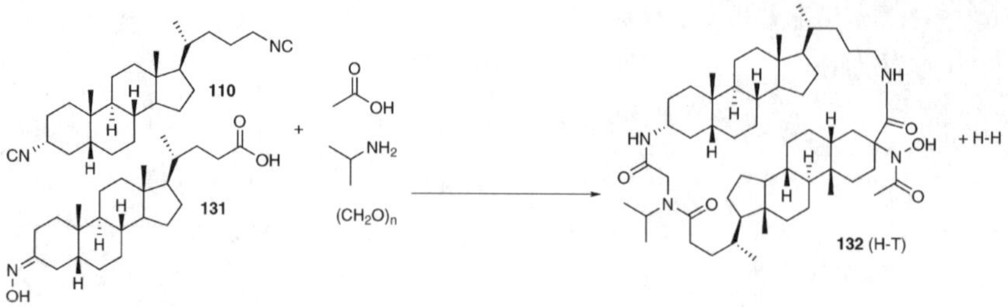

**Scheme 28**

### 3.3.4
### Conclusion

The synthesis of a small library of very large (up to 60-membered) steroid/peptide hybrid macrocycles has been achieved using double and fourfold Ugi reactions. This type of compound has not previously been described in literature. Neither have multicomponent reactions been used so far to form directly macrocycles of this size. In fact, synthetic macrocycles of this size with this structural complexity are very rare.

## 4
## Postmodification Studies of Macrocycles

Post-modification has already been described above as the final stage in the biosynthesis of natural products, concerning mainly oxidation/reduction, elimination/addition, and side chain attachment or removal after construction of the macrocycle, and combination of metabolic pathways. In a synthetic context, the term refers to the same set of minor modifications, performed on a known active natural product. A striking example for the usefulness of chemical post-modification is the discovery of ketolides as post-modified erythromycin derivatives. In these compounds, the erythromycin C-3 cladinose sugar is cleaved off and the remaining alcohol is oxidized to the corresponding ketone. The sugar, which was formerly thought to be essential for the biological activity, could be omitted, and the resulting ketolides show higher stability, while retaining high antibiotic activity, even against erythromycin-resistant strains. New active derivatives based on these variations continue to be found [77].

A promising approach to this topic is the development of biocompatible solid phase attachment systems for macrocycles that allow on-bead enzymatic and chemical modification [4]. While making use of recent developments in polymeric support for resins, we endeavored in constructing a new linker system, which allows easy attachment of macrocycles to the solid phase, simple organic or enzymatic reactions, and cleavage from solid support under mild conditions [98].

We chose to employ PEGA$_{1900}$ and PEG (polyacrylamide backbone with PEG spacer and amine functionalization) because its higher polarity allows water or buffer as solvent and better enzyme permeation with respect to classical polystyrene-based resins [99–100], while it is still compatible with a wide range of organic transformations and solvents.

The model macrocycle for this project was decided to be the rifamycin SV family, because of its high (antibiotic) activity and the presence of many (partially sensitive) functional groups.

Rifamycin SV (**133**, Fig. 23) is a naturally occurring macrocycle isolated from *Nocardia mediterranei* by Senti, Greco and Ballotta in 1959. It shows a high in vitro antibiotic activity through inhibition of DNA-dependent RNA polymerase, but bioavailability is low due to its poor water solubility. The semisynthetic derivative rifampicin (**135**) displays markedly higher water solubility and in vivo activity. We chose 3-formylrifamycin (**134**) as model macrocycle for the possibility of binding it to solid phase by hydrazone bond formation in a manner similar to rifampicin (Fig. 24).

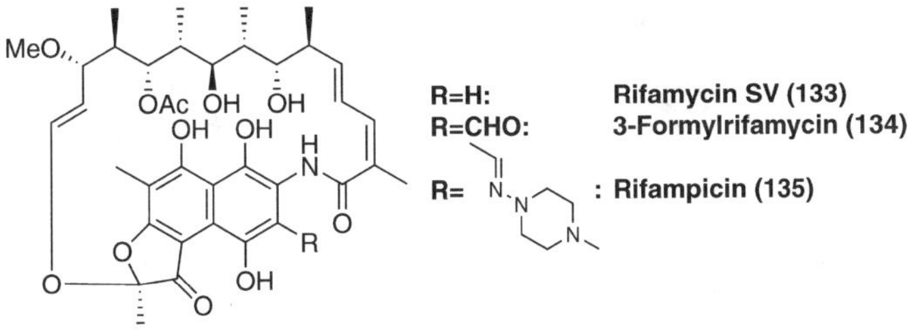

**Fig. 23** Rifamycin SV and derivatives

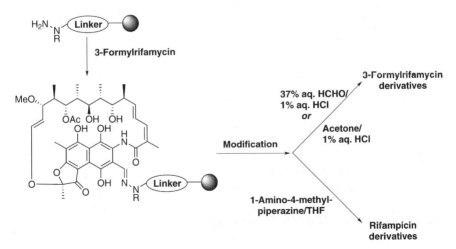

**Fig. 24** Solid phase attachment of 3-formylrifamycin, modification and cleavage

The primary objective was to perform a systematic chemical screening of the positions available for modifications and later, selection of those available positions, the modification of which retains or even enhances or modifies the biological activity.

The first task was to design a linker between the primary amine functionality on the resin and the hydrazine function that ultimately binds the macrocyclic aldehyde. As mentioned above, the employed resin (PEGA$_{1900}$) features a primary amino group at the end of a PEG spacer connected to the polyacrylamide backbone. To assure monofunctionalization, the amino group was monobenzylated, thereby preventing possible proximity problems. Subsequently, the resulting secondary amine was provided with a handle for the introduction of the hydrazine functionality (see Scheme 29).

**Scheme 29** *a* PhCHO, EtOH; *b* NaBH$_4$, MeOH; *c* α,α'-dibromo-*p*-xylene, DMF; *d* hydrazine monohydrate, DMF; *e* 3-formylrifamycin (**134**, =RifCHO), DMF; *f* 37% aq. HCHO/1% aq. HCl, 85%; *g* acetone/1% aq. HCl, 80%; *h* 1-amino-4-methylpiperazine, THF, 75%

3-Formylrifamycin was then attached to the solid support through hydrazone formation. Under several cleavage conditions, it proved possible to obtain either 3-formylrifamycin or rifampicin in good to excellent yield. For example, in case of cleavage with acetone/1% HCl, the product is obtained in 80% overall yield over linker synthesis and connection/ disconnection (total six steps) [98].

To determine the scope of this linker system, the connection/disconnection sequence was also performed with a variety of other aromatic aldehydes (see Table 2). Overall yields usually vary from reasonable to good, with some exceptions, where little or no cleavage product was isolated.

**Table 2** Attachment to and cleavage from PEGA-hydrazine linker of selected aromatic aldehydes (Isolated % over attachment and cleavage)

| Aldehyde | Total Yield (%) |
|---|---|
| p-Bromobenzaldehyde | 84 |
| 3,5-Di-*tert*-butylsalicaldehyde | 83 |
| 3-Formylrifamycin | 80 |
| 3-Formylsalicylic acid | 55 |
| 2,4-Dihydroxybenzaldehyde | 45 |
| Cinnamaldehyde | 43 |
| p-Cyanobenzaldehyde | <10 |
| Benzophenone | <10 |

Although some specific functional groups in these molecules are incompatible with the current method, the fact that 3-formylrifamycin is bound and cleaved in excellent yield demonstrates that several sensitive functionalities (acetal, enol ether, ketone, dienoate) are well tolerated. Various positions of the antibiotic could be acylated selectively.

## References

1. Newman DJ, Cragg GM, Snader KM (2003) J Nat Prod 66:1022
2. Lipinski CA, Lombardo F, Dominy BW, Feeney PJ (1997) Adv Drug Delivery Rev 46:3
3. Lee M-L, Schneider G (2001) J Comb Chem 3:284
4. Wessjohann LA, Ruijter E, Garcia-Rivera O, Brandt W (2005) Molecular Diversity 9(1): (in press)
5. Wessjohann LA, Wild H, Schrekker HS (2004) Tetrahedron Lett, in press
6. Hoffmann RW (1987) Angew Chem 99:503; Angew Chem Int Ed 26:489
7. Brautaset T, Sekurova ON, Sletta HE, T.E., Strøm AR, Valla S, Zotchev SB (2003) Chem Biol 7:395
8. Nakatsuka M, Ragan JA, Sammakia T, Smith DB, Uehling DE, Schreiber SL (1990) J Am Chem Soc 112:5583
9. Nicolaou KC, Roschangar F, Vourloumis D (1998) Angew Chem Int Ed Engl 37:2014; 10:2120
10. Mulzer J (2000) Monatsh Chem 131:205
11. Altmann K-H (2001) Curr Opin Chem Biol 5:424; (2004) Org. Biomol. Chem. 2:2137
12. Wessjohann LA (1997) Angew Chem 109:738; Angew Chem Int Ed 36:715
13. Wessjohann LA, Scheid GO (2000) In: Schmalz H-G (ed) Organic synthesis highlights, vol IV. Wiley-VCH, Weinheim, p 251
14. Schreiber SL (1998) Bioorg Med Chem 6:1127
15. Schreiber SL (2000) Science 287:1964
16. Lokey RS (2003) Curr Opin Chem Biol 7:91
17. Hotha S, Yarrow JC, Yang JG, Garrett S, Renduchintala KV, Mayer TU, Kapoor TM (2003) Angew Chem 115:2481; Angew Chem Int Ed 42:2379
18. Pelish HE, Westwood NJ, Feng Y, Kirchhausen T, Shair MD (2001) J Am Chem Soc 123:6740

19. Nielsen J (2003) Curr Opin Chem Biol 6:297
20. Wessjohann LA (2000) Curr Opin Chem Biol 4:303
21. Henkel T, Brunne RM, Mueller H, Reichel F (1999) Angew Chem 111:688; Angew Chem Int Ed 38:643
22. Pojer F, Li S-M, Heide L (2002) Microbiology 148:3901; Galm U, Dessoy M-A, Schmidt J, Wessjohann LA, Heide L (2004) Chem. Biol. 11:173
23. Breinbauer R, Vetter IR, Waldmann H (2002) Angew Chem 114:3002; Angew Chem Int Ed 41:2878
24. Breinbauer R, Manger M, Scheck M, Waldmann H (2002) Curr Med Chem 9:2129
25. Stahl P, Kissau L, Mazitschek R, Huwe A, Furet P, Giannis A, Waldmann H (2001) J Am Chem Soc 123:11586
26. Stahl P, Kissau L, Mazitschek R, Giannis A, Waldmann H (2002) Angew Chem 114:1222; Angew Chem Int Ed 41:1174
27. Ghosh AK, Wang Y (2001) Tetrahedron Lett 42:3399
28. Crimmins MT, Al-awar RS, Vallin IM, Hollis WGJ, O'Mahony R, Lever JG, Bankaitis-Davis DM (1996) J Am Chem Soc 118:7513
29. Mitsunobu O (1981) Synthesis 1
30. Kruizinga WH, Kellogg RM (1981) J Am Chem Soc 103:5183
31. Mori K, Tomioka H (1992) Liebigs Ann 1011
32. Pawar AS, Chattopadhyay S, Chattopadhyay A, Mamdapur VR (1993) J Org Chem 58:7535
33. Kohli RM, Walsh CT, Burkart MD (2002) Nature 418:658
34. Martín MT, Plou FJ, Alcalde M, Ballesteros A (2003) J Mol Cat B Enzym 21:299
35. Yokokawa F, Sameshima H, In Y, Minoura K, Ishida T, Shioiri T (2002) Tetrahedron 58:8127
36. Erickson SD, Simon JA, Still WC (1993) J Org Chem 58:1305
37. Meutermans WDF, Bourne GT, Golding SW, Horton DA, Campitelli MR, Craik D, Scanlon M, Smythe ML (2003) Org Lett 5:2711
38. Schuster M, Blechert S (1997) Angew Chem 109:2124; Angew Chem Int Ed 36:2036
39. Smith AB, Zheng J (2002) Tetrahedron 58:6455
40. Fürstner A, Mathes C, Lehmann CW (2001) Chemistry 7:5299
41. Beck B, Larbig G, Magnin-Lachaux M, Picard A, Herdtweck E, Dömling A (2003) Org Lett 5:1047
42. van Maarseveen JH (1998) Comb Chem High Throughput Screening 1:185
43. Park K-H, Kurth MJ (2000) Drugs Fut 25:1265
44. van Maarseveen JH, den Hartog JAJ, Engelen V, Finner E, Visser G, Kruse CG (1996) Tetrahedron Lett 37:8249
45. Nicolaou KC, Vourloumis D, Li T, Pastor J, Winssinger N, He Y, Ninkovic S, Sarabia F, Vallberg H, Roschangar F, King NP, Finlay MRV, Giannakakou P, Verdierpinard P, Hamel E (1997) Angew Chem 109:2181; Angew Chem Int Ed 36:2097
46. Nicolaou KC, Winssinger N, Pastor J, Ninkovic S, Sarabia F, He Y, Vourloumis D, Yang Z, Li T, Giannakakou P, Hamel E (1997) Nature 387:268
47. Brohm D, Philipe N, Metzger S, Bhargava A, Müller O, Lieb F, Waldmann H (2002) J Am Chem Soc 124:13171
48. Brohm D, Metzger S, Bhargava A, Müller O, Lieb F, Waldmann H (2002) Angew Chem 114:319; Angew Chem Int Ed 41:307
49. Balog A, Meng D, Kamenecka T, Bertinato P, Su D-S, Sorensen EJ, Danishefsky SJ (1996) Angew Chem 108:2976; Angew Chem Int Ed 35:2801

50. Eichelberger U, Scheid GO, Wessjohann LA (2003) (unpublished results); see also: Wessjohann LA, Scheid GO (2000) German patent DE 10051136 (16.10.2000); CA 136:325358
51. Pattenden G, Sinclair DJ (2002) J Organomet Chem 653:261
52. Stocks MJ, Harrison RP, Teague SJ (1995) Tetrahedron Lett 36:6555
53. Sellès P, Lett R (2002) Tetrahedron Lett 43:4627
54. Spring DR, Krishnan S, Blackwell HE, Schreiber SL (2001) J Am Chem Soc 124:1354
55. Laib T, Zhu J (1999) Tetrahedron Lett 40:83
56. Temal-Laib T, Chastanet J, Zhu J (2002) J Am Chem Soc 124:583
57. Boisnard S, Zhu J (2002) Tetrahedron Lett 43:2577
58. Venkatraman S, Njoroge FG, Girijavallabhan V (2002) Tetrahedron Lett 58:5453
59. Pearson AJ, Bignan G (1996) Tetrahedron Lett 37:735
60. Pearson AJ, Heo J-N (2000) Tetrahedron Lett 41:5991
61. Pearson AJ, Zigmantas S (2001) Tetrahedron Lett 42:8765
62. Otto S, Furlan RLE, Sanders JKM (2003) Curr Opin Chem Biol 6:321
63. Huc I, Nguyen R (2001) Comb Chem High Throughput Screening 4:53
64. Furlan RLE, Ng Y-F, Otto S, Sanders JKM (2001) J Am Chem Soc 123:8876
65. Weber L (2002) Curr Med Chem 9:1241
66. Weber L (2002) Drug Disc Today 7:143
67. Beck B, Magnin-Lachaux M, Dömling A (2001) Org Lett 3:2875
68. Gámez-Montaño R, González-Zamora E, Potier P, Zhu J (2002) Tetrahedron 58:6351
69. Kolb J, Beck B, Dömling A (2002) Tetrahedron Lett 43:6897
70. Hebach C, Kazmaier U (2003) Chem Commun 596
71. Lee D, Sello JK, Schreiber SL (1999) J Am Chem Soc 121:10648
72. Kolb HC, Finn MG, Sharpless KB (2001) Angew Chem 113:2056; Angew Chem Int Ed 40:2004
73. Merrifield RB (1963) J Am Chem Soc 85:2149
74. Blackwell HE, Clemons PA, Schreiber SL (2001) Org Lett 3:1185
75. Tan DS, Foley MA, Stockwell BR, Shair MD, Schreiber SL (1999) J Am Chem Soc 121:9073
76. Tan DS, Foley MA, Shair MD, Schreiber SL (1998) J Am Chem Soc 120:8565
77. Akritopoulou-Zanze I, Sowin TJ (2001) J Comb Chem 3:301
78. Wessjohann LA, Ruijter E (2005) Molecular Diversity 9(1): (in print); 2nd International Conference on Multi Component Reactions, Combinatorial and Related Chemistry (2003) Genova, April 14–16, 2003
79. Lee D, Sello JK, Schreiber SL (2000) Org Lett 2:709
80. Gerth K, Bedorf N, Höfle G, Irschik H, Reichenbach H (1996) J Antibiotics 49:560
81. Höfle G, Bedorf N, Steinmetz H, Schomburg D, Gerth K, Reichenbach H (1996) Angew Chem 108:1671; Angew Chem Int Ed Engl 35:1567
82. Bollag DM, McQueney PA, Zhu J, Hensens O, Koupal L, Liesch J, Goetz M, Lazarides E, Woods CM (1995) Cancer Res 55:2325
83. Nicolaou KC, Ninkovic S, Sarabia F, Vourloumis D, He Y, Vallberg H, Finlay MRV, Yang Y (1997) J Am Chem Soc 119:7974
84. Harris CR, Danishefsky SJ (1999) J Org Chem 64:8434
85. Mulzer J, Karig G, Pojarliev P (2000) Tetrahedron Lett 41:7635
86. Scheid GO, Kuit W, Ruijter E, Orru RVA, Henke E, Bornscheuer U, Wessjohann LA (2004) Eur J Org Chem: 1063
87. Scheid GO, Ruijter E, Konarzycka-Bessler M, Bornscheuer U, Wessjohann LA (2004) Tetrahedron Asymm. 15:2861

88. Wessjohann LA, Scheid GO, Bornscheuer U, Henke E, Kuit W, Orru RVA (2001) German patent DE 10134172 A1 (13.7.2001); CA 136:340534; (2002) International patent, PCT WO 02/32844 A2 (25.4.2002); CA 136:340534
89. Braun M (1987) Angew Chem 99:24; Angew Chem Int Ed 26:24
90. Wessjohann LA, Scheid GO (1999) Synthesis 1
91. Gabriel T, Wessjohann LA (1997) Tetrahedron Lett 38:4387
92. Gabriel T, Wessjohann LA (1997) Tetrahedron Lett 38:1363
93. Hardt IH, Steinmetz H, Gerth K, Sasse F, Reichenbach H, Höfle G (2001) J Nat Prod 64:847
94. de Greef M, Abeln S, Belkasmi K, Dömling A, Orru RVA, Wessjohann LA (2004): synlett submitted
95. Rodrigues O, Braga AL, Wessjohann LA (2004): submitted
96. Jauch J (2001) J Org Chem 66:609
97. Wessjohann LA, Voigt B, Garcia-Rivera D (2004) Angew Chem: submitted
98. Zhu M, Ruijter E, Wessjohann LA (2004) Org Lett: in print
99. Ulijn RV, Baragana B, Halling PJ, Flitsch SL (2002) J Am Chem Soc 124:10988
100. Basso A, De Martin L, Gardossi L, Margetts G, Brazendale I, Bosma AY, Ulijn RV, Flitsch SL (2003) Chem Commun 1296
101. Basso A, Ebert C, Gardossi L, Linda P, Tran Phuong T, Zhu M, Wessjohann L (2004) Chem Comm: submitted

# Enantioselective Synthesis of C(8)-Hydroxylated Lignans: Early Approaches and Recent Advances

Michael Sefkow

Institut für Chemie, Universität Potsdam, Karl-Liebknecht-Strasse 24–25, 14476 Golm, Germany
*sefkow@rz.uni-potsdam.de*

| 1 | Introduction | 186 |
|---|---|---|
| 2 | Lignans by Chiral Auxiliary-Based Biaryl Coupling | 187 |
| 3 | Synthesis by $\alpha$-Oxygenation of C(8)-Unfunctionalized Lactone Lignans | 191 |
| 3.1 | $\alpha$-Hydroxylated Lactone Lignans by $\alpha$-Hydroxylation of Lignan Enolates | 193 |
| 3.2 | $\alpha$-Hydroxylated Biaryl Lignans by Addition of Water to Benzylidene Lignans | 194 |
| 4 | Synthesis from Chiral Natural Products | 198 |
| 4.1 | $\alpha$-Hydroxylated Lignans from Sugars | 198 |
| 4.1.1 | $\alpha$-Hydroxylated Lactone Lignans from L-Arabinose | 199 |
| 4.1.2 | Tetrahydrofuran Lignans from D-Xylose or L-Arabinose | 202 |
| 4.2 | $\alpha$-Hydroxylated Lignans from Malic Acid | 205 |
| 4.2.1 | Furofuran Lignans from Diethyl Malate | 206 |
| 4.2.2 | $\alpha$-Hydroxylated Lactone Lignans from Dialkyl Malates | 208 |
| References | | 222 |

**Abstract** $\alpha$-Hydroxylated lignans, dimers of phenylpropene with important biological properties, have been enantioselectively prepared following different synthetic strategies. The early approaches to optically active $\alpha$-hydroxylated lignans involved the use of chiral auxiliaries or were based on the stereoselective $\alpha$-oxygenation of lactone lignan enolates. Most recently compounds from the pool of chiral building blocks, such as sugars or malic acid, became important starting materials for the asymmetric construction of $\alpha$-hydroxylated lignans. The biaryl lignans schizandrin and isoschizandrin were obtained either by oxazoline-based biaryl coupling as key reaction or by asymmetric hydrogenation followed by stereoselective water addition. The preparation of furofuran lignans was achieved from dialkyl malates and tetrahydrofuran lignans were accessed from arabinose and xylose, respectively. Arabinose and malic acid, as well as unfunctionalized lactone lignans, prepared from chiral $\beta$-benzyl-$\gamma$-butyrolactones, were employed for the synthesis of $\alpha$-hydroxylated lactone lignans. Further details of the strategies and an outlook for future investigations into methodology and interesting targets are provided in this review.

**Keywords** $\alpha$-Hydroxylated lignans · Natural product synthesis · Enantioselective hydrogenation · Chiral building blocks · Malic acid

**List of Abbreviations**

| | |
|---|---|
| 2,6-lut | 2,6-Lutidine |
| cmpd | Compound |
| DMP | 2,2-Dimethoxypropane |
| HB | Hünig's base |
| im | Imidazole |
| MBA | (+)-Methylbenzylamine |
| MIBK | Methylisobutylketone |
| NMO | N-Methylmorpholinoxide |
| TES | Triethylsilyl |
| TMB | 3,4,5-Trimethoxybenzaldehyde |

# 1
# Introduction

Lignans are secondary plant metabolites possessing a variety of biological activities [1, 2]. They are dimers of phenylpropenes, which are by definition connected between C(8) and C(8′) [3]. Lignans are of great structural variety due to numerous potential oxidation states at the C(7)/C(7′) and C(9)/C(9′) positions, and to the possibility of aryl-aryl bond formation [1–3].

Some lignans are additionally hydroxylated at C(8). These C(8), or so-called α-hydroxylated, lignans have also attracted the interest of synthetic chemists, since they exhibit important pharmacological properties or act as insecticide enhancers. Some of the biologically most active α-hydroxylated lignans with different core structures are displayed in Fig. 1. The biphenyl gomisin A (**1**) is hepatoprotective [4], the tetrahydrofuran olivil (**2**) is used as radical scavenger [5], the lactone wikstromol (**3**) was found to be antileukemic in vivo [6] and also hepatoprotective [7], and the furofuran paulownin (**4**) exhibit insecticide-enhancing activity [8].

The first diastereoselective synthesis of α-hydroxylated lignans appeared in the literature over two decades ago [9]. The first synthesis of optically active compounds, however, was achieved a decade later [10]. It is the aim of this chapter to provide an overview of the efforts made towards the *enantioselective* synthesis of α-hydroxylated lignans. Despite their attractive properties, only about 20 articles appeared in the literature describing the synthesis of optically pure α-hydroxylated lignans. Three general strategies have been followed: 1. chiral auxiliary-based synthesis; 2. α-hydroxylation of optically active lactone lignans, unfunctionalized at C(8); 3. synthesis from the pool of chiral building blocks (sugars or malic acid). All of these approaches will be discussed in the following sections.

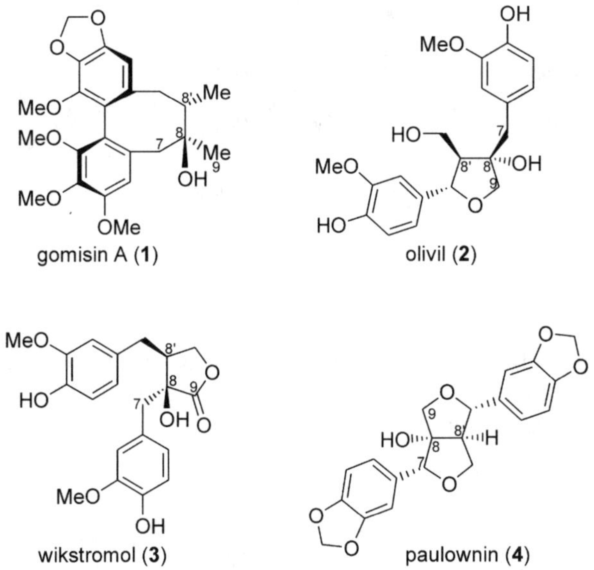

gomisin A (1)

olivil (2)

wikstromol (3)

paulownin (4)

**Fig. 1**

## 2
## Lignans by Chiral Auxiliary-Based Biaryl Coupling

One of the first asymmetric syntheses of an $\alpha$-hydroxylated lignan was published in 1990, from Meyers and co-workers [11]. The target molecules were schizandrin (5) and iso-schizandrin (6), possessing similar structures and biological properties to their congener gomisin A (1). At this time, the absolute configuration of the naturally occurring (+)-enantiomer of schizandrin, as well as the relative configuration of isoschizandrin, was uncertain and one motivation for the synthesis of 5 and 6 was to provide proof for the previously determined stereochemistry.

Retrosynthetic analysis of 5 and 6 yielded ketone 7 as pivotal intermediate. A key reaction for the synthesis of 7 was the auxiliary controlled, diastereoselective biaryl coupling of a phenyl magnesium bromide (from 8 or 9) with aryloxazoline 10 (Scheme 1). This coupling strategy was developed in the Meyers laboratory [12] and previously applied to the enantioselective synthesis of other naturally occurring biaryl lignans, such as steganone [13].

The coupling of aryl bromide 8 or 9 with aryloxazoline 10 was conducted by in situ formation of the Grignard reagent in refluxing THF. The diastereoselectivity of this reaction was highly dependent on the substituent R at the oxomethylene group. The best (S)-selective aryl coupling in combination with a good yield (dr (S)-11:(R)-11=6.2:1, yield 68%) was achieved with a TBDMS protecting group (bromide 8). In contrast, the opposite diastereoselectivity (1:5.2, (S)-12:(R)-12) was observed in the coupling reaction of bro-

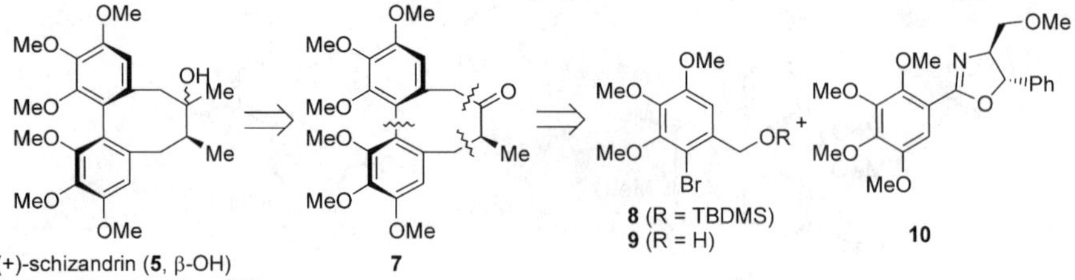


(+)-schizandrin (**5**, β-OH)
(+)-isoschizandrin (**6**, α-OH)
**7**
**8** (R = TBDMS)
**9** (R = H)
**10**

**Scheme 1**

mide **9**, bearing a free hydroxy group, with oxazoline **10** (Scheme 2), but the yield was low (26%) and could not be substantially improved upon by varying the reaction conditions.

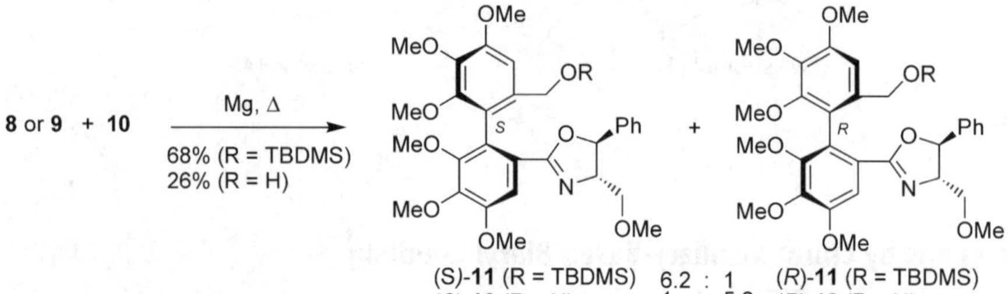


(*S*)-**11** (R = TBDMS) 6.2 : 1 (*R*)-**11** (R = TBDMS)
(*S*)-**12** (R = H) 1 : 5.2 (*R*)-**12** (R = H)

**Scheme 2**

The oxazoline moiety and the TBDMS group were removed to afford the enantiomerically pure hydroxy acid **13** in 75% yield. Reprotection of the hydroxy group with TBDMSCl, reduction of the carboxyl group with DIBALH, and bromide formation ($Ph_3P$, NBS), provided benzyl bromide **14** (70%) alongside the corresponding dibromide (7%). Extension by a three-carbon unit was achieved by alkylation of the lithium enolate of *N*-propylidenecyclo-hexaneamine with benzyl bromide **14**. No diastereoselectivity was observed in this step, furnishing the two C(8)-diastereoisomers as a 1:1 mixture. Acid-catalyzed hydrolysis of the imino group and the TBDMS ether afforded the hydroxy aldehyde, which was converted to the bromo aldehyde **15** in 66% overall yield. Reductive cyclization of **15** to the corresponding cyclooctanol-epimers was achieved with $SmI_2$ and with HMPA as co-solvent. The presence of the carcinogenic HMPA was essential for acceptable yields. In the absence of HMPA, a Meerwein-Pondorf-type oxidation of the resulting alcohol to ketones (*S*)-**16**/(*R*)-**16** occurred at the expense of aldehyde **15**.

The alcohol epimers were oxidized with PCC to ketones (*S*)-**16**/(*R*)-**16** in 71% combined yield. The separation of the diastereoisomers (*S*)-**16** and

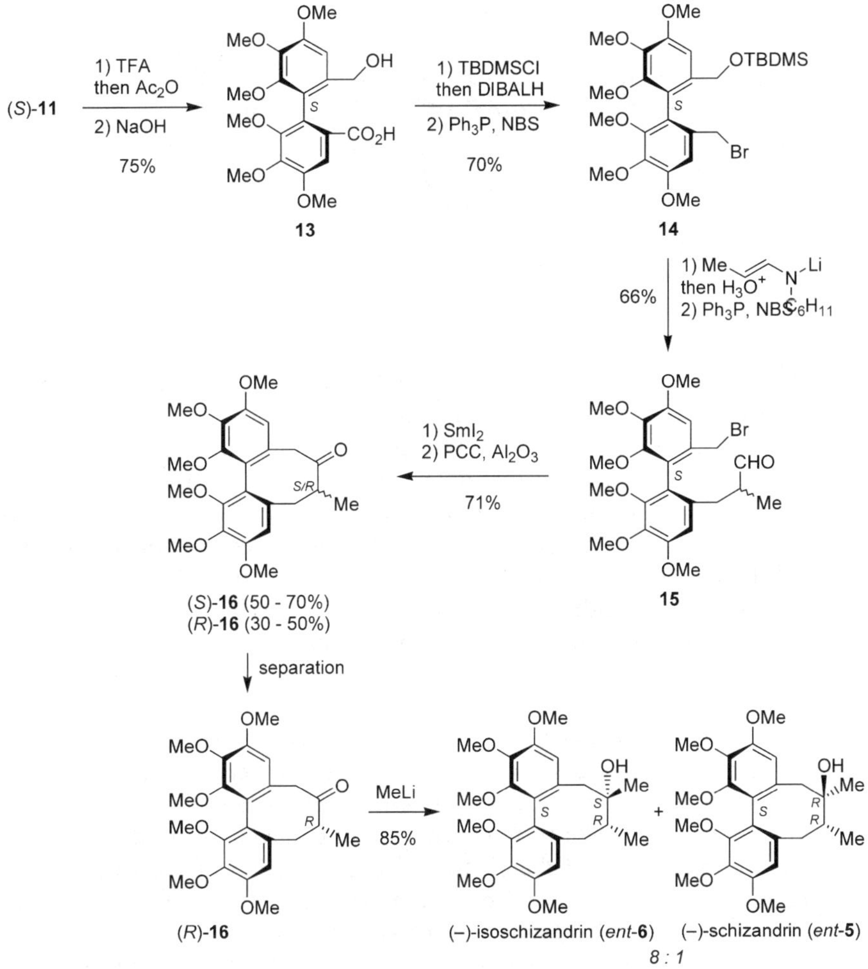

**Scheme 3**

(R)-**16** was difficult at this point, but it was readily achieved at the alcohol stage, despite the additional stereocenter at C(8). As described, both ketones, (S)-**16** and (R)-**16**, were obtained in almost equal amounts. Acid or base-mediated equilibration of the stereocenter at C(8) produced a 7:3 mixture of (S)-**16** and (R)-**16** (Scheme 3). This experimental result was in accordance with Dreiding models of the cyclooctane diastereoisomers.

As mentioned, the relative configuration of isoschizandrin (**6**) was uncertain, and structure (S)-**17** or structure (R)-**17** were proposed for this lignan (see Fig. 2) [14]. Addition of MeLi to ketone (S)-**16** yielded compound (R)-**17** as a single isomer (96%, dr>100:1). Epimer (S)-**17** was obtained in three steps by olefination (CH$_2$Br$_2$, Zn, TiCl$_4$), epoxidation (MCPBA), and

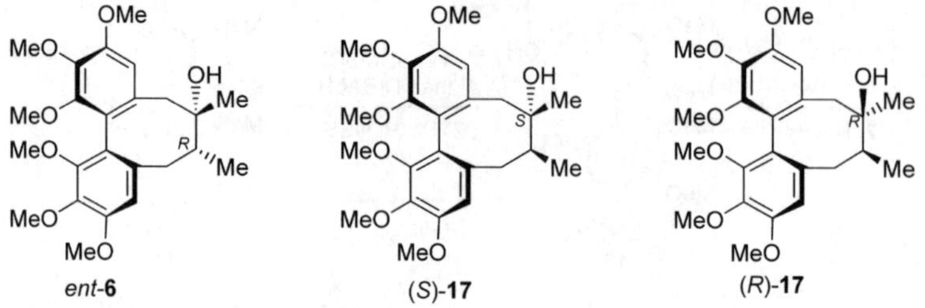

**Fig. 2**

reductive opening of the oxirane (LiBEt₃H), in 60% yield (dr 14:1). Comparison of the NMR spectra of compounds (S)-**17** and (R)-**17** with that of isoschizandrin (**6**) indicated several differences. Consequently, the structure of isoschizandrin was neither that of (S)-**17** nor that of (R)-**17**.

Upon treatment of diastereoisomerically pure ketone (R)-**16** with MeLi an 8:1 mixture of compounds *ent*-**6** and *ent*-**5** was obtained in 85% yield (Scheme 3). The synthetic isoschzandrin was in all respects identical to (+)-isoschizandrin (**6**). However, both, (−)-schizandrin (*ent*-**5**) and (−)-isoschizandrin (*ent*-**6**) have the opposite sign of optical rotation from those of the natural products, establishing the absolute configuration of the schizandra lignans.

Since ketone (R)-**16** was prepared in a non-selective way when an achiral imino enolate was alkylated, it was considered whether alkylation of chiral enolates, such as that of oxazoline **18**, with benzyl bromide **14**, would provide stereoselective access to the corresponding alkylation product **19** with R-configuration at C(8) (Scheme 4). Indeed, alkylation of **18** with **14** gave the biaryl **19** and its diastereoisomer almost quantitatively, in a 14:1 ratio. However, reductive hydrolysis using the sequence 1. MeOTf, 2. NaBH₄, and 3. H₃O⁺, afforded hydroxy aldehyde **20** in 25% yield at best. Furthermore, partial epimerization at C(8) occurred (dr 7.7:1). An alternative route, using chiral hydrazones, was even less successful.

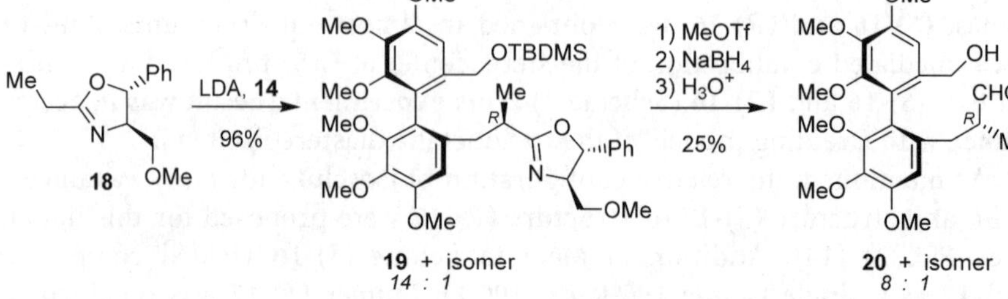

**Scheme 4**

The enantiomers of the naturally occurring lignans, schizandrin (**5**) and isoschizandrin (**6**), have been prepared from oxazoline **10** in 11 steps with 0.7% and 5.5% overall yield, respectively. Although both natural products are accessible by this strategy, the reported synthetic approach is basically a route to isoschizandrin (**6**). Schizandrin (**5**) was obtained only as minor congener and a selective synthesis of **5** has not been accomplished by the authors.

## 3
## Synthesis by $\alpha$-Oxygenation of C(8)-Unfunctionalized Lactone Lignans

An obvious approach to $\alpha$-hydroxylated lignans is the oxygenation at C(8) of $\alpha$-unfunctionalized lactone lignans. The major advantages are 1. a plethora of chemical syntheses of optically pure lactone lignans [1, 2] and 2. the availability of enantiomerically pure lignans from natural resources [15]. The most frequently used protocol for the enantioselective synthesis of lignans employ $\beta$-benzyl-$\gamma$-butyrolactones as chiral intermediates. Diastereoselective alkylation of these key intermediates and further functionalization produce a variety of lignans with different core structures.

Both, $\alpha$-hydroxylated lignans of the dibenzyllactone-type and of the biarylcyclooctane-type have been enantioselectively prepared from the corresponding $\beta$-benzyl-$\gamma$-butyrolactones via $\alpha$-alkylation followed by $\alpha$-oxygenation (Scheme 5). The hydroxy group was introduced in two different

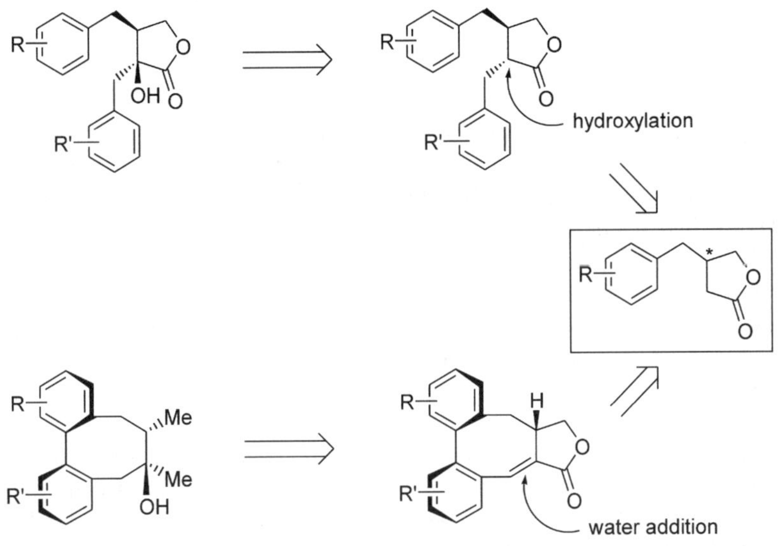

**Scheme 5**

ways: either by the α-hydroxylation of a lactone enolate or by the addition of water to an α,β-unsaturated lactone.

As mentioned above, several syntheses of optically active β-benzyl-γ-butyrolactones have been reported in the literature [1, 2]. A comprehensive overview of the methods available for the synthesis of β-benzyl-γ-butyrolactones is displayed in Scheme 6. Asymmetry was introduced involving either

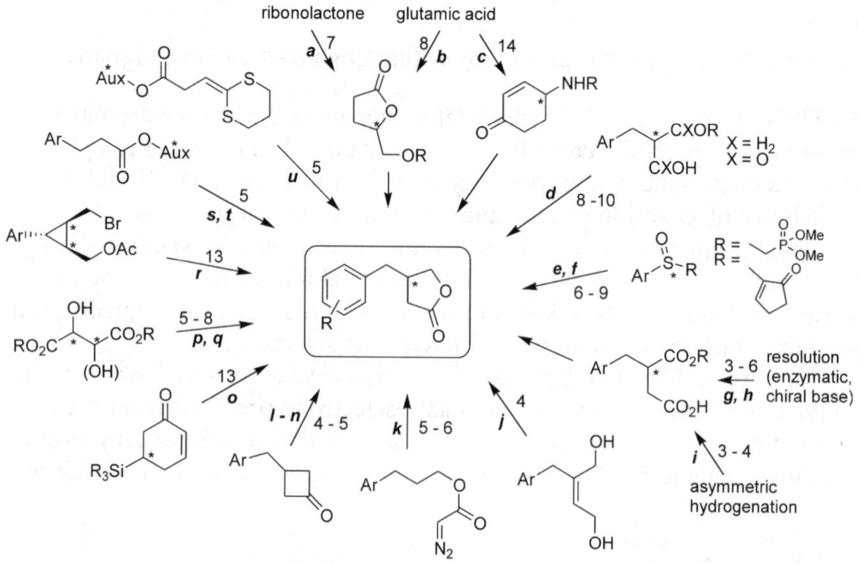

**Scheme 6**

starting materials from the pool of chiral building blocks (a–c [16–18] and p [19], q [20]) or auxiliary-controlled diastereoselective syntheses (s–u [21–23]). Transition metal-catalyzed reactions, such as the carbene insertion (k [24]) or hydrogenation (i [25]), have also been applied. Furthermore, chiral sulfoxides (e [26], f [27]) and silanes (o [28]), as well as enzyme-catalyzed desymmetrization of achiral compounds (d [29], m [30]), have been successfully employed for the synthesis of β-benzyl-γ-butyrolactones. Desymmetrization of achiral cyclobutanones (l [31]) or resolution of racemic succinates (h [32, 33]), was also possible with chiral amine bases (e.g., methylbenzylamine or ephedrine).

In the past few years, new approaches for the enantioselective synthesis of β-benzyl-γ-butyrolactones appeared in the literature. Some of these approaches involve the asymmetric hydrogenation of 2-benzyl-2-butenediols (j [34]), the radical mediated rearrangement of chiral cyclopropanes (r [35]), the transition metal catalyzed asymmetric Bayer-Villiger oxidation of cyclobutanones (n [36]), or the enzymatic resolution of racemic succinates (g [37]).

Two classes of α-hydroxylated lignans have been enantioselectively prepared: a) wikstromol (**3**) [10, 38] and related natural products [39] and b) gomisin A (**1**) and congeners [40, 41]. In both cases, chiral, non-racemic itaconic acid derivatives have been synthesized as key compounds for the preparation of β-benzyl-γ-butyrolactones (either by resolution (*g* [32]) or by asymmetric hydrogenation (*h* [25])).

## 3.1
### α-Hydroxylated Lactone Lignans by α-Hydroxylation of Lignan Enolates

The initial step in the synthesis of wikstromol (**3**) was the Stobbe condensation of dimethyl succinate **21** with vanilline. Interestingly, a high yield was only obtained when LiOMe was employed as base (ca. 90%), otherwise the yield was in the range of 20% [32a]. Hydrogenation of the Stobbe condensation product quantitatively yielded the racemic itaconic acid ester *rac*-**22**. Resolution of *rac*-**22** with (+)-methylbenzylamine (MBA) afforded the enantiomers (S)-**22** and (R)-**22** in 23% and 26% overall yield, respectively [33]. The selective reduction of the ester moiety in **22** was achieved with $Ca(BH_4)_2$, generated from $CaCl_2$ and $NaBH_4$ (ca. 85% yield). Benzylation of the phenol moiety and α-alkylation with LDA/benzyl bromide **23** provided di-*O*-benzyl-matairesinol **24** in 76% yield (from (R)-**22**). The introduction of oxygen was achieved by the reaction of the lithium enolate of **24** with molecular oxygen (Scheme 7), but this reaction was of low selectivity. In the best case, a 2:1 mixture of the *cis*-isomer **26** (desired) and the *trans*-product **25**

**Scheme 7** MBA=(R)-(+)-methylbenzylamine

was obtained. Furthermore, the combined yield of the oxidation reaction did not exceed 47%. Other methods of enolate oxidation (e.g., Mo-peroxides [42]) produced only minor improvements. The removal of the benzyl groups from compound **26** provided wikstromol (**3**) in 3% overall yield over eight steps. The low overall yield and the non-selective synthesis of the key intermediates **26** are crucial drawbacks of this route to optically pure α-hydroxylated lactone lignans.

From di-*O*-benzylwikstromol **26**, other classes of naturally occurring α-hydroxylated lignans, such as lactols or diols, were prepared as well. Thus, the reduction of **26** with a large excess of LiAlH$_4$, followed by hydrogenation, afforded carinol **27** in 53% yield [39]. The partial reduction of the carbonyl group with DIBALH and hydrogenation provided carissanol **28** in 66% yield (Scheme 8).

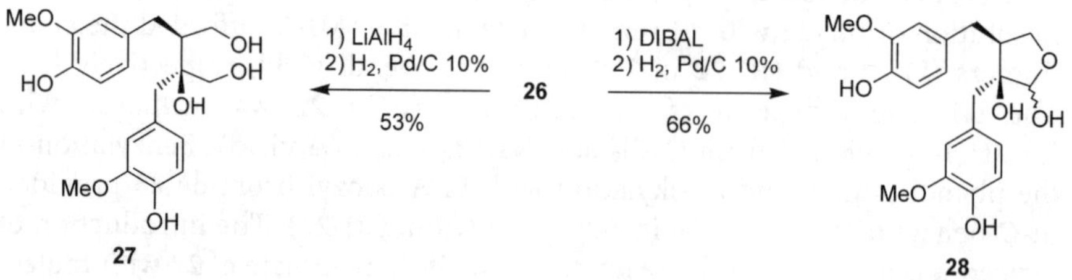

**Scheme 8**

## 3.2
### α-Hydroxylated Biaryl Lignans by Addition of Water to Benzylidene Lignans

The initial step for the synthesis of α-hydroxylated biaryl lignans, such as gomisin A (**1**), was again a Stobbe-condensation, as shown in Scheme 7. The chirality was introduced by the asymmetric hydrogenation of the double bonds in compounds **29** and **30** (Scheme 9) [25, 40]. This reaction was achieved with Rh(cod)BF$_4$ and (*R*,*R*)-MOD-DIOP as chiral ligand, a modified DIOP ligand, originally developed by Achiwa and co-workers [43]. With this catalyst, the (*S*)-configured products **31** and **32** have been obtained quantitatively and in 94% enantiomeric excess. Recently, new chiral ligands for the asymmetric hydrogenation of 2-benzylidene succinic acid mono esters have been introduced by Burk and co-workers ((*R*,*R*)-Et-DuPHOS, [44]) and by Zhang et al. ((*S*,*S*)-TangPhos, [45]), with which better ees have been realized (97–98% EtDuPHOS and 95–99% TangPhos). The latter ligand produced the (*R*)-enantiomer.

Conversion of the monoesters **31** and **32** to the lactones **33** and **34** was achieved in 91% and 78% yield, respectively. Optically pure lactones **33** and

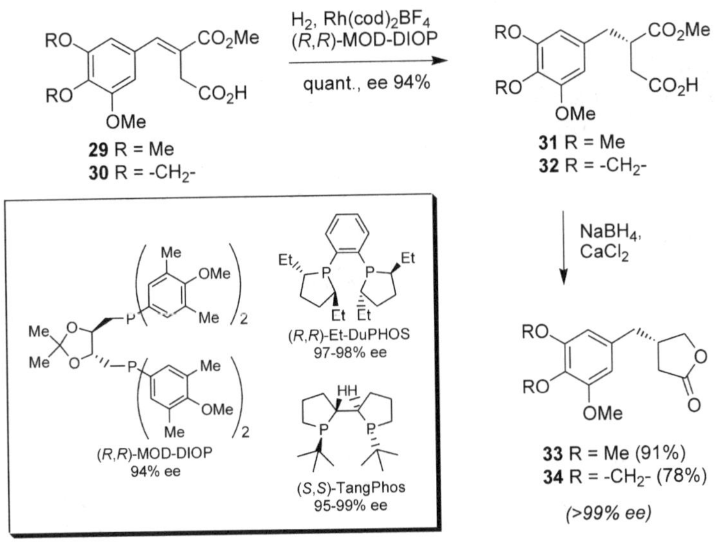

## Scheme 9

34 have been obtained by either the recrystallization of the lactone (33) or the methanol extraction of the monoester (32).

The condensation of lactones 33 and 34 with 3,4,5-trimethoxy-benzaldehyde (TMB) was carried out in a three step sequence (1. LDA, 2. Ac$_2$O, 3. DBU), affording the bisbenzyllactones 35 (91%) and 36 (94%). The oxidative biaryl coupling of 35 and 36 to lactones 37 and 38 was examined with different metal reagents with TFA as the proton source (see the Table in Scheme 10). The highest yields were realized with Ru(O$_2$CCF$_3$)$_4$ and with Fe(ClO$_4$)$_3$, the latter in the presence of dichloromethane as co-solvent. The cost factor and the reaction time favor the use of Fe(ClO$_4$)$_3$. The reaction was diastereoselective and the newly-formed (axial) stereocenter had the (R)-configuration. Lactone 37 was produced in 91% yield. The asymmetric lactone 38, on the other hand, was obtained in only 46% yield. The major side-product was the regioisomeric coupling product (7%) [40a]. The yield was improved when the methylendioxy group was cleaved prior to the coupling reaction. After re-installation of the methylendioxy group, lactone 38 was obtained in 59% overall yield from 36. DIBALH reduction of the carbonyl group followed by acetalization yielded bis-acetals 39 (96%) and 40 (92%). The introduction of oxygen at C(8) was achieved after a palladium catalyzed allylic rearrangement of the double bond. Dihydroxylation of compounds 41 and 42 with OsO$_4$ followed by dimesylation and epoxide formation with NaH, gave the compounds 43/44 (85%) and 45/46 (96%), each mixture of epimers in a ca. 1:4 ratio. Lithium aluminum hydride reduction of the isomerically-pure compounds 44 and 46 provided (+)-schizandrin (5)

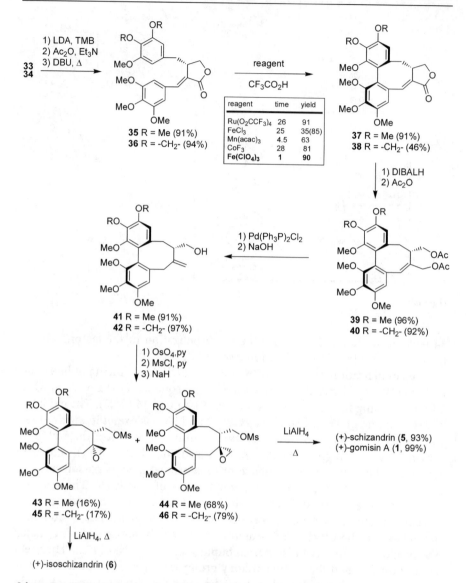

**Scheme 10** TMB=3,4,5-trimethoxybenzaldehyde

and (+)-gomisin A (**1**) in almost quantitative yield (Scheme 10). The synthesis of schizandrin (**5**) and gomisin A (**1**) was very effective. From itaconic acid derivatives **29** and **30**, 13 and 15 steps, respectively, were necessary to afford **1** and **5** in overall yields of 41% and 31%, respectively.

Since the synthetic sequence presented in Scheme 10 is lengthy, and the oxygenation of the endo double bond in compounds **39** or **40** seemed to be a more straight forward strategy, shorter routes towards **1** and **5** have been examined. However, better overall yields of schizandrin **5** were realized neither by dihydroxylation nor by epoxidation, though only 3–4 steps were required. On the other hand, formal addition of water to the α,β-unsaturated lactones **37** and **38** according to the Mukaiyama protocol (O$_2$, Mn(acac)$_2$, PhSiH$_3$ [46]), produced the diastereoisomers **47/48** and **49/50**, each isomer in a 1:6 ratio (~80% overall yield). Reduction of the lactone moiety according to the sequence shown in Scheme 11 furnished **5** and **1** in 63% and 48% yield, respectively. Although this route was shorter and more elegant, the overall yields were not satisfactory (31% for **5** and 14% for **1**).

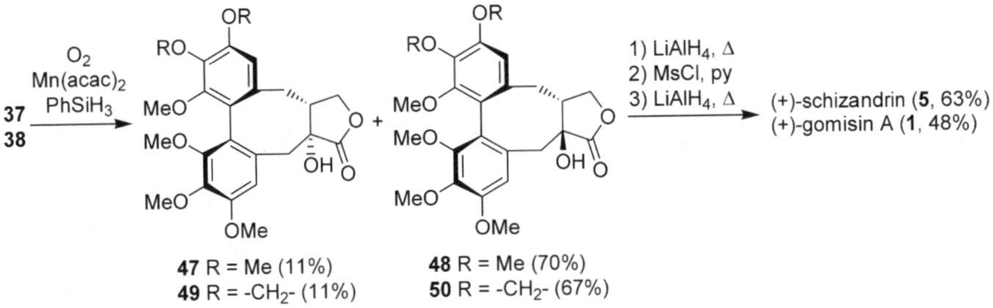

**Scheme 11**

(+)-Isoschizandrin (**6**) could be prepared from epoxide **43**; however, since **43** was obtained only as minor diastereoisomer, a stereoselective synthesis for **6** was developed. The starting material was diol **51** (from the reduction of lactone **37**), which was stereoselectively epoxidized to compound **52** using VO(acac)$_2$. Nucleophilic displacement with iodide, and reductive epoxide opening, provided cyclooctane **53** containing an exo double bond and a benzylic hydroxy group with (R)-configuration. Due to the preferred conformation of the cyclooctane ring (supported by molecular mechanics calculations), the benzylic hydroxy group directs the epoxidation reagent (VO(acac)$_2$) to the si-face, affording the (S)-configured epoxide as a single isomer. After mesylation, compound **54** was obtained in 92% yield. Sodium borohydride reduction of **54** finally provided (+)-isoschizandrin (**6**) in 49% yield (Scheme 12).

In this chapter two syntheses of optically active α-hydroxylated lignans from unhydroxylated lactone lignans have been described. Both synthetic strategies used optically pure β-benzyl-γ-butyrolactones as key precursors. These were prepared either from the corresponding racemic mixture by resolution with chiral amine bases or by asymmetric hydrogenation of the corresponding benzylidene succinates. Both synthetic sequences suffer from lengthy syntheses and from oxygenation reactions, which proceed with rela-

**Scheme 12** MIBK=methylisobutylketone

tively low diastereoselectivity. Despite the number of necessary steps, the synthesis of the biaryl lignans was very effective (average yield for each step 92–93%).

# 4
# Synthesis from Chiral Natural Products

## 4.1
## α-Hydroxylated Lignans from Sugars

Sugars are often used as chiral precursors for the synthesis of optically active compounds, because they are readily available in large quantities and they are relatively inexpensive. The major restriction is that only the D-series of sugars is usually available. An exception is arabinose, which is an attractive chiral source since both enantiomers are commercially available.

Two different classes of optically active α-hydroxylated lignans have been prepared from sugars by Yamauchi and co-workers: lactone lignans [47, 48] and tetrahydrofuran lignans [49, 50]. In both cases, only model compounds were synthesized, such as lactone lignan 55 (Scheme 13), as a result of the synthetic strategy. The key intermediate for the synthesis of lactone lignan

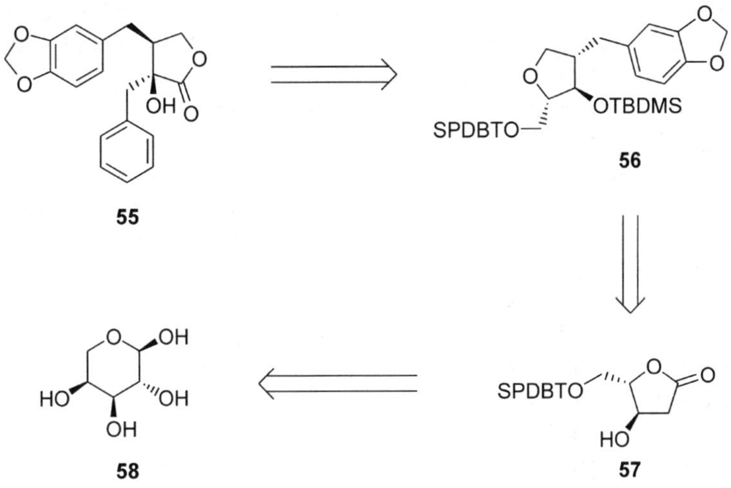

**Scheme 13**

55 is tetrahydrofuran 56. This precursor is accessible from lactone 57, which is prepared in 4 steps from L-arabinose (58) (Scheme 13).

The strategy was chosen because compound 57 is readily prepared from 58 and is easily transformed to 56 by α-alkylation followed by reduction of the lactone moiety. Additionally, both *cis*- and *trans*-lactone lignans can be synthesized by one common intermediate.

### 4.1.1
#### α-Hydroxylated Lactone Lignans from L-Arabinose

The transformation of L-arabinose (58) to lactone 57 was based on a route developed by Marquez and Sharma [51]: Selective protection of the primary hydroxy group with TBDPSCl and oxidation of the lactol moiety with bromine afforded lactone 59. Subsequent selective deoxygenation α to the carbonyl group proceeded under Barton-McCombie conditions providing lactone 57 in 21% yield (Scheme 14).

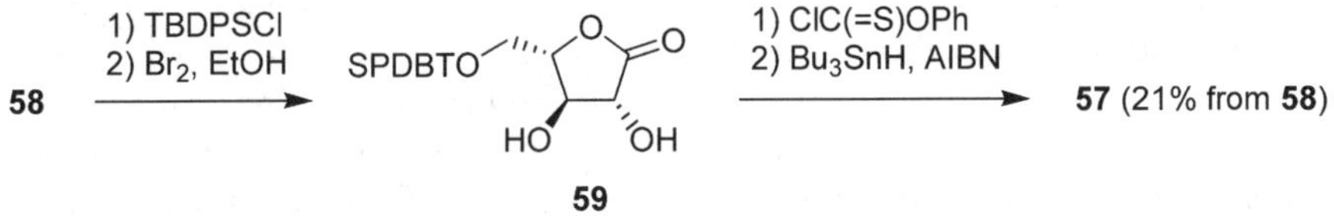

**Scheme 14**

Addition of piperonal (**60**) to the lithium enolate of **57**, followed by protection of the hydroxy groups, gave lactone **61** in 43% yield and with high stereoselection at C(2). The diastereoselectivity at the benzylic carbon was 4:1 (*S:R*), which is not important since this stereocenter will be removed in a later stage of the synthesis. Reduction of the lactone moiety and regioselective removal of the benzylic silylether was achieved by treatment of **60** with DIBALH and $Et_3SiH/BF_3$, furnishing tetrahydrofuran **56** in 71% yield [47]. Selective deprotection of the TBDMS ether and oxidation provided furanone **62** in 81% yield. The second benzyl group was introduced by nucleophilic addition of benzyl magnesium bromide to the carbonyl group of **62**. The reaction proceeded with excellent stereoselectivity but moderate yield (59%). Completion of the synthesis of the lignan analog **55** required the conversion of the silyloxymethyl group into a carbonyl group. TBAF mediated removal of the TBDPS group, followed by iodide formation ($I_2/Ph_3P$, imidazole) afforded tetrahydrofuran **63**. Reductive ring opening followed by protection produced olefin **64** (63% yield from **63**). The additional carbon was removed by $OsO_4$/*N*-methylmorpholinoxide (NMO) dihydroxylation and cleavage of the vicinal diol with $NaIO_4$. After treatment with potassium carbonate, lactol **65** was obtained in 13% yield (from **64**). Both oxidation steps proceeded in low yields. For example, the yield for the dihydroxylation was 24% (50% based on recovered starting material). No reasons are given for the low conversion in the dihydroxylation step, and improvements of the yield failed. Silver carbonate-mediated oxidation of the lactol group of **65** afforded lactone **55** in 51% yield (Scheme 15). In total, 20 steps are required for the transformation of L-arabinose (**58**) into *trans*-lactone **55** and the overall yield over all steps was only 0.1%. Furthermore, since the synthetic strategy required the use of a benzyl nucleophile which is hardly available from highly functionalized benzyl halides, the generality of this route to α-hydroxylated lactone lignans is arguable. Additionally, the synthesis is based on several key reactions with a low yield, such as the aldol addition and the dihydroxylation, which could not be improved.

On the other hand, the flexibility of the synthetic strategy was demonstrated by the synthesis of the corresponding *cis*-lactone **70** from intermediate **62** [48]. In order to obtain the epimer at C(3) it was necessary to introduce the "benzylic" carbon first and the oxygen second. The extension by one carbon at C(3) was effected by olefination using the Tebbe reagent (which will provide the "benzylic" carbon). The oxygen was then stereoselectively introduced at C(3) by dihydroxylation with $OsO_4$/NMO (*R:S*=97:3). The regioselective transformation of the silyloxymethylene group of compound **66** into a carbonyl group was achieved after protection of the diol moiety. Thus, the silyl ether was converted via iodide **67** to lactone **68** using essentially the same sequence as described above (**66**→**67**→**68**, 75% overall yield). Cleavage of the dioxolane, iodide formation, and ring closure gave oxirane **69** in 80% yield. The key reaction of the synthesis was the introduction of the second

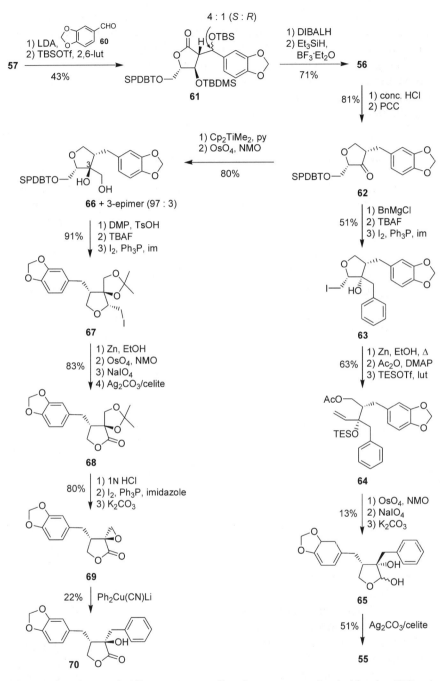

**Scheme 15** 2,6-lut=2,6-lutidine; DMP=2,2-dimethoxypropane; im=imidazole; TES=triethylsilyl, NMO=N-methylmorpholinoxide

aromatic moiety. This was accomplished by addition of the higher order cuprate Ph$_2$Cu(CN)Li to the oxirane to provide the *trans*-lactone lignan 70, albeit in low yield (22%). The yield could not be improved by varying the reaction conditions. The synthesis of an α-hydroxylated *trans*-lactone lignan was achieved in a lengthy sequence (23 steps) and in low overall yield (0.5%). The major restriction of the synthesis is, in analogy to the *cis*-isomer, the introduction of the second aryl group by an aryl nucleophile.

### 4.1.2
### Tetrahydrofuran Lignans from D-Xylose or L-Arabinose

The Yamauchi group used two pentoses, D-xylose (59) [49] and L-arabinose (58) [50], for the enantioselective construction of olivil-type lignans (Scheme 16). Depending on the pentose employed (58 or 59), they have de-

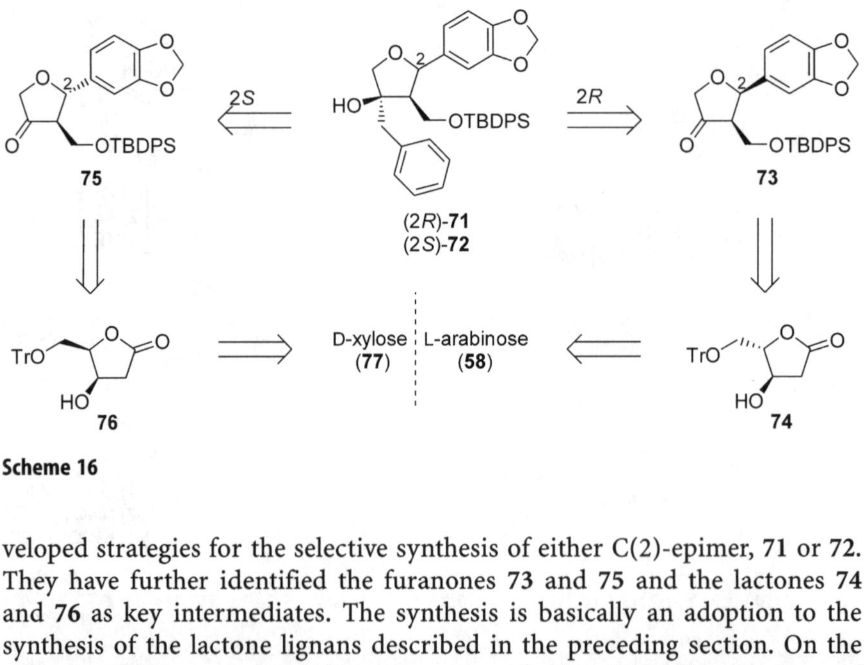

**Scheme 16**

veloped strategies for the selective synthesis of either C(2)-epimer, 71 or 72. They have further identified the furanones 73 and 75 and the lactones 74 and 76 as key intermediates. The synthesis is basically an adoption to the synthesis of the lactone lignans described in the preceding section. On the other hand, comparison of all stereocenters of either 71 or 72 with those of olivil (4) revealed that neither of the synthetic derivatives have the exact relative configuration of olivil-type lignans. This belongs to the strategy by which the stereocenter at C(4) can only be synthesized with the configuration opposite to that of olivil (4).

According to the procedure described in the preceding section, D-xylose (77) was converted in four steps and 30% overall yield to the trityl ether 76. Aldol addition to piperonal (60) and protection of the hydroxy groups as

MOM ethers provided compound **78** as a 1:1 mixture of epimers at the benzylic carbon. These epimers were separable by chromatography and the subsequent reactions were carried out with either of the separated isomers (the yields shown in Scheme 17 are average yields for the transformations of

**Scheme 17** HB=Hünig's base

each epimers). LiAlH$_4$ reduction followed by regioselective protection of the primary hydroxy group using pivaloyl chloride afforded **79** in 86% yield. The trityl ether was selectively removed with formic acid. Subsequent diol cleavage with NaIO$_4$ produced the corresponding aldehyde. The selective deprotection of the benzylic MOM ether was effected with PPTS in $t$-BuOH. Reduction of the carbonyl group gave compound **80** in 50% yield from **79**. The cyclization of the diol to the tetrahydrofuran was accomplished with TsOH. This reaction proceeded with high stereocontrol, no matter which epimer was used. Consequently, a separation of the two C(2)-epimers, ($R$)-**78** and ($S$)-**78**, was not necessary. DIBALH reductive removal of the pivaloyl group, cleavage of the second MOM ether and selective protection of the primary hydroxy group (as TBDPS ether) afforded compound **81** as single isomer in 25% overall yield from **80**. Swern oxidation of the secondary alcohol gave furanone **75** in 87% yield. The second benzyl group was introduced as benzyl Grignard reagent, similarly to the synthesis of the *trans*-lactone lignans (see above). TBAF mediated removal of the silyl group afforded the (2S)-epimer **72** in 52% yield. The difficulties in preparation of functionalized benzyl magnesium halides restricts this routes to the synthesis of olivil analogs.

By the same strategy applied for **72**, the 2R-epimer **71** was synthesized from L-arabinose (**58**) (Scheme 18). Aldol adduct **82** was prepared from **58**

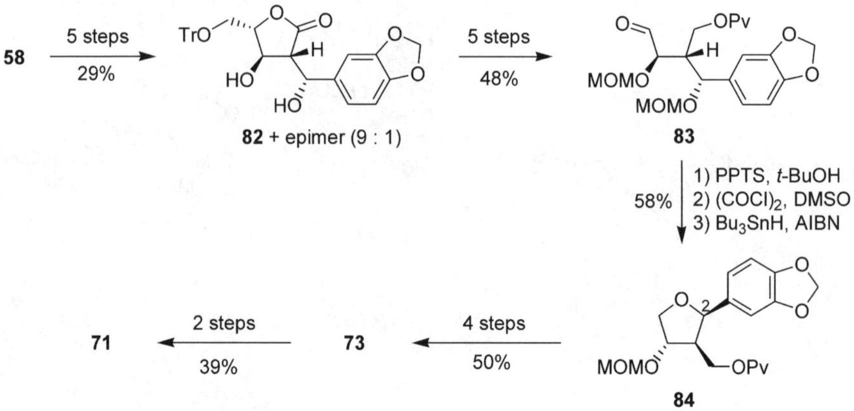

**Scheme 18**

in 5 steps and 29% overall yield. Interestingly, the stereochemistry of the aldol addition of lactone **74** to piperonal (**60**) was significantly different to that of lactone **76** to **60**. Whereas the latter aldol reaction produced a 1:1 mixture of epimers at the benzylic carbon, the aldol adduct **82** and its epimer were obtained in a 9:1 ratio. The different ratios can be explained by modeling the approach of the carbonyl group of **60** to the enolates of **74** and **76**. Both lactone enolates approach the carbonyl group from their *Re*-face, guided by the adjacent oxo anion. Consequently, the newly formed stereocenter at C(2) has *R*-configuration. The carbonyl group of piperonal (**60**), on the other hand, can be approached from either the *Si*- or the *Re*-face. In the case of compound **76**, having the appendages of both stereocenters on the opposite site of the ring system, the enolate can approach the carbonyl group from either the *Si*- or the *Re*-face (Fig. 3, structures **A** and **B**) with no difference in steric repulsion. Consequently, aldol adduct **78** was formed as a 1:1 mixture of epimers at the benzylic carbon. The enolate of lactone **74** contains an

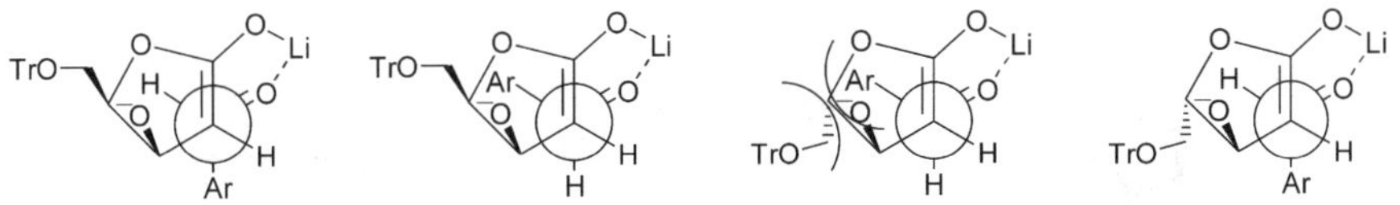

**Fig. 3**

oxo anion, determining the stereochemistry of the nucleophile, and a trityloxymethylene group, which occupies the space on the opposite site of the ring system. The latter substituent strongly interacts with the aryl moiety when the carbonyl group was approached from the *Si*-face whereas the steric repulsion is minimized when the nucleophile attacks from the *Re*-face (Fig. 3, structures **C** and **D**). Therefore, compound **82** with *R*-configuration (*threo*) at the benzylic carbon is favored over the *erythro* product (dr *RR*:*RS*=9:1).

Aldehyde **83** was obtained from lactone **82** in 5 steps and in 48% overall yield, according to the sequence described for its epimer in Scheme 17. Selective cleavage of the benzylic MOM ether using PPTS in *t*-BuOH produced the lactol, which upon chloride formation under Swern conditions and dehalogenation (Bu$_3$SnH, AIBN) furnished tetrahydrofuran **84** with *R*-configuration at C(2). The synthetic strategy underlines the importance of good stereocontrol in the aldol reaction because the stereocenter at the benzylic carbon is not removed as it is in the acid catalyzed cyclization of **80**. From tetrahydrofuran **84**, furanone **73** was prepared in 50% yield using the same sequence as described for the epimer. Finally, benzyl Grignard addition, followed by TBAF mediated silyl ether cleavage gave the 2*R*-configurated olivil analog **71** in 39% overall yield.

In conclusion, analogs of olivil-type lignans with different configurations at C(3), compounds **71** and **72**, have been prepared in 19 steps and 1.6% and 1.0% overall yield, respectively. The lengthy sequence, the low yield of key reactions, and the restrictions in the functionality of one of the two aryl groups are the major drawbacks derived from the use of sugars as starting materials. Additionally, derivatives with the correct olivil-type stereochemistry cannot be produced by this strategy.

## 4.2
### α-Hydroxylated Lignans from Malic Acid

Although the use of starting materials from the pool of chiral building blocks is an attractive approach to the synthesis of α-hydroxylated lignans, sugars are of limited value. Having a look at the structures of α-hydroxylated lignans, another chiral natural product, malic acid, seems to be a reasonable starting material. This inexpensive α-hydroxy acid is available in both enan-

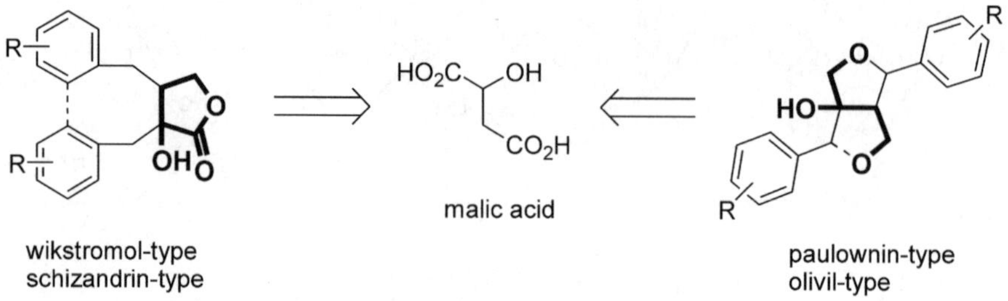

**Scheme 19**

tiomeric forms and has already a properly functionalized core unit (Scheme 19, emboldened substructures). All that has to be done is the stereoselective introduction of the two benzyl groups. For these reasons, malic acid is the most frequently used chiral starting material for the construction of optically active α-hydroxylated lignans. Depending on the electrophiles employed in the stereoselective C-C-bond forming reactions, several classes of α-hydroxylated lignans, such as lactone (wikstromol-) or furofuran (paulownin-type) lignans, were prepared from malic acid, which will be discussed in this chapter.

### 4.2.1
### Furofuran Lignans from Diethyl Malate

(*R*)-Diethyl malate (**85**) was used as the starting material for the synthesis of paulownin (**2**) [51]. Malate **85** was converted to lactone **86** using the protocol of Saito et al. [52]. Aldol addition of the lithium enolate of **86** to piperonal (**60**) gave compound **87** in 80% yield and in a 2:1 mixture of the *S*- and *R*-epimer at the benzylic carbon. The stereoselection at the benzylic carbon is in agreement with that obtained for the reaction of **76** with **60** (see above). The lactone moiety was transformed in three steps into the tetrahydrofuran **88** as a single isomer in 84% overall yield. The structure of compound **88** is similar to that of the intermediates **73** and **75** (see Scheme 16), although it was realized in a more efficient way than these precursors. From diol **88** two routes to paulownin (**2**) have been examined (Scheme 20) [51, 53]. The initial synthesis involved eight steps: functionalization at C(4) with appropriate stereochemistry was conducted in four steps by selective protection of the primary hydroxy group, oxidation of the secondary, olefination using the Tebbe reagent, and dihydroxylation with $OsO_4$/NMO (44% overall yield). This sequence for the stereoselective introduction of a carbon at C(4) is matching that for the synthesis of *cis*-lactone lignan **70**. The dihydroxylation proceeded with 6:1 stereoselection at C(4) in favor of diol **89** over its C(4)-epimer (Scheme 20). Oxidation of the primary hydroxy group, addition of an aryl Grignard reagent, and deprotection afforded the triol **90** in 22%

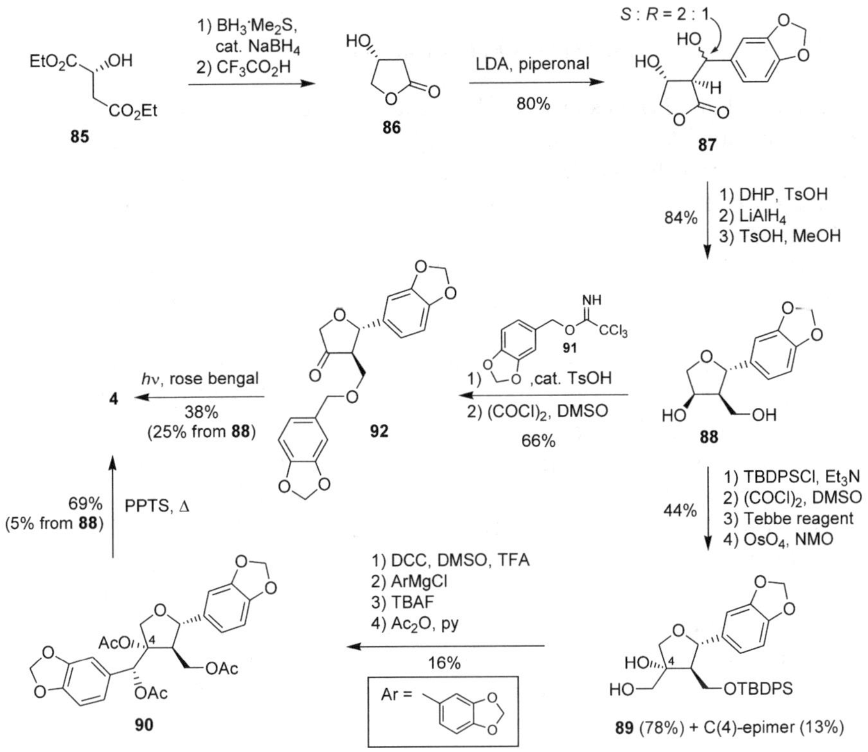

**Scheme 20**

overall yield as a single isomer. The good stereoselectivity of the Grignard addition to the carbonyl group can be explained by the Cram chelate model. However, the Grignard addition proved to be a reaction with a low yield.

PPTS-catalyzed cyclization of triol **90** provided paulownin (**2**) in 69% yield. The overall yield for the synthesis of paulownin (**2**) from the key intermediate **88** was only 5%. Therefore, a shorter and more efficient route involving a photocyclization was examined [53]. Etherification of the primary hydroxy group of compound **88** with benzyl trichloroacetimidate **91**, followed by Swern oxidation afforded furanone **92** in 66% yield. Photocyclization according to the Kraus procedure [54], using a mercury lamp and rose Bengal as sensitizer, gave paulownin (**2**) in 38% yield. Following this strategy, **2** was prepared in three steps from intermediate **88** and in an overall yield of 25% (Scheme 20). The synthesis of paulownin (**2**) was the first preparation of an optically-active furofuran lignan. The modified procedure involving a photocyclization [53] allows the preparation of **2** in 8 steps and 13% overall yield from diethyl malate.

## 4.2.2
### α-Hydroxylated Lactone Lignans from Dialkyl Malates

Diethyl malate (85) was also used for the construction of meridinol (97), an α-hydroxylated lactone lignan with sudirific and sedative properties, and its diastereoisomer 98 [55]. α-Alkylation of 85 with benzyl iodide 93, according to the procedure developed by Seebach and Wasmuth [56], gave compound 94 with *anti*-configuration in 78% yield. The amount of *syn*-isomer was not reported. The carboxyl group at C(4) of compound 94 was converted to the alcohol as follows: LiAlH$_4$ reduction and acetalization with anisaldehyde produced stereo- and regioselectively the C(2)-C(4) acetal. Re-oxidation of the remaining primary hydroxy group at C(1) and esterification with *t*-BuOH/Ac$_2$O provided the dioxane 95 (28% overall yield from 94). Subsequent hydrolysis of the acetal and the ester followed by etherification gave lactone 96 in 83% yield. The alkylation of lactone 96 was achieved with LDA and benzyl iodide 93. However, this reaction proceeded neither with high yield nor with acceptable stereoselectivity, thus providing after TMSBr-mediated removal of the MOM group meridinol (97) and its epimer 98 in 19% and 15% yield (from 96), respectively. As a result, the combined overall yield for both isomers was only 5% (Scheme 21). Two crucial disadvantages are associated with this strategy: 1. the non-selective reduction of the carboxy group at C(4), along with an unavoidable reduction-oxidation sequence, and 2. the low-yielding, non-selective α-alkylation of the γ-butyrolactone. The

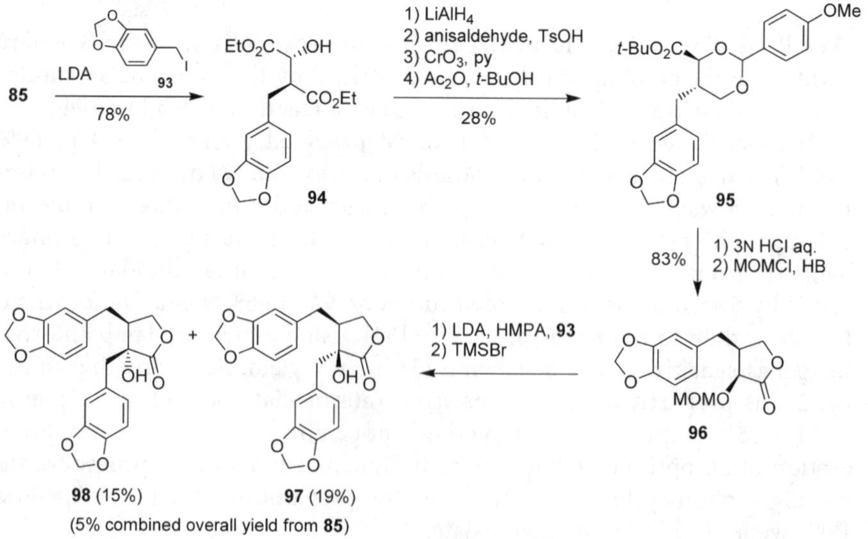

**Scheme 21**

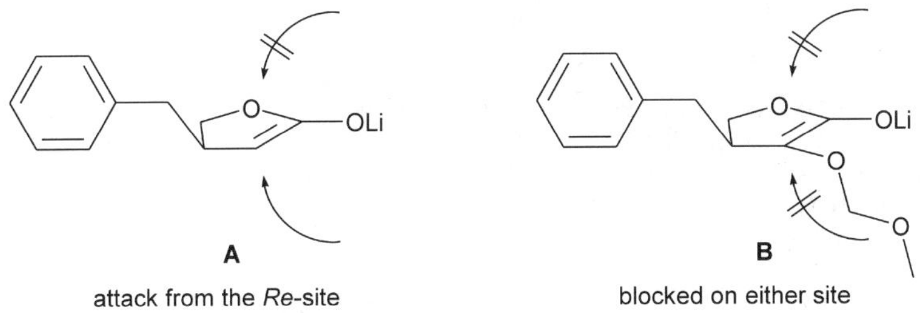

**Fig. 4**

A — attack from the Re-site
B — blocked on either site

non-selective alkylation is surprising, when compared with the highly stereoselective alkylations of 2-unsubstituted γ-butyrolactones (see above). This result can be explained as follows (supported by molecular modeling calculations). α-Unsubstituted β-benzyl-γ-butyrolactone enolates are stereoselectively trans-alkylated due to the difference in the steric repulsion of the Re- and the Si-face (Fig. 4, structure A). In α-substituted β-benzyl-γ-butyrolactone enolates, the stereocenter at C(3) forces the MOM group in a position opposite to the stereocenter. Thus, the lowest energy conformation of the enolate of **96** is that in which the methoxymethyl moiety and the stereocenter occupying both half-spaces (Fig. 4, structure B) and the approaching electrophile is partially blocked by either the stereocenter at C(3) or the MOM group at C(2) (Fig. 4). A similar observation has been made in the alkylation of chiral dioxanones [57].

The stereoselective introduction of both benzyl groups simultaneously in one step seemed to be particularly attractive for a short synthesis of α-hydroxylated lactone lignans from malic acid (**99**). Such a simultaneous double alkylation requires the formation of a chiral 1,3-diene-1,4-diolate, which was not known. On the other hand, achiral 1,3-diene-1,4-diolates (di-enolates) have been previously prepared by Garrett et al. [58] and subsequently employed for the synthesis of racemic lignans by Snieckus [59] and Pohmakotr [60]. With knowledge of the synthesis and reactivity of di-enolates, we planned to prepare chiral di-enolates from dioxolanones and to alkylate these di-enolates in a stereocontrolled manner (Scheme 22). For the development of the described double deprotonation/alkylation strategy, tert-butyl

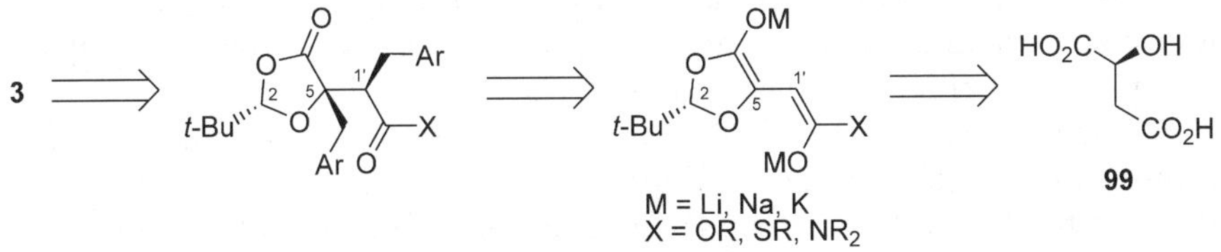

M = Li, Na, K
X = OR, SR, NR$_2$

**Scheme 22**

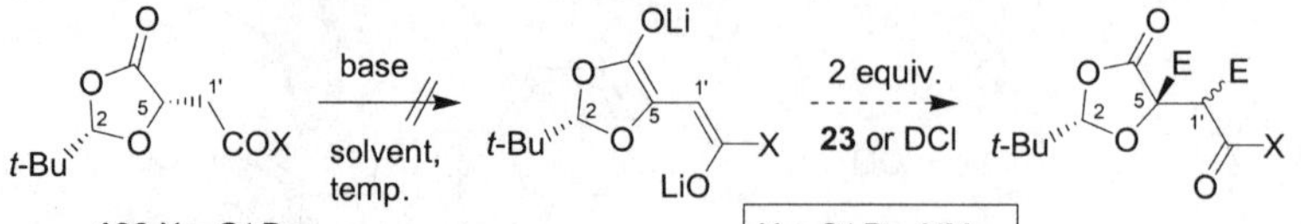

**Scheme 23**

ester **100** [61] and dimethyl-amide **101** (Scheme 23) were synthesized as substrates from (S)-(−)-malic acid (**99**) [62]. Ester **100** was chosen because of its steric hindrance and selective removal under mild conditions. On the other hand, it was shown that 1,4-bis-dialkylamides provide more stable di-enolates than the corresponding esters, for which reason dimethylamide **101** was prepared [63].

Dioxolanones **100** and **101** were treated with two equivalents of lithium amide bases, whereupon various electrophiles were added in excess (Scheme 23 and Table 1). Unfortunately, addition of two equivalents of LDA

**Table 1** Conditions for the attempted double deprotonation of dioxolanones **100** and **101**

| Entry | Cmpd | Base (equiv.) | $T$ [°C] | $t$ [min]$^a$ | E$^+$ | Products |
|---|---|---|---|---|---|---|
| 1 | 100 | LDA (2) | −78 | 0 | 23 | Decomp. |
| 2 | 100 | LHMDS (2) | −78 | 10 | DCl | 102 (20%) |
| 3 | 101 | LDA (2) | −78 | 10 | 23 | 103 (25%) |
| 4 | 101 | LHMDS (2) | −78 | 15 | DCl | 103 (41%) |
| 5 | 101 | LHMDS (2) | −105 | 0 | 23 | 103 (35%) |

$^a$ Addition of the electrophile after $x$ min; 0 min means metallation of the dioxolanones in the presence of the electrophile

to dioxolanone **100** at −71 °C, even in the presence of benzyl bromide **23**, furnished complete decomposition (Table 1, entry 1). Metallation of **100** with two equivalents of LHMDS gave spirolactone **102** (Scheme 24) in 20% yield and >20:1 diastereoselectivity (entry 2). An analogous aldol product, alcohol **103** (Scheme 24), was obtained when amide **101** was reacted with two equivalents of LDA or LHMDS and treated with DCl or benzyl bromide **23** (entries 3–5). Neither deuterated nor alkylated dioxolanones were formed under any examined conditions. Aldol product **103** was isolated in up to 40% yield and with good diastereoselectivity (>10:1).

The formation of **102** and **103** can be explained if one assumes that deprotonation at C(5) or at C(1′) occurred with similar reaction rates to afford the enolates **A** and **B** (Scheme 24). Evidence for this assumption is the finding that upon treatment with one equivalent of base (data not shown), the

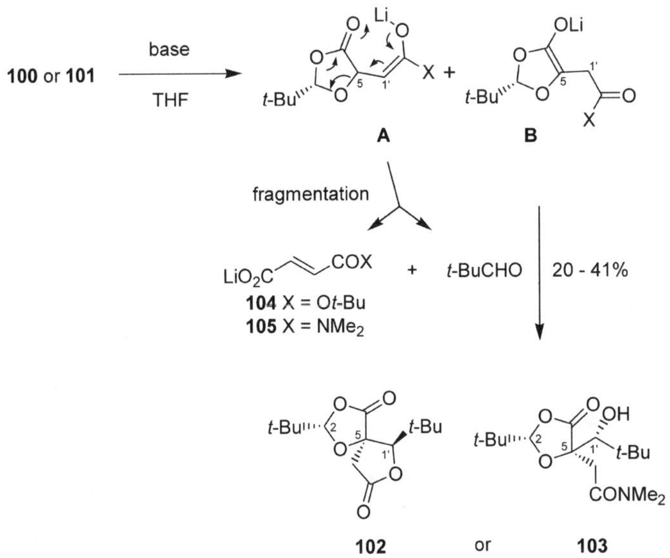

**Scheme 24**

same aldol products, **102** and **103**, were formed. Instead of a subsequent α-deprotonation of the second available proton, enolate **A** underwent a rapid β-elimination, even at −105 °C, producing the lithium salts of fumaric acids **104** and **105** (depending on the dioxolanone employed), and pivaldehyde. The fumaric acid salts were isolated as their corresponding acids after work-up. The pivaldehyde then reacted, as soon as it was liberated, with enolate **B** to give aldol products **102** or **103** in good yields, keeping in mind that the maximum yield is 50% for this reaction.

The absolute configuration of the major isomers of spirolactone **102** and amide **103** was 2S,5R,1′R, based on the known configuration of malic acid and determination of the relative configuration by an X-ray structure analysis of amide **103** [62].

Since the formation of optically active, dioxolanone-based di-enolates was not successful, a consecutive alkylation strategy was developed for a short synthesis of (−)-wikstromol (*ent*-**3**) from (−)-malic acid (**99**) (Scheme 25). The first alkylation reaction was analogous to that reported for the enantioselective total synthesis of (−)-meridinol (**97**). In order to avoid a reduction/re-oxidation sequence and an almost unselective second alkylation, two disadvantages of the synthesis of meridinol (**97**) [55], we planned to use a different strategy for the second alkylation. Therefore, we have focused our strategy on two stereoselective alkylation reactions, one of dialkyl malates and one of a dioxolanone prepared thereof. Both alkylation reactions were previously described by Seebach and coworker [56, 63, 64]. The

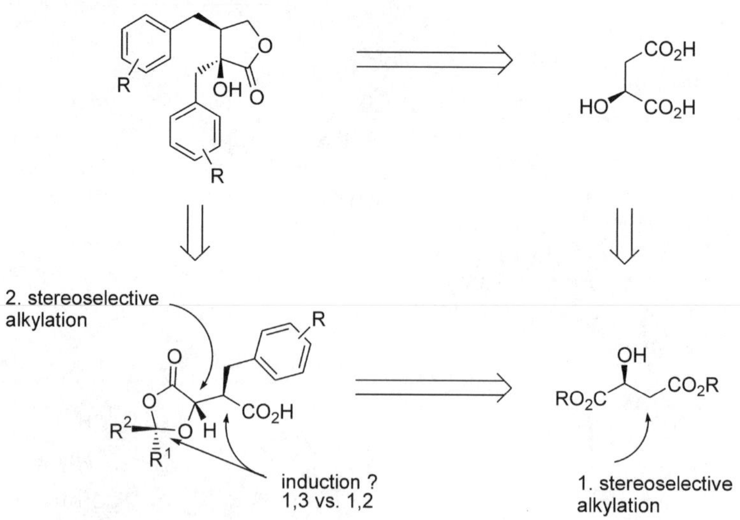

**Scheme 25**

major challenge of our consecutive alkylation strategy was the stereoselective synthesis of a C(1')-benzyl substituted dioxolanone and its diastereoselective alkylation at C(5). The preparation and alkylation of C(1')-substituted dioxolanones has not been reported and the stereochemical outcome was uncertain, due to the additional stereocenter in the side chain (Scheme 25).

The diastereoselective alkylation of dialkyl malates has been frequently used in the past [65]. However, according to the original procedure [63] (dialkyl malate, base, −78→−20 °C, then −78 °C, electrophile, then −78→0 °C, 16 h), the alkylation proceeded in average yields of about 50–60% and in diastereoselectivities in the range of 9:1 *anti / syn*. In our hands, application of this procedure to the reaction of benzyl bromide **23** with dimethyl malate **106** produced the alkylated compounds in only 20% yield. The yield of the alkylation was easily improved (>75%) when the ester was deprotonated with LHMDS *in the presence of the electrophile* at −78 °C and the reaction mixture was allowed to warm to 10 °C (Scheme 26 and Table 2).

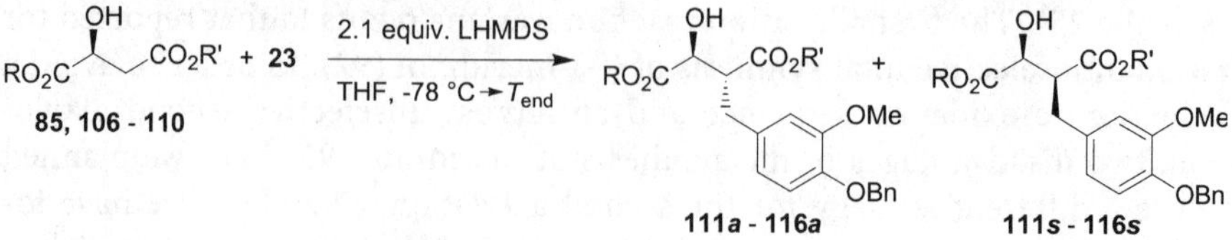

**Scheme 26**

**Table 2** Yields and stereoselectivities of the alkylation of diesters 85, 106–110

| Entry | cmpd | R | R' | $T_{end}$ [°C] | Products | dr | Yield [%] | Remarks |
|---|---|---|---|---|---|---|---|---|
| 1 | 106 | Me | Me | 14 | 111a: 111s | 8:1 | 75 | |
| 2 | 85 | Et | Et | 14 | 112a: 112s | 9:1 | 91 | |
| 3 | 107 | i-Pr | i-Pr | 9 | 113a: 113s | 19:1 | 80 | (17:1; 78%)[a] |
| 4 | 108 | t-Bu | t-Bu | 9 | 114a: 114s | 7:1 | 94 | (9:1; 90%)[a] |
| 5 | 109 | i-Pr | t-Bu | 16 | 115a: 115s | 9:2 | 90 | |
| 6 | 110 | t-Bu | i-Pr | 16 | 116a: 116s | 40:1 | 78 | |

[a] Results of a repetition experiment ($T_{end}$=16 °C)

Furthermore, a significant simplification of the original procedure was realized. Yield and ratio of *syn*- and *anti*-product was dependent on the steric effects of the ester groups. Commercially available dimethyl, diethyl, and diisopropyl malate **106**, **85**, and **107** afforded the alkylation products **111a–113a** and **111s–113s** in 75%, 92%, and 80% combined yield, respectively. In the case of diisopropyl malate **107**, the reaction was incomplete and ca. 10% of starting material was recovered. The diastereomeric ratio increased in the order Me, Et, i-Pr ester from 8:1 to 19:1 (*anti-:syn-*isomer). It was assumed that the diastereomeric ratio could be increased even more, when the bulkier *tert*-butyl group ("bigger is better") was used for both ester moieties. Di-*tert*-butyl malate **108** was prepared [66] and alkylated using our modified reaction conditions [67]. Surprisingly, the two diastereomers **114a** and **114s** were obtained in a 7:1 ratio (!) but in almost quantitative yield. We reasoned that only one of the *tert*-butyl groups of malate **108** was responsible for the low stereoselectivity and prepared the mixed esters, isopropyl-*tert*-butyl ester **109** and *tert*-butyl-isopropyl ester **110**.

The hitherto unknown esters **109** and **110** were obtained from malic acid **99** in 85% and 20% overall yield, respectively (Scheme 27). Esterification of

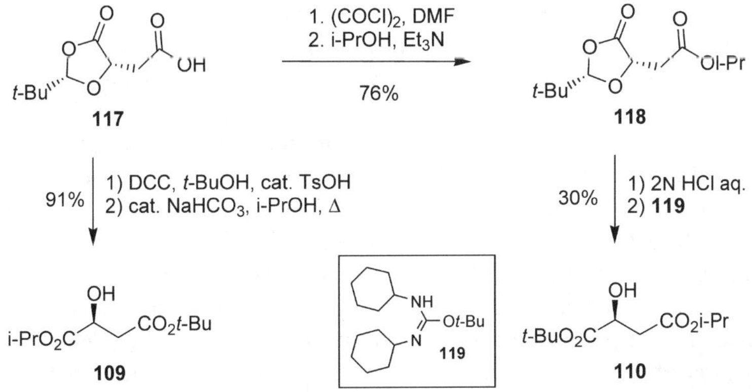

**Scheme 27**

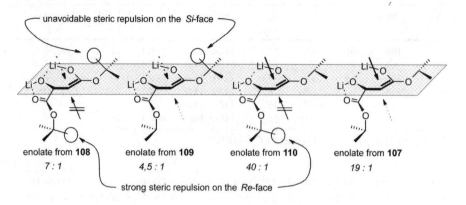

**Fig. 5**

dioxolanone 117 (DCC, *t*-BuOH), followed by base-catalyzed trans-esterification in hot i-PrOH in the presence of a catalytic amount of $NaHCO_3$ [61], afforded the ester 109 in 91% yield. Ester 110 was prepared by the esterification of the carboxyl group of dioxolanone 117 (→118, 76%), followed by the selective hydrolysis of the acetal moiety with 2N HCl and esterification of the C(1) carboxyl group with isourea 119 [68] (→110, 30% yield).

Alkylation of 109 and 110 with benzyl bromide 23 provided the following results [69]: succinates 115*a* and 115*s* were formed in 90% yield and in a ratio of 9:2, whereas succinates 116*a* and 116*s* were obtained in a ratio of 40:1, although the yield was only 78% and ca. 10% of the starting material remained. Thus, malate 109 and di-*tert*-butyl malate 108 reacted comparably with respect to both yield and stereoselectivity, as did malates 110 and 107.

Molecular mechanics calculations of the intermediate enolates provided a rationale for these findings, which is displayed in Fig. 5. Since the electrophile approaches the enolate perpendicular to the C=C double bond [70], steric hindrance caused by the alkyl groups is pivotal for either the *Re*- or the *Si*-site attack. The *anti*-selectivity was enhanced by a factor of two when the isopropyl ester at C(1) (malate 107) was substituted by a *tert*-butyl ester (malate 110). In this case, the *Re*-face is more effectively blocked because one of the methyl substituents of the *tert*-butyl group is positioned exactly in the trajectory of the electrophile. On the other hand, steric repulsion on the opposite *Si*-face should decrease the *anti*-selectivity. In fact, this is the case with malate 108 having a *tert*-butyl group at C(4). Again, the difference between an i-Pr group and a *t*-Bu group becomes apparent. Whereas both Me groups of the i-Pr group are positioned away from the enolate moiety

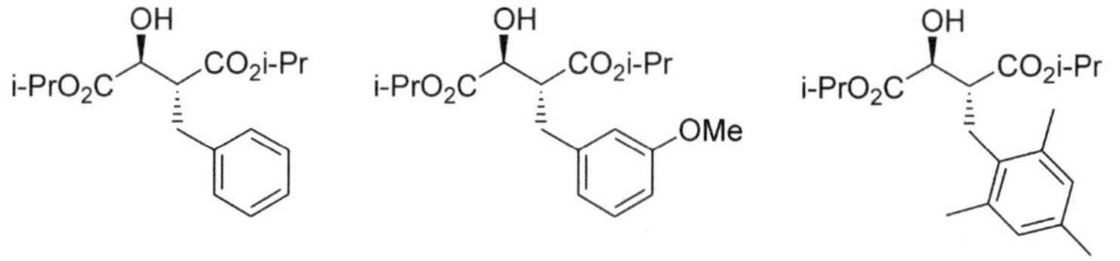

**120** (53%, dr 94:6)   **121** (65%, dr 97:3)   **122** (56%, dr 83:17)

**Fig. 6**

this is not possible with the *t*-Bu group, with which one Me group inevitably shield the *Si*-face. As a result, malate **108** was alkylated with lower stereoselectivity than malate **107**. Consequently, the lowest *anti*-selectivity was achieved with malate **109**, in which the bulky *tert*-butyl group is at C(4) and the comparably smaller isopropyl group is at C(1). Therefore, increased steric hindrance at C(1) increases the *anti*-selectivity, whereas increased steric hindrance at C(4) decreases it. *In conclusion, dialkyl malate **110** may be described as a matched and dialkyl malate **109** as a mismatched pair of differently encumbered alkyl esters.*

The best compromise with respect to reactivity *and* availability of the starting material was the use of diisopropyl malate **107**. This malic acid ester is easy to prepare and its alkylation with various benzyl bromides can be achieved with good yields (53–67%, not optimized) and high stereoselectivities (dr~95:5 for **120** and **121**). An exception with respect to the stereoselectivity was the 2,4,6-trimethylbenzyl substituted succinate **122**, which was obtained in a dr of only 83:17 (Fig. 6) [71].

Saponification of the succinates **113**, **120–122** was carried out in ethanolic KOH. The reaction proceeded smoothly under these conditions. However, even if diastereomerically pure starting material was employed, mixtures of *anti*- and *syn*-isomer of diacids **123–125** (ratio ca. 10:1) were formed in quantitative yields [71]. The more crowded succinate **122** was saponified to its diacid **126a** without epimerization at C(2). Diacid **123** could be recrystallized to afford the pure *anti*-isomer, but the recrystallization was not reliable. Therefore, the acetalizations with pivaldehyde were carried out with the mixtures of diastereomers in 72–83% combined yield. This reaction was best achieved in benzene as solvent, affording the *cis*-dioxolanones **127a**–**130a** and the *trans*-dioxolanones **131a–134a** (from the *anti*-diacids), alongside the *cis*-dioxolanones **127s–129s** (from the *syn*-diacids), in a ca. 5:1:1 ratio (Scheme 28) [72]. The fourth possible isomer, the *trans*-dioxolanones from the *anti*-diacids, having (2$S$,2'$R$,4'$S$)-configuration, was not detected in any case.

Isomerically pure dioxolanone **127a** was obtained from the crude reaction mixture after recrystallization. Again, this process was irreproducible, and the yield for stereoisomerically pure **127a** was low (50–70%). For this rea-

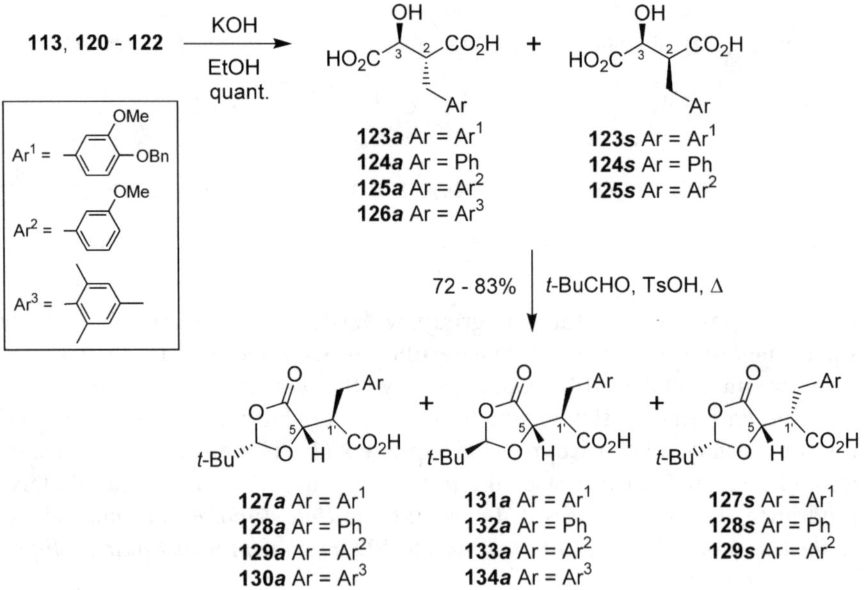

**Scheme 28**

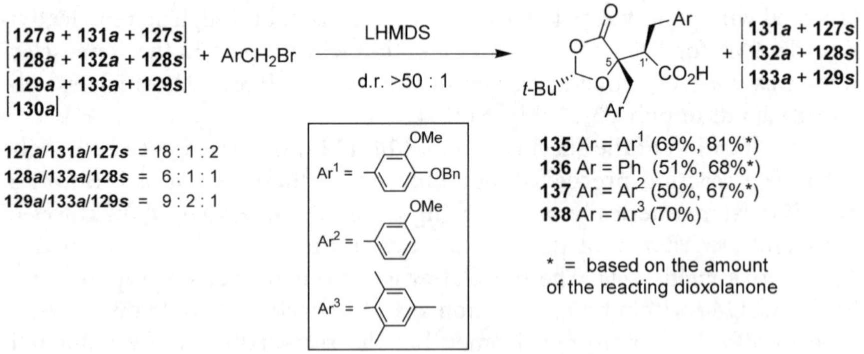

**Scheme 29**

son, a mixture of the isomers was allowed to react with the benzyl bromides at −78 °C, using LHMDS as the base (Scheme 29). Surprisingly, only the *cis*-dioxolanones **127a–130a** were alkylated under these conditions, providing diastereomerically pure dioxolanones **135–138** according to the $^1$H NMR spectra of the crude reaction mixtures [69]. *Cis*-dioxolanones **127s–129s**, as well as *trans*-dioxolanones **131a–133a**, were recovered unchanged. The alkylation products **135–138** were isolated in 50–70% yield (67–81% yield based on the amount of the reacting *cis*-dioxolanones **127a–130a**).

The different alkylation ability of dioxolanone **127a** vs dioxolanones **131a** and **127s** may be explained by either of the following mechanisms (Scheme 30) [69]. Initially, the lithium carboxylates of the dioxolanones **127a**, **131a**, and **127s** (**127a·Li**, **131a·Li**, **127s·Li**) are formed with one equivalent of LHMDS. This reaction is very exothermic and could be monitored easily with an internal thermometer. Then the carboxylates may be deprotonated by a second equivalent of LHMDS to afford the enolates **127a·2Li**, **131a·2Li**, **127s·2Li** (Scheme 30, path A). The latter two are enantiomers and have the same relative configuration. This reduces the problem to the comparison of the enolate structures of **127a·2Li** and **131a·2Li**. If the enolates were formed in all cases, then the steric interactions of the enolates with the incoming electrophile must be responsible for the different reactivity (Scheme 30, path A).

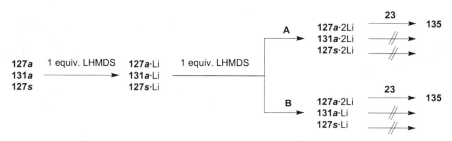

**Scheme 30**

On the other hand, LHMDS is a bulky base and, more likely, the kinetic deprotonation at C(4′) occurs selectively with carboxylate **127a·Li**, providing the enolate **127a·2Li**, which is alkylated, while the isomeric carboxylates **131a·Li** and **127s·Li** remained unchanged (Scheme 6, path B).

The assumption that from a mixtures of dioxolanones only one diastereomer was alkylated, was independently confirmed by alkylation experiments involving stereoisomerically-pure dioxolanones **127a** and **131a** with benzyl bromide **23** as the electrophile and LHMDS as the base. Dioxolanone **127a** was alkylated within 5 h at −78 °C to afford product **135**, whereas dioxolanone **131a** was recovered unchanged (no alkylation or isomerization products were detected) in over 90% yield under identical reaction conditions. If the enolate **131a·2Li** would have been formed, then reprotonation should occur from both stereofaces. In fact, this was observed when the reaction of diastereomerically pure dioxolanone **127a·2Li** with benzyl bromide **23** was quenched prior to the complete consumption of the nucleophile. In this case, product **135** was accompanied by *cis*-dioxolanone **127a** and the corresponding *trans*-dioxolanone, which were formed in a 2:1 ratio!

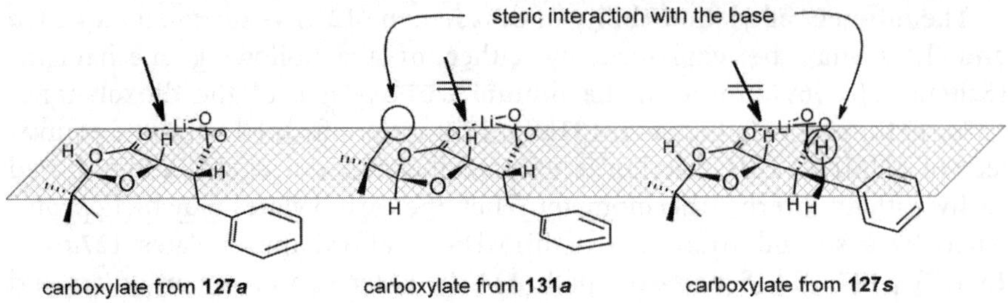

steric interaction with the base

carboxylate from **127a**    carboxylate from **131a**    carboxylate from **127s**

**Fig. 7**

Further evidence was provided by molecular mechanics calculations of the three carboxylates **127a**·Li, **131a**·Li, **127s**·Li, whose minimum energy conformations are depicted in Fig. 7. The calculated structures of the lithium carboxylates **131a**·Li and **127s**·Li indicated that the proton at C(4′) is shielded by either the *tert*-butyl group of the dioxolanone moiety (**131a**·Li, $d_{H-C(4')...H-C(tert)}=3.15$ Å) or the $CH_2$ group of the side chain (synaxial interaction between H-C(4′) and H-C(3) in **127s**·Li, $d_{H-C(4')...H-C(3)}=2.43$ Å). In the case of dioxolanone **127a**·Li, neither of the stereocenters interfere with the base, and deprotonation to the enolate is possible even at −78 °C (Fig. 7).

Dioxolanones **135–138** were converted to the corresponding lactone lignans in a one-pot procedure by regioselective reduction of the carboxyl group using borane dimethylsulfide complex followed by acid hydrolysis of the dioxolanone moiety. In the reduction step, an appropriate solvent is crucial for good yields. Dioxolanone **135** upon treatment of borane dimethylsulfide in THF gave, after hydrolysis, lactone **139** and lactol **140** in 19% and 20%, respectively. Several by-products have been formed, some of which were structurally characterized [70]. Much better results were obtained when the reaction was carried out in $Et_2O$. The yields of the lactone lignans **141–143** were generally over 80% (Scheme 31).

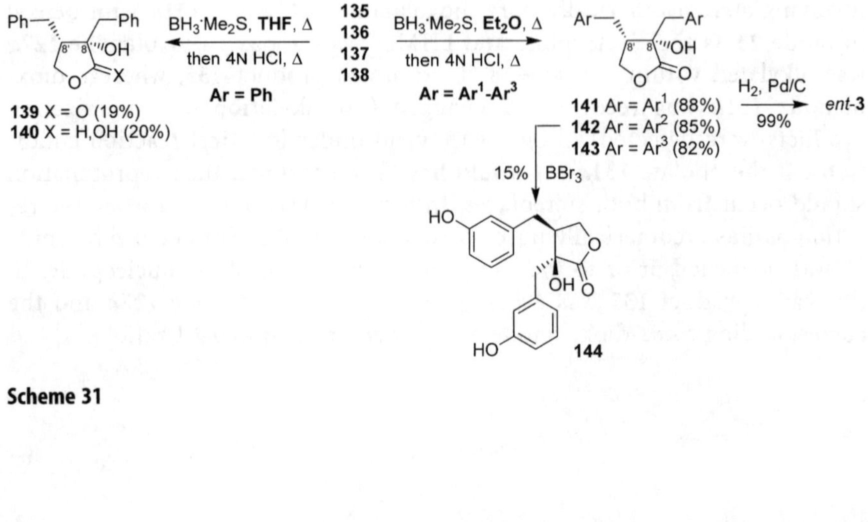

**139** X = O (19%)
**140** X = H,OH (20%)

$BH_3·Me_2S$, THF, Δ
then 4N HCl, Δ
Ar = Ph

**135**
**136**
**137**
**138**

$BH_3·Me_2S$, $Et_2O$, Δ
then 4N HCl, Δ
Ar = Ar¹-Ar³

**141** Ar = Ar¹ (88%)
**142** Ar = Ar² (85%)
**143** Ar = Ar³ (82%)

$H_2$, Pd/C
99%
*ent*-**3**

15% | $BBr_3$

**144**

**Scheme 31**

Hydrogenolysis of the benzylether of lactone **141** (H$_2$, 10% Pd/C) quantitatively afforded (−)-wikstromol (*ent*-**3**). The overall yield of *ent*-**3** from diisopropyl malate **107** over six steps was 30% [62]. This strategy for the construction of optically-active lactone lignans provided a short and stereoselective synthesis with comparably good overall yield.

Demethylation of the methoxy groups in lactone **142** was achieved with BBr$_3$ in dichloromethane. The reaction conditions were not optimized, and the yield of α-hydroxylated enterolactone **144** was low (15%). Several byproducts, such as both regioisomers of the mono-methylether, were obtained.

The amount of borane is also important in the regioselective reduction of the dioxolanones (Scheme 31). The use of a large excess of borane resulted in the partial reduction of both carboxyl moieties. Completion of the reduction with LiAlH$_4$, and further functional group manipulations, gave (−)-carinol (*ent*-**26**) in 35% overall yield (Scheme 32) [62].

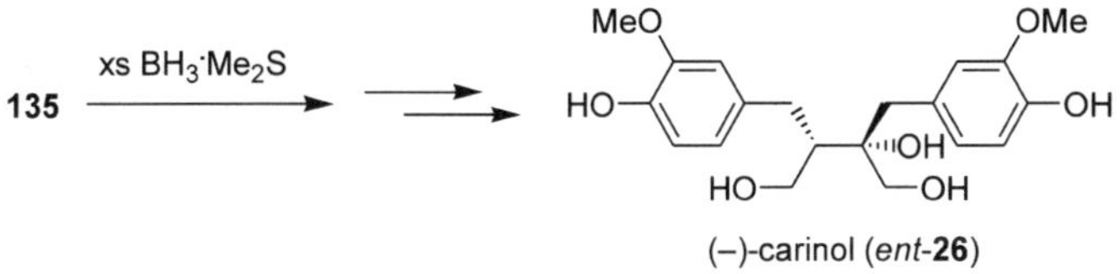

**Scheme 32**

The synthesis of the biologically more active (+)-isomer of wikstromol (**3**) was achieved using the non-naturally-occurring (*R*)-malic acid (*ent*-**99**) in 20% overall yield utilizing the strategy described above [67]. However, acetalization of substituted succinates, such as diacid **123a** (Scheme 28), with pivaldehyde under thermodynamic control afforded the diastereoisomeric *trans*- and *cis*-dioxolanones in a ratio of ca. 5:1, in agreement with previous observations [72]. Much better stereoselectivities have been realized, when malic acid was converted to the dioxolanone using Noyori's kinetic acetalization [73, 74]. In an attempt to prepare dioxolanone *ent*-**127u** from the per-silylated succinate **145**, only the corresponding diacid was recovered from the reaction mixture (Scheme 33).

The inhibition of proliferation of several cancer cells was investigated with the compounds shown in Fig. 8. One of these lignans, compound **143**, was active, having an IC$_{50}$ of ca. 35 μmol/l against HT 29 colon cancer cells [75] and others [76]. This result was in agreement with previously observed cytotoxicities of α-unfunctionalized lactone lignans [77].

Our strategy for the synthesis of α-hydroxylated lactone lignans might be extended to the general synthesis of α-hydroxylated lignans, when either of the alkylation steps is changed to an aldol addition. For example, berchemol

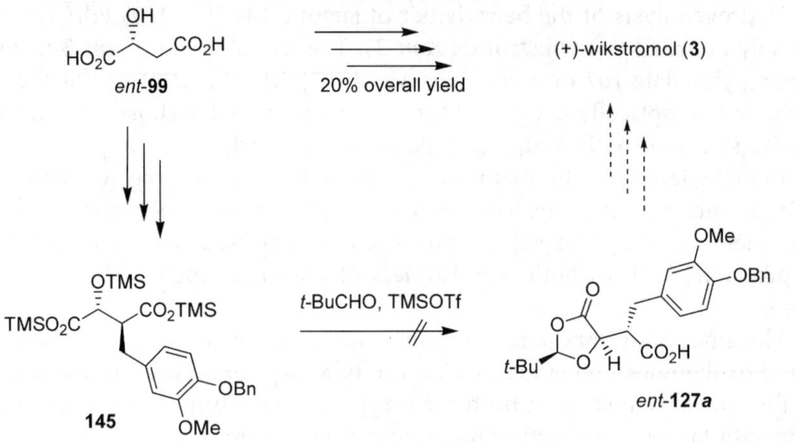

**Scheme 33**

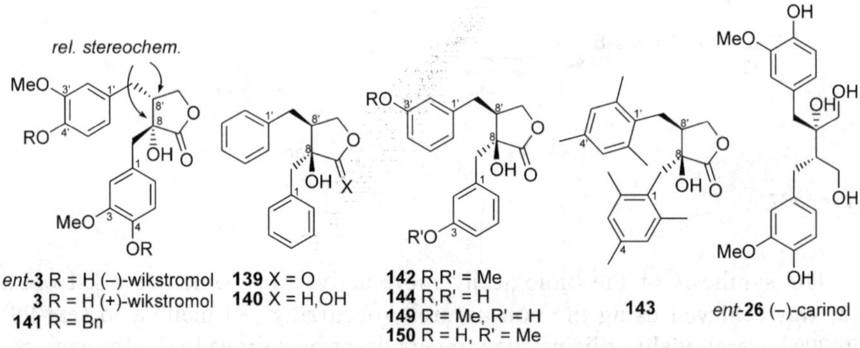

**Fig. 8**

(148) [78], a tetrahydrofuran lignan containing a vicinal diol subunit, should be produced when the second alkylation is replaced by an aldol reaction. Thus, the dioxolanone **127a**, after enolization with LHMDS and treatment with O-benzyl-vanilline (**146**), furnished the aldol product **147** in 80% yield as a mixture of diastereomeric benzylic alcohols (ratio ca. 1.5:1). Unfortunately, the initial strategy for the conversion of dioxolanone **147** to berchemol (**148**), which has not yet been synthesized, was unsuccessful (Scheme 34). We are currently examining alternative routes for this conversion.

Upon supplementing the first alkylation by an aldol reaction, other types of α-hydroxylated lignans should be attainable as well. In the case in which both C-C bond formations are aldol reactions, furofuran lignans, such as paulownin or analogs, should be obtained. On the other hand, aldol addition

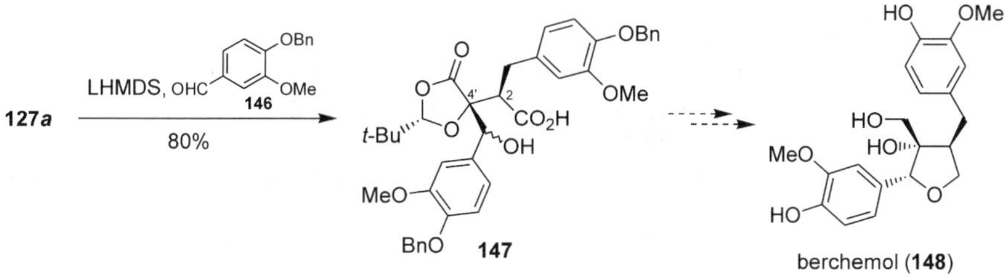

**Scheme 34**

followed by alkylation should yield lignans of the olivil-type. As mentioned above, the latter class of lignans has not been prepared with natural occurring configuration and substitution pattern. The only change in the synthetic strategy is the appropriate protection of the additional benzylic hydroxy group (Scheme 35). The application of these considerations towards a short synthesis of paulownin (4) and olivil (2) is underway.

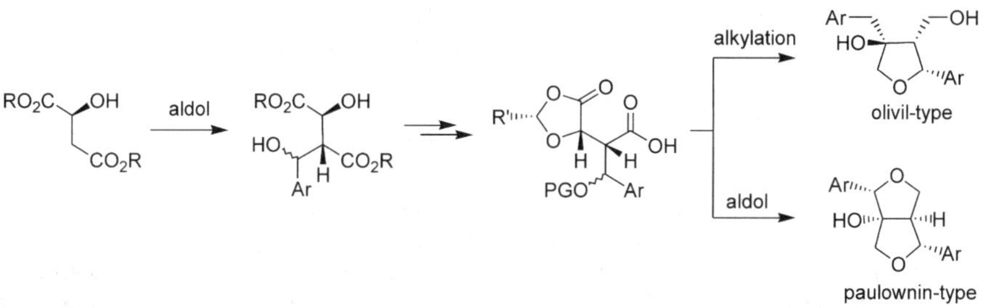

**Scheme 35**

## References

1. Ayres DC, Loike JD (1990) Lignans: chemical, biological and clinical properties. Cambridge University Press, Cambridge
2. Ward RS (1999) Nat Prod Rep 16:75
3. Moss GP (2000) Pure Appl Chem 72:1493
4. a) Shiota G, Yamada S, Kawasaki H (1996) Res Commun Mol Pathol Pharm 94:141; b) Yamada S, Murawaki Y, Kawasaki H (1993) Biochem Pharmacol 46:1081
5. Rao J, Madhusudana T, Ashok K, Upparapalli S, Yadav JS, Raghavan KV (2003) WO 0311279
6. Lee KH, Tagahara K, Suzuki H, Wu RY, Haruna M, Hall IH, Huang HC, Ito K, Iida T, Lai JS (1981) J Nat Prod 44:530
7. Rao J, Madhusudana T, Ashok K, Srinivas PV, Yadav JS, Raghavan KV (2002) US 2002 094351
8. a) Kurozumi A, Kobayashi S, Masui A, Taniguchi E (1989) JP 01160982; b) Matsubara H (1972) Bull Inst Chem Res 50:197

9. Ghera E, Ben-David Y (1978) J Chem Soc Chem Commun 480
10. Khamlach K, Dhal R, Brown E (1989) Tetrahedron Lett 30:2221
11. Warshawsky AM, Meyers AI (1990) J Am Chem Soc 112:8090
12. Meyers AI, Himmelsbach RJ (1985) J Am Chem Soc 107:682
13. Meyers AI, Flisak RJ, Aitken RA (1987) J Am Chem Soc 109:5446
14. Ikeya Y, Taguchi H, Mitsuhashi H, Takeda S, Kase Y, Aburada M (1988) Phytochemistry 27:569
15. Eklund P, Lindholm A, Mikkola JP, Smeds A, lehtilae R, Sjoeholm R (2003) Org Lett 5:491
16. Tomioka K, Koga K (1979) Tetrahedron Lett 3315
17. Camps P, Font J, Ponsati O (1981) Tetrahedron Lett 22:1471
18. Yoda H, Naito S, Takabe K, Tanaka N, Hosoya K (1990) Tetrahedron Lett 31:7623
19. Takano S, Ohashi K, Sugihara T, Ogasawara K (1991) Chem Lett 203
20. Yoda H, Kitayama H, Katagiri T, Takabe K (1992) Tetrahedron 48:3313
21. a) Costa PRR, Ferreira VF, Filho HCA, Pinheiro S (1996) J Brazil Chem Soc 7:67; b) Filho HCA, Filho UFL, Pinheiro S, Vasconcellos MLAA, Costa PRR (1994) Tetrahedron Asymmetry 5:1219
22. a) Sibi MP, Liu P, Johnson MD (2000) Can J Chem 78:133; b) de L Vanderlei JM, Coelho F, Almeida WP (1998) Synth Commun 28:3047
23. Canan Koch SS, Chamberlin AR (1993) J Org Chem 58:2725
24. a) Bode JW, Doyle MP, Protopopova MN, Zhou QL (1996) J Org Chem 61:9146; b) Doyle MP, Protopopova MN, Zhou QL, Bode JW (1995) J Org Chem 60:6654
25. a) Morimoto T, Chiba M, Achiwa K (1993) Tetrahedron 49:1793; b) Morimoto T, Chiba M, Achiwa K (1992) Heterocycles 33:435
26. Posner GH, Kogan TP, Haines SR, Frye LL (1984) Tetrahedron Lett 25:2627
27. Kosugi H, Tagami K, Takehashi A, Kanna H, Uda H (1989) J Chem Soc Perkin Trans 1 935
28. Asaoka M, Fujii N, Shima K, Takei H (1988) Chem Lett 805
29. a) Chęnevert R, Mohammadi-Zirani G, Caron D, Dasser M (1999) Can J Chem 77:223; b) Shiotani S, Okada H, Yamamoto T, Nakamata K, Adachi J, Nakamoto H (1996) Heterocycles 43:113; c) Yang LM, Lin SJ, Yang TH, Lee KH (1996) Bioorg Med Chem Lett 6:941; d) Itoh T, Chika J, Takagi Y, Nishiyama S (1993) J Org Chem 58:5717
30. a) Alphand V, Mazzani C, Lebreton J, Furstoss R (1998) J Mol Catal B 5:219; b) Gagnon R, Grogan G, Groussain E, Pedragosa-Moreau S, Richardson PF, Roberts SM, Willets AJ, Alphand V. Lebreton J, Furstoss R (1995) J Chem Soc Perkin Trans 1 2527
31. Honda T, Kimura N, Sato S, Kato D, Tominaga H (1994) J Chem Soc Perkin Trans 1 1043
32. a) Daugan A, Brown E (1991) J Nat Prod 54:110; b) Brown E, Daugan A (1986) Tetrahedron Lett 27:3719
33. a) Lalami K, Dhal R, Brown E (1991) J Nat Prod 54:119; b) Lalami K, Dhal R, Brown E (1988) Heterocycles 27:1131
34. Kamlage S, Sefkow M, Zimmermann N, Peter MG (2002) Synlett 77
35. a) Takekawa Y, Shishido K (2001) J Org Chem 66:8490; b) Takekawa Y, Shishido K (1999) Tetrahedron Lett 40:6817
36. a) Bolm C, Beckmann O, Cosp A, Palazzi C (2001) Synlett 1461; b) Bolm C, Beckmann O, Palazzi C (2001) Can J Chem 79:1593
37. Caro Y, Masaguer CF, Raviña E (2001) Tetrahedron Asymmetry 12:1723; see also Brinksma J, van der Deen H, van Oeveren A, Feringa BL (1998) J Chem Soc, Perkin Trans 1 4159; Barnier JP, Blanco L, Guibé-Jampel E, Rousseau G (1989) Tetrahedron 45:5051

38. Kenza Khamlach M, Dhal R, Brown E (1992) Tetrahedron 48:10115
39. Khamlach K, Dhal R, Brown E (1990) Heterocycles 31:2195
40. a) Tanaka M, Mukaiyama C, Mitsuhashi H, Maruno M, Wakamatsu T (1995) J Org Chem 60:4339; b) Tanaka M, Itoh H, Mitsuhashi H, Maruno M, Wakamatsu T (1993) Tetrahedron Asymmetry 4:605; c) Tanaka M, Mukaiyama C, Mitsuhashi H, Wakamatsu T (1992) Tetrahedron Lett 33:4165
41. a) Tanaka M, Mitsuhashi H, Maruno M, Wakamatsu T (1996) Heterocycles 42:359; b) Tanaka M, Ikeya Y, Mitsuhashi H, Maruno M, Wakamatsu T (1995) Tetrahedron 51:11703; c) Tanaka M, Mitsuhashi H, Maruno M, Wakamatsu T (1994) Synlett 604
42. a) Moritani Y, Fukushima C, Ukita T, Miyagishima T, Ohmizu H, Iwasaki T (1996) J Org Chem 61:6922; b) Belletire JL, Fry DF (1988) J Org Chem 53:4724
43. Morimoto T, Chiba M, Achiwa K (1989) Tetrahedron Lett 30:735
44. Burk MJ, Bienewald F, Harris M, Canotti-Gerosa A (1998) Angew Chem Intern Ed Engl 37:1931
45. Tang W, Liu D, Zhang X (2003) Org Lett 5:205
46. a) Inoki S, Kato K, Isayama S, Mukaiyama T (1991) Chem Lett 1869; b) Mukaiyama T, Isayama S, Inoki S, Kato K, Yamada T, Takei T (1989) Chem Lett 449
47. Yamauchi S, Kinoshita Y (1998) Biosci Biotechnol Biochem 62:521
48. Yamauchi S, Kinoshita Y (1998) Biosci Biotechnol Biochem 62:1726
49. Yamauchi S, Kinoshita Y (2000) Biosci Biotechnol Biochem 64:1563
50. Yamauchi S, Kinoshita Y (2000) Biosci Biotechnol Biochem 64:2320
51. Sharma R, Marquez VE (1994) Synth Commun 24:1937; see also Yamauchi S, Kinoshita Y (1997) Biosci Biotechnol Biochem 61:1342
52. a) Ishibashi F, Hayashita M, Okazaki M, Shuto Y (2001) Biosci Biotechnol Biochem 65:29; b) Okazaki M, Ishibashi F, Shuto Y, Taniguchi E (1997) Biosci Biotechnol Biochem 61:743
53. Saito S, Hasegawa T, Inaba M, Nishida R, Fujii T, Nomizu S, Moriwake T (1984) Chem Lett 1389
54. Kraus GA, Chen L (1990) J Am Chem Soc 112:3464
55. Takano D, Matsumi Y, Yoshihara K, Kinoshita T (1999) J Heterocycl Chem 36:221
56. Seebach D, Wasmuth D (1980) Helv Chim Acta 63:197
57. a) Scherowsky G, Sefkow M (1993) Chimia 47:19; b) Pietzonka T, Seebach D (1991) Chem Ber 124:1837
58. Garratt PJ, Zahler R (1978) J Am Chem Soc 100:7753
59. Mahalanabis KK, Mumtaz M, Snieckus V (1982) Tetrahedron Lett 23:3975
60. Pohmakotr M, Reutrakul V, Prongpradit T, Chansri A (1982) Chem Lett 687
61. Tain ZQ, Brown BB, Mack DP, Hutton CA, Bartlett PA (1997) J Org Chem 62:514
62. Sefkow, M (2001) J Org Chem 66:2343
63. Seebach D, Aebi J, Wasmuth D (1990) Org Synth, Col Vol VII. Wiley, p 153
64. Seebach D, Naef R, Calderari G (1984) Tetrahedron 40:1313
65. Gawroński J, Gawrońska K (1999) Tartaric and malic acid in synthesis. Wiley, New York
66. Kuo CH, Robichaud AJ, Rew DJ, Bergstrom JD, Berger GD (1994) Bioorg Med Chem Lett 4:1591
67. Sefkow M (2001) Tetrahedron Asymmetry 12:987
68. Mathias LJ (1976) Synthesis 561
69. Sefkow M, Koch A, Kleinpeter E (2002) Helv Chim Acta 85:4216
70. a) Baldwin JE, Kruse LI (1977) J Chem Soc Chem Commun 233; b) House HO, Phillips WV, Sayer TSB, Yau CC (1978) J Org Chem 43:700
71. Sefkow M, Kelling A, Schilde U (2001) Tetrahedron Lett 42:5101

72. Gabrielsen MV (1975) Acta Chem Scand A 29:7
73. a) Noyori R, Murato S, Suzuki M (1981) Tetrahedron 37:3899; b) Tsunoda T, Suzuki M, Noyori R (1980) Tetrahedron Lett 21:1357
74. Hoye TR, Peterson BH, Miller JD (1987) J Org Chem 52:1351
75. Sefkow M, Raschke M, Steiner C (2003) Pure Appl Chem 75:273
76. Results of the NCI-test panel of 60 different cancer cell lines (not published)
77. a) Kamlage S, Sefkow M, Peter MG, Pool-Zobel BL (2001) Chem Commun 331; b) Kamlage S, Sefkow M, Peter MG, Pool-Zobel BL (2001) Chem Commun 507
78. a) Sakurai N, Nagashima S, Kawai K, Inoue T (1989) Chem Pharm Bull 37:3311; b) Sakurai N, Nagashima S, Kawai K, Inoue T (1990) Chem Pharm Bull 38:1801

# Author Index Volumes 201–243

Author Index Vols. 26–50 see Vol. 50
Author Index Vols. 51–100 see Vol. 100
Author Index Vols. 101–150 see Vol. 150
Author Index Vols. 151–200 see Vol. 200

*The volume numbers are printed in italics*

Achilefu S, Dorshow RB (2002) Dynamic and Continuous Monitoring of Renal and Hepatic Functions with Exogenous Markers. *222*: 31–72
Albert M, see Dax K (2001) *215*: 193–275
Angelov D, see Douki T (2004) *236*: 1–25
Angyal SJ (2001) The Lobry de Bruyn-Alberda van Ekenstein Transformation and Related Reactions. *215*: 1–14
Armentrout PB (2003) Threshold Collision-Induced Dissociations for the Determination of Accurate Gas-Phase Binding Energies and Reaction Barriers. *225*: 227–256
Astruc D, Blais J-C, Cloutet E, Djakovitch L, Rigaut S, Ruiz J, Sartor V, Valério C (2000) The First Organometallic Dendrimers: Design and Redox Functions. *210*: 229–259
Augé J, see Lubineau A (1999) *206*: 1–39
Baars MWPL, Meijer EW (2000) Host-Guest Chemistry of Dendritic Molecules. *210*: 131–182
Bach T, see Basler B (2005) *243*: 1–42
Balazs G, Johnson BP, Scheer M (2003) Complexes with a Metal-Phosphorus Triple Bond. *232*: 1-23
Balczewski P, see Mikoloajczyk M (2003) *223*: 161–214
Ballauff M (2001) Structure of Dendrimers in Dilute Solution. *212*: 177–194
Baltzer L (1999) Functionalization and Properties of Designed Folded Polypeptides. *202*: 39–76
Balzani V, Ceroni P, Maestri M, Saudan C, Vicinelli V (2003) Luminescent Dendrimers. Recent Advances. *228*: 159–191
Balazs G, Johnson BP, Scheer M (2003) Complexes with a Metal-Phosphorus Triple Bond. *232*: 1-23
Bandichhor R, Nosse B, Reiser O (2005): Paraconic Acids – The Natural Products from *Lichen Symbiont*. *243*: 43–72
Barré L, see Lasne M-C (2002) *222*: 201–258
Bartlett RJ, see Sun J-Q (1999) *203*: 121–145
Barton JK, see O'Neill MA (2004) *236*: 67–115
Basler B, Brandes S, Spiegel A, Bach T (2005) Total Syntheses of Kelsoene and Preussin. *243*: 1–42
Beaulac R, see Bussière G (2004) *241*: 97–118
Behrens C, Cichon MK, Grolle F, Hennecke U, Carell T (2004) Excess Electron Transfer in Defined Donor-Nucleobase and Donor-DNA-Acceptor Systems. *236*: 187-204
Bélisle H, see Bussière G (2004) *241*: 97–118
Beratan D, see Berlin YA (2004) *237*: 1–36
Berlin YA, Kurnikov IV, Beratan D, Ratner MA, Burin AL (2004) DNA Electron Transfer Processes: Some Theoretical Notions. *237*: 1–36
Bertrand G, Bourissou D (2002) Diphosphorus-Containing Unsaturated Three-Menbered Rings: Comparison of Carbon, Nitrogen, and Phosphorus Chemistry. *220*: 1–25
Betzemeier B, Knochel P (1999) Perfluorinated Solvents – a Novel Reaction Medium in Organic Chemistry. *206*: 61–78
Bibette J, see Schmitt V (2003) *227*: 195–215
Blais J-C, see Astruc D (2000) *210*: 229–259
Blomberg MRA, see Lundberg M (2004) *238*: 79–112
Bogár F, see Pipek J (1999) *203*: 43–61

Bohme DK, see Petrie S (2003) 225: 35-73
Bourissou D, see Bertrand G (2002) 220: 1-25
Bowers MT, see Wyttenbach T (2003) 225: 201-226
Brand SC, see Haley MM (1999) 201: 81-129
Brandes S, see Basler B (2005) 243: 1-42
Bray KL (2001) High Pressure Probes of Electronic Structure and Luminescence Properties of Transition Metal and Lanthanide Systems. 213: 1-94
Bronstein LM (2003) Nanoparticles Made in Mesoporous Solids. 226: 55-89
Brönstrup M (2003) High Throughput Mass Spectrometry for Compound Characterization in Drug Discovery. 225: 275-294
Brücher E (2002) Kinetic Stabilities of Gadolinium(III) Chelates Used as MRI Contrast Agents. 221: 103-122
Brunel JM, Buono G (2002) New Chiral Organophosphorus atalysts in Asymmetric Synthesis. 220: 79-106
Buchwald SL, see Muci A R (2002) 219: 131-209
Bunz UHF (1999) Carbon-Rich Molecular Objects from Multiply Ethynylated p-Complexes. 201: 131-161
Buono G, see Brunel JM (2002) 220: 79-106
Burin AL, see Berlin YA (2004) 237: 1-36
Bussière G, Beaulac R, Bélisle H, Lescop C, Luneau D, Rey P, Reber C (2004) Excited States and Optical Spectroscopy of Nitronyl Nitroxides and Their Lanthanide and Transition Metal Complexes. 241: 97-118
Byun KL, see Gao J (2004) 238: 113-136
Cadet J, see Douki T (2004) 236: 1-25
Cadierno V, see Majoral J-P (2002) 220: 53-77
Cai Z, Sevilla MD (2004) Studies of Excess Electron and Hole Transfer in DNA at Low Temperatures. 237: 103-127
Caminade A-M, see Majoral J-P (2003) 223: 111-159
Carell T, see Behrens C (2004) 236: 187-204
Carmichael D, Mathey F (2002) New Trends in Phosphametallocene Chemistry. 220: 27-51
Caruso F (2003) Hollow Inorganic Capsules via Colloid-Templated Layer-by-Layer Electrostatic Assembly. 227: 145-168
Caruso RA (2003) Nanocasting and Nanocoating. 226: 91-118
Ceroni P, see Balzani V (2003) 228: 159-191
Chamberlin AR, see Gilmore MA (1999) 202: 77-99
Che C-M, see Lai S-W (2004) 241: 27-63
Chivers T (2003) Imido Analogues of Phosphorus Oxo and Chalcogenido Anions. 229: 143-159
Chow H-F, Leung C-F, Wang G-X, Zhang J (2001) Dendritic Oligoethers. 217: 1-50
Cichon MK, see Behrens C (2004) 236: 187-204
Clarkson RB (2002) Blood-Pool MRI Contrast Agents: Properties and Characterization. 221: 201-235
Cloutet E, see Astruc D (2000) 210: 229-259
Co CC, see Hentze H-P (2003) 226: 197-223
Conwell E (2004) Polarons and Transport in DNA. 237: 73-101
Cooper DL, see Raimondi M (1999) 203: 105-120
Cornils B (1999) Modern Solvent Systems in Industrial Homogeneous Catalysis. 206: 133-152
Corot C, see Idee J-M (2002) 222: 151-171
Crépy KVL, Imamoto T (2003) New P-Chirogenic Phosphine Ligands and Their Use in Catalytic Asymmetric Reactions. 229: 1-40
Cristau H-J, see Taillefer M (2003) 229: 41-73
Crooks RM, Lemon III BI, Yeung LK, Zhao M (2001) Dendrimer-Encapsulated Metals and Semiconductors: Synthesis, Characterization, and Applications. 212: 81-135
Croteau R, see Davis EM (2000) 209: 53-95
Crouzel C, see Lasne M-C (2002) 222: 201-258
Cuniberti G, see Porath D (2004) 237: 183-227
Curran DP, see Maul JJ (1999) 206: 79-105
Currie F, see Häger M (2003) 227: 53-74

Dabkowski W, see Michalski J (2003) *232*: 93-144
Daniel C (2004) Electronic Spectroscopy and Photoreactivity of Transition Metal Complexes: Quantum Chemistry and Wave Packet Dynamics. *241*: 119-165
Davidson P, see Gabriel J-C P (2003) *226*: 119-172
Davis EM, Croteau R (2000) Cyclization Enzymes in the Biosynthesis of Monoterpenes, Sesquiterpenes and Diterpenes. *209*: 53-95
Davies JA, see Schwert DD (2002) *221*: 165-200
Dax K, Albert M (2001) Rearrangements in the Course of Nucleophilic Substitution Reactions. *215*: 193-275
de Keizer A, see Kleinjan WE (2003) *230:* 167-188
de la Plata BC, see Ruano JLG (1999) *204*: 1-126
de Meijere A, Kozhushkov SI (1999) Macrocyclic Structurally Homoconjugated Oligoacetylenes: Acetylene- and Diacetylene-Expanded Cycloalkanes and Rotanes. *201*: 1-42
de Meijere A, Kozhushkov SI, Khlebnikov AF (2000) Bicyclopropylidene – A Unique Tetrasubstituted Alkene and a Versatile $C_6$-Building Block. *207*: 89-147
de Meijere A, Kozhushkov SI, Hadjiaraoglou LP (2000) Alkyl 2-Chloro-2-cyclopropylidene-acetates – Remarkably Versatile Building Blocks for Organic Synthesis. *207*: 149-227
Dennig J (2003) Gene Transfer in Eukaryotic Cells Using Activated Dendrimers. *228*: 227-236
de Raadt A, Fechter MH (2001) Miscellaneous. *215*: 327-345
Desreux JF, see Jacques V (2002) *221*: 123-164
Diederich F, Gobbi L (1999) Cyclic and Linear Acetylenic Molecular Scaffolding. *201*: 43-79
Diederich F, see Smith DK (2000) *210*: 183-227
Di Felice, R, see Porath D (2004) *237*: 183-227
Djakovitch L, see Astruc D (2000) *210*: 229-259
Dolle F, see Lasne M-C (2002) *222*: 201-258
Donges D, see Yersin H (2001) *214*: 81-186
Dormán G (2000) Photoaffinity Labeling in Biological Signal Transduction. *211*: 169-225
Dorn H, see McWilliams AR (2002) *220*: 141-167
Dorshow RB, see Achilefu S (2002) *222*: 31-72
Douki T, Ravanat J-L, Angelov D, Wagner JR, Cadet J (2004) Effects of Duplex Stability on Charge-Transfer Efficiency within DNA. *236*: 1-25
Drabowicz J, Mikołajczyk M (2000) Selenium at Higher Oxidation States. *208*: 143-176
Dutasta J-P (2003) New Phosphorylated Hosts for the Design of New Supramolecular Assemblies. *232*: 55-91
Eckert B, see Steudel R (2003) *230:* 1-79
Eckert B, Steudel R (2003) Molecular Spectra of Sulfur Molecules and Solid Sulfur Allotropes. *231:* 31-97
Ehses M, Romerosa A, Peruzzini M (2002) Metal-Mediated Degradation and Reaggregation of White Phosphorus. *220*: 107-140
Eder B, see Wrodnigg TM (2001) The Amadori and Heyns Rearrangements: Landmarks in the History of Carbohydrate Chemistry or Unrecognized Synthetic Opportunities? *215*: 115-175
Edwards DS, see Liu S (2002) *222*: 259-278
Elaissari A, Ganachaud F, Pichot C (2003) Biorelevant Latexes and Microgels for the Interaction with Nucleic Acids. *227*: 169-193
Esumi K (2003) Dendrimers for Nanoparticle Synthesis and Dispersion Stabilization. *227*: 31-52
Famulok M, Jenne A (1999) Catalysis Based on Nucleid Acid Structures. *202*: 101-131
Fechter MH, see de Raadt A (2001) *215*: 327-345
Ferrier RJ (2001) Substitution-with-Allylic-Rearrangement Reactions of Glycal Derivatives. *215*: 153-175
Ferrier RJ (2001) Direct Conversion of 5,6-Unsaturated Hexopyranosyl Compounds to Functionalized Glycohexanones. *215*: 277-291
Frey H, Schlenk C (2000) Silicon-Based Dendrimers. *210*: 69-129
Förster S (2003) Amphiphilic Block Copolymers for Templating Applications. *226*: 1-28
Frullano L, Rohovec J, Peters JA, Geraldes CFGC (2002) Structures of MRI Contrast Agents in Solution. *221*: 25-60
Fugami K, Kosugi M (2002) Organotin Compounds. *219*: 87-130

Fuhrhop J-H, see Li G (2002) *218*: 133–158

Furukawa N, Sato S (1999) New Aspects of Hypervalent Organosulfur Compounds. *205*: 89–129

Gabriel J-C P, Davidson P (2003) Mineral Liquid Crystals from Self-Assembly of Anisotropic Nanosystems. *226*: 119–172

Gamelin DR, Güdel HU (2001) Upconversion Processes in Transition Metal and Rare Earth Metal Systems. *214*: 1–56

Ganachaud F, see Elaissari A (2003) *227*: 169–193

Gao J, Byun KL, Kluge R (2004) Catalysis by Enzyme Conformational Change. *238*: 113-136

García R, see Tromas C (2002) *218*: 115–132

Geacintov NE, see Shafirovich V (2004) *237*: 129-157

Geraldes CFGC, see Frullano L (2002) *221*: 25–60

Giese B (2004) Hole Injection and Hole Transfer through DNA : The Hopping Mechanism. *236*: 27–44

Gilmore MA, Steward LE, Chamberlin AR (1999) Incorporation of Noncoded Amino Acids by In Vitro Protein Biosynthesis. *202*: 77–99

Glasbeek M (2001) Excited State Spectroscopy and Excited State Dynamics of Rh(III) and Pd(II) Chelates as Studied by Optically Detected Magnetic Resonance Techniques. *213*: 95–142

Glass RS (1999) Sulfur Radical Cations. *205*: 1–87

Gobbi L, see Diederich F (1999) *201*: 43–129

Göltner-Spickermann C (2003) Nanocasting of Lyotropic Liquid Crystal Phases for Metals and Ceramics. *226*: 29–54

Gouzy M-F, see Li G (2002) *218*: 133–158

Gries H (2002) Extracellular MRI Contrast Agents Based on Gadolinium. *221*: 1–24

Grolle F, see Behrens C (2004) *236*: 187-204

Gruber C, see Tovar GEM (2003) *227*: 125–144

Gudat D (2003): Zwitterionic Phospholide Derivatives – New Ambiphilic Ligands. *232*: 175-212

Güdel HU, see Gamelin DR (2001) *214*: 1–56

Guga P, Okruszek A, Stec WJ (2002) Recent Advances in Stereocontrolled Synthesis of P-Chiral Analogues of Biophosphates. *220*: 169–200

Gulea M, Masson S (2003) Recent Advances in the Chemistry of Difunctionalized Organo-Phosphorus and -Sulfur Compounds. *229*: 161–198

Hackmann-Schlichter N, see Krause W (2000) *210*: 261–308

Hadjiaraoglou LP, see de Meijere A (2000) *207*: 149–227

Häger M, Currie F, Holmberg K (2003) Organic Reactions in Microemulsions. *227*: 53–74

Häusler H, Stütz AE (2001) d-Xylose (d-Glucose) Isomerase and Related Enzymes in Carbohydrate Synthesis. 215: 77–114

Haley MM, Pak JJ, Brand SC (1999) Macrocyclic Oligo(phenylacetylenes) and Oligo(phenyldiacetylenes). *201*: 81–129

Harada A, see Yamaguchi H (2003) *228*: 237–258

Hartmann T, Ober D (2000) Biosynthesis and Metabolism of Pyrrolizidine Alkaloids in Plants and Specialized Insect Herbivores. *209*: 207–243

Haseley SR, Kamerling JP, Vliegenthart JFG (2002) Unravelling Carbohydrate Interactions with Biosensors Using Surface Plasmon Resonance (SPR) Detection. *218*: 93–114

Hassner A, see Namboothiri INN (2001) *216*: 1–49

Hauser A, von Arx ME, Langford VS, Oetliker U, Kairouani S, Pillonnet A (2004) Photophysical Properties of Three-Dimensional Transition Metal Tris-Oxalate Network Structures. *241*: 65-96

Helm L, see Tóth E (2002) *221*: 61–101

Helmboldt H, see Hiersemann M (2005) *243*: 73-136

Hemscheidt T (2000) Tropane and Related Alkaloids. *209*: 175–206

Hennecke U, see Behrens C (2004) *236*: 187-204

Hentze H-P, Co CC, McKelvey CA, Kaler EW (2003) Templating Vesicles, Microemulsions and Lyotropic Mesophases by Organic Polymerization Processes. *226*: 197–223

Hergenrother PJ, Martin SF (2000) Phosphatidylcholine-Preferring Phospholipase C from *B. cereus*. Function, Structure, and Mechanism. *211*: 131–167

Hermann C, see Kuhlmann J (2000) *211*: 61–116

Heydt H (2003) The Fascinating Chemistry of Triphosphabenzenes and Valence Isomers. 223: 215-249
Hiersemann M, Helmboldt H (2005): Recent Progress in the Total Synthesis of Dolabellane and Dolastane Diterpenes. 243: 73-136
Hirsch A, Vostrowsky O (2001) Dendrimers with Carbon Rich-Cores. 217: 51-93
Hiyama T, Shirakawa E (2002) Organosilicon Compounds. 219: 61-85
Holmberg K, see Häger M (2003) 227: 53-74
Houk KN, Tantillo DJ, Stanton C, Hu Y (2004) What have Theory and Crystallography Revealed About the Mechanism of Catalysis by Orotidine Monophosphate Decarboxylase? 238: 1-22
Houseman BT, Mrksich M (2002) Model Systems for Studying Polyvalent Carbohydrate Binding Interactions. 218: 1-44
Hricovíniová Z, see Petruš L (2001) 215: 15-41
Hu Y, see Houk KN (2004) 238: 1-22
Idee J-M, Tichkowsky I, Port M, Petta M, Le Lem G, Le Greneur S, Meyer D, Corot C (2002) Iodiated Contrast Media: from Non-Specific to Blood-Pool Agents. 222: 151-171
Igau A, see Majoral J-P (2002) 220: 53-77
Ikeda Y, see Takagi Y (2003) 232: 213-251
Imamoto T, see Crépy KVL (2003) 229: 1-40
Iwaoka M, Tomoda S (2000) Nucleophilic Selenium. 208: 55-80
Iwasawa N, Narasaka K (2000) Transition Metal Promated Ring Expansion of Alkynyl- and Propadienylcyclopropanes. 207: 69-88
Imperiali B, McDonnell KA, Shogren-Knaak M (1999) Design and Construction of Novel Peptides and Proteins by Tailored Incorparation of Coenzyme Functionality. 202: 1-38
Ito S, see Yoshifuji M (2003) 223: 67-89
Jacques V, Desreux JF (2002) New Classes of MRI Contrast Agents. 221: 123-164
James TD, Shinkai S (2002) Artificial Receptors as Chemosensors for Carbohydrates. 218: 159-200
Janssen AJH, see Kleinjan WE (2003) 230: 167-188
Jenne A, see Famulok M (1999) 202: 101-131
Johnson BP, see Balazs G (2003) 232: 1-23
Junker T, see Trauger SA (2003) 225: 257-274
Kairouani S, see Hauser A (2004) 241: 65-96
Kaler EW, see Hentze H-P (2003) 226: 197-223
Kamerling JP, see Haseley SR (2002) 218: 93-114
Kashemirov BA, see Mc Kenna CE (2002) 220: 201-238
Kato S, see Murai T (2000) 208: 177-199
Katti KV, Pillarsetty N, Raghuraman K (2003) New Vistas in Chemistry and Applications of Primary Phosphines. 229: 121-141
Kawa M (2003) Antenna Effects of Aromatic Dendrons and Their Luminescene Applications. 228: 193-204
Kawai K, Majima T (2004) Hole Transfer in DNA by Monitoring the Transient Absorption of Radical Cations of Organic Molecules Conjugated to DNA. 236: 117-137
Kee TP, Nixon TD (2003) The Asymmetric Phospho-Aldol Reaction. Past, Present, and Future. 223: 45-65
Khlebnikov AF, see de Meijere A (2000) 207: 89-147
Kim K, see Lee JW (2003) 228: 111-140
Kirtman B (1999) Local Space Approximation Methods for Correlated Electronic Structure Calculations in Large Delocalized Systems that are Locally Perturbed. 203: 147-166
Kita Y, see Tohma H (2003) 224: 209-248
Kleij AW, see Kreiter R (2001) 217: 163-199
Klein Gebbink RJM, see Kreiter R (2001) 217: 163-199
Kleinjan WE, de Keizer A, Janssen AJH (2003) Biologically Produced Sulfur. 230: 167-188
Klibanov AL (2002) Ultrasound Contrast Agents: Development of the Field and Current Status. 222: 73-106
Klopper W, Kutzelnigg W, Müller H, Noga J, Vogtner S (1999) Extremal Electron Pairs - Application to Electron Correlation, Especially the R12 Method. 203: 21-42

Kluge R, see Gao J (2004) *238*: 113-136
Knochel P, see Betzemeier B (1999) *206*: 61-78
Koser GF (2003) C-Heteroatom-Bond Forming Reactions. *224*: 137-172
Koser GF (2003) Heteroatom-Heteroatom-Bond Forming Reactions. *224*: 173-183
Kosugi M, see Fugami K (2002) *219*: 87-130
Kozhushkov SI, see de Meijere A (1999) *201*: 1-42
Kozhushkov SI, see de Meijere A (2000) *207*: 89-147
Kozhushkov SI, see de Meijere A (2000) *207*: 149-227
Krause W (2002) Liver-Specific X-Ray Contrast Agents. *222*: 173-200
Krause W, Hackmann-Schlichter N, Maier FK, Müller R (2000) Dendrimers in Diagnostics. *210*: 261-308
Krause W, Schneider PW (2002) Chemistry of X-Ray Contrast Agents. *222*: 107-150
Kräuter I, see Tovar GEM (2003) *227*: 125-144
Kreiter R, Kleij AW, Klein Gebbink RJM, van Koten G (2001) Dendritic Catalysts. *217*: 163-199
Krossing I (2003) Homoatomic Sulfur Cations. *230*: 135-152
Kuhlmann J, Herrmann C (2000) Biophysical Characterization of the Ras Protein. *211*: 61-116
Kunkely H, see Vogler A (2001) *213*: 143-182
Kurnikov IV, see Berlin YA (2004) *237*: 1-36
Kutzelnigg W, see Klopper W (1999) *203*: 21-42
Lai S-W, Che C-M (2004) Luminescent Cyclometalated Diimine Platinum(II) Complexes. Photophysical Studies and Applications. *241*: 27-63
Lammertsma K (2003) Phosphinidenes. *229*: 95-119
Landfester K (2003) Miniemulsions for Nanoparticle Synthesis. *227*: 75-123
Landman U, see Schuster GB (2004) *236*: 139-161
Langford VS, see Hauser A (2004) *241*: 65-96
Lasne M-C, Perrio C, Rouden J, Barré L, Roeda D, Dolle F, Crouzel C (2002) Chemistry of $b^+$-Emitting Compounds Based on Fluorine-18. *222*: 201-258
Lawless LJ, see Zimmermann SC (2001) *217*: 95-120
Leal-Calderon F, see Schmitt V (2003) *227*: 195-215
Lee JW, Kim K (2003) Rotaxane Dendrimers. *228*: 111-140
Le Bideau, see Vioux A (2003) *232*: 145-174
Le Greneur S, see Idee J-M (2002) *222*: 151-171
Le Lem G, see Idee J-M (2002) *222*: 151-171
Leclercq D, see Vioux A (2003) *232*: 145-174
Leitner W (1999) Reactions in Supercritical Carbon Dioxide (scCO$_2$). *206*: 107-132
Lemon III BI, see Crooks RM (2001) *212*: 81-135
Lescop C, see Bussière G (2004) *241*: 97-118
Leung C-F, see Chow H-F (2001) *217*: 1-50
Levitzki A (2000) Protein Tyrosine Kinase Inhibitors as Therapeutic Agents. *211*: 1-15
Lewis, FD, Wasielewski MR (2004) Dynamics and Equilibrium for Single Step Hole Transport Processes in Duplex DNA. *236*: 45-65
Li G, Gouzy M-F, Fuhrhop J-H (2002) Recognition Processes with Amphiphilic Carbohydrates in Water. *218*: 133-158
Li X, see Paldus J (1999) *203*: 1-20
Licha K (2002) Contrast Agents for Optical Imaging. *222*: 1-29
Linclau B, see Maul JJ (1999) *206*: 79-105
Lindhorst TK (2002) Artificial Multivalent Sugar Ligands to Understand and Manipulate Carbohydrate-Protein Interactions. *218*: 201-235
Lindhorst TK, see Röckendorf N (2001) *217*: 201-238
Liu S, Edwards DS (2002) Fundamentals of Receptor-Based Diagnostic Metalloradiopharmaceuticals. *222*: 259-278
Liz-Marzán L, see Mulvaney P (2003) *226*: 225-246
Loudet JC, Poulin P (2003) Monodisperse Aligned Emulsions from Demixing in Bulk Liquid Crystals. *226*: 173-196
Loupy A (1999) Solvent-Free Reactions. *206*: 153-207

Lubineau A, Augé J (1999) Water as Solvent in Organic Synthesis. 206: 1–39
Lundberg M, Blomberg MRA, Siegbahn PEM (2004) Developing Active Site Models of ODCase – from Large Quantum Models to a QM/MM Approach. 238: 79-112
Lundt I, Madsen R (2001) Synthetically Useful Base Induced Rearrangements of Aldonolactones. 215: 177–191
Luneau D, see Bussière G (2004) 241: 97-118
Madsen R, see Lundt I (2001) 215: 177–191
Maestri M, see Balzani V (2003) 228: 159–191
Maier FK, see Krause W (2000) 210: 261–308
Majima T, see Kawai K (2004) 236: 117-137
Majoral J-P, Caminade A-M (2003) What to do with Phosphorus in Dendrimer Chemistry. 223: 111-159
Majoral J-P, Igau A, Cadierno V, Zablocka M (2002) Benzyne-Zirconocene Reagents as Tools in Phosphorus Chemistry. 220: 53–77
Manners I (2002), see McWilliams AR (2002) 220: 141–167
March NH (1999) Localization via Density Functionals. 203: 201–230
Martin SF, see Hergenrother PJ (2000) 211: 131–167
Mashiko S, see Yokoyama S (2003) 228: 205–226
Masson S, see Gulea M (2003) 229: 161-198
Mathey F, see Carmichael D (2002) 220: 27–51
Maul JJ, Ostrowski PJ, Ublacker GA, Linclau B, Curran DP (1999) Benzotrifluoride and Derivates: Useful Solvents for Organic Synthesis and Fluorous Synthesis. 206: 79–105
McDonnell KA, see Imperiali B (1999) 202: 1–38
McKelvey CA, see Hentze H-P (2003) 226: 197-223
Mc Kenna CE, Kashemirov BA (2002) Recent Progress in Carbonylphosphonate Chemistry. 220: 201–238
McWilliams AR, Dorn H, Manners I (2002) New Inorganic Polymers Containing Phosphorus. 220: 141–167
Meijer EW, see Baars MWPL (2000) 210: 131–182
Merbach AE, see Tóth E (2002) 221: 61–101
Metzner P (1999) Thiocarbonyl Compounds as Specific Tools for Organic Synthesis. 204:127–181
Meyer D, see Idee J-M (2002) 222: 151–171
Mezey PG (1999) Local Electron Densities and Functional Groups in Quantum Chemistry. 203: 167–186
Michalski J, Dabkowski W (2003) State of the Art. Chemical Synthesis of Biophosphates and Their Analogues via $P^{III}$ Derivatives. 232: 93-144
Miller BG (2004) Insight into the Catalytic Mechanism of Orotidine 5′-Phosphate Decarboxylase from Crystallography and Mutagenesis. 238: 43-62
Mikołajczyk M, Balczewski P (2003) Phosphonate Chemistry and Reagents in the Synthesis of Biologically Active and Natural Products. 223: 161–214
Mikołajczyk M, see Drabowicz J (2000) 208: 143–176
Miura M, Nomura M (2002) Direct Arylation via Cleavage of Activated and Unactivated C-H Bonds. 219: 211-241
Miyaura N (2002) Organoboron Compounds. 219: 11–59
Miyaura N, see Tamao K (2002) 219: 1–9
Möller M, see Sheiko SS (2001) 212: 137–175
Morales JC, see Rojo J (2002) 218: 45–92
Mori H, Müller A (2003) Hyperbranched (Meth)acrylates in Solution, in the Melt, and Grafted From Surfaces. 228: 1–37
Mrksich M, see Houseman BT (2002) 218:1–44
Muci AR, Buchwald SL (2002) Practical Palladium Catalysts for C-N and C-O Bond Formation. 219: 131–209
Müllen K, see Wiesler U-M (2001) 212: 1–40
Müller A, see Mori H (2003) 228: 1–37
Müller G (2000) Peptidomimetic SH2 Domain Antagonists for Targeting Signal Transduction. 211: 17–59

Müller H, see Klopper W (1999) *203*: 21–42
Müller R, see Krause W (2000) *210*: 261–308
Mulvaney P, Liz-Marzán L (2003) Rational Material Design Using Au Core-Shell Nanocrystals. *226*: 225–246
Murai T, Kato S (2000) Selenocarbonyls. *208*: 177–199
Muscat D, van Benthem RATM (2001) Hyperbranched Polyesteramides – New Dendritic Polymers. *212*: 41–80
Mutin PH, see Vioux A (2003) *232*: 145-174
Naka K (2003) Effect of Dendrimers on the Crystallization of Calcium Carbonate in Aqueous Solution. *228*: 141–158
Nakahama T, see Yokoyama S (2003) *228*: 205–226
Nakatani K, Saito I (2004) Charge Transport in Duplex DNA Containing Modified Nucleotide Bases. 236: 163-186
Nakayama J, Sugihara Y (1999) Chemistry of Thiophene 1,1-Dioxides. *205*: 131–195
Namboothiri INN, Hassner A (2001) Stereoselective Intramolecular 1,3-Dipolar Cycloadditions. *216*: 1–49
Narasaka K, see Iwasawa N (2000) *207*: 69–88
Nierengarten J-F (2003) Fullerodendrimers: Fullerene-Containing Macromolecules with Intriguing Properties. *228*: 87–110
Nishibayashi Y, Uemura S (2000) Selenoxide Elimination and [2,3] Sigmatropic Rearrangements. *208*: 201–233
Nishibayashi Y, Uemura S (2000) Selenium Compounds as Ligands and Catalysts. *208:* 235–255
Nixon TD, see Kee TP (2003) *223*: 45–65
Noga J, see Klopper W (1999) *203*: 21–42
Nomura M, see Miura M (2002) *219*: 211–241
Nosse B, see Bandichhor R (2005) *243*: 43-72
Nubbemeyer U (2001) Synthesis of Medium-Sized Ring Lactams. *216*: 125–196
Nummelin S, Skrifvars M, Rissanen K (2000) Polyester and Ester Functionalized Dendrimers. *210*: 1–67
Ober D, see Hemscheidt T (2000) *209*: 175–206
Ochiai M (2003) Reactivities, Properties and Structures. *224*: 5–68
Oetliker U, see Hauser A (2004) *241*: 65-96
Okazaki R, see Takeda N (2003) *231*:153-202
Okruszek A, see Guga P (2002) *220*: 169–200
Okuno Y, see Yokoyama S (2003) *228*: 205–226
O'Neill MA, Barton JK (2004) DNA-Mediated Charge Transport Chemistry and Biology. *236*: 67-115
Onitsuka K, Takahashi S (2003) Metallodendrimers Composed of Organometallic Building Blocks. *228*: 39–63
Osanai S (2001) Nickel (II) Catalyzed Rearrangements of Free Sugars. *215*: 43–76
Ostrowski PJ, see Maul JJ (1999) *206*: 79–105
Otomo A, see Yokoyama S (2003) *228*: 205–226
Pai EP, see Wu N (2004) *238*: 23-42
Pak JJ, see Haley MM (1999) *201*: 81–129
Paldus J, Li X (1999) Electron Correlation in Small Molecules: Grafting CI onto CC. *203*: 1–20
Paleos CM, Tsiourvas D (2003) Molecular Recognition and Hydrogen-Bonded Amphiphilies. *227*: 1–29
Paulmier C, see Ponthieux S (2000) *208*: 113–142
Penadés S, see Rojo J (2002) *218*: 45–92
Perrio C, see Lasne M-C (2002) *222*: 201–258
Peruzzini M, see Ehses M (2002) *220*: 107–140
Peters JA, see Frullano L (2002) *221*: 25–60
Petrie S, Bohme DK (2003) Mass Spectrometric Approaches to Interstellar Chemistry. *225*: 35–73
Petruš L, Petrušová M, Hricovíniová (2001) The Bílik Reaction. *215*: 15–41
Petrušová M, see Petruš L (2001) *215*: 15–41

Petta M, see Idee J-M (2002) *222*: 151-171
Pichot C, see Elaissari A (2003) *227*: 169-193
Pillarsetty N, see Katti KV (2003) *229*: 121-141
Pillonnet A, see Hauser A (2004) *241*: 65-96
Pipek J, Bogár F (1999) Many-Body Perturbation Theory with Localized Orbitals – Kapuy's Approach. *203*: 43-61
Plattner DA (2003) Metalorganic Chemistry in the Gas Phase: Insight into Catalysis. *225*: 149-199
Ponthieux S, Paulmier C (2000) Selenium-Stabilized Carbanions. *208*: 113-142
Porath D, Cuniberti G, Di Felice, R (2004) Charge Transport in DNA-Based Devices. *237*: 183-227
Port M, see Idee J-M (2002) *222*: 151-171
Poulin P, see Loudet JC (2003) *226*: 173-196
Raghuraman K, see Katti KV (2003) *229*: 121-141
Raimondi M, Cooper DL (1999) Ab Initio Modern Valence Bond Theory. *203*: 105-120
Ratner MA, see Berlin YA (2004) *237*: 1-36
Ravanat J-L, see Douki T (2004) *236*: 1-25
Reber C, see Bussière G (2004) *241*: 97-118
Reinhoudt DN, see van Manen H-J (2001) *217*: 121-162
Reiser O, see Bandichhor R (2005) *243*: 43-72
Renaud P (2000) Radical Reactions Using Selenium Precursors. *208*: 81-112
Rey P, see Bussière G (2004) *241*: 97-118
Richardson N, see Schwert DD (2002) *221*: 165-200
Rigaut S, see Astruc D (2000) *210*: 229-259
Riley MJ (2001) Geometric and Electronic Information From the Spectroscopy of Six-Coordinate Copper(II) Compounds. *214*: 57-80
Rissanen K, see Nummelin S (2000) *210*: 1-67
Røeggen I (1999) Extended Geminal Models. *203*: 89-103
Röckendorf N, Lindhorst TK (2001) Glycodendrimers. *217*: 201-238
Roeda D, see Lasne M-C (2002) *222*: 201-258
Rösch N, Voityuk AA (2004) Quantum Chemical Calculation of Donor-Acceptor Coupling for Charge Transfer in DNA. *237*: 37-72
Rohovec J, see Frullano L (2002) *221*: 25-60
Rojo J, Morales JC, Penadés S (2002) Carbohydrate-Carbohydrate Interactions in Biological and Model Systems. *218*: 45-92
Romerosa A, see Ehses M (2002) *220*: 107-140
Rouden J, see Lasne M-C (2002) *222*: 201258
Ruano JLG, de la Plata BC (1999) Asymmetric [4+2] Cycloadditions Mediated by Sulfoxides. *204*: 1-126
Ruijter E, see Wessjohann LA (2005) *243*: 137-184
Ruiz J, see Astruc D (2000) *210*: 229-259
Rychnovsky SD, see Sinz CJ (2001) *216*: 51-92
Saito I, see Nakatani K (2004) *236*: 163-186
Salaün J (2000) Cyclopropane Derivates and their Diverse Biological Activities. *207*: 1-67
Sanz-Cervera JF, see Williams RM (2000) *209*: 97-173
Sartor V, see Astruc D (2000) *210*: 229-259
Sato S, see Furukawa N (1999) *205*: 89-129
Saudan C, see Balzani V (2003) *228*: 159-191
Scheer M, see Balazs G (2003) *232*: 1-23
Scherf U (1999) Oligo- and Polyarylenes, Oligo- and Polyarylenevinylenes. *201*: 163-222
Schlenk C, see Frey H (2000) *210*: 69-129
Schmitt V, Leal-Calderon F, Bibette J (2003) Preparation of Monodisperse Particles and Emulsions by Controlled Shear. *227*: 195-215
Schoeller WW (2003) Donor-Acceptor Complexes of Low-Coordinated Cationic p-Bonded Phosphorus Systems. *229*: 75-94
Schröder D, Schwarz H (2003) Diastereoselective Effects in Gas-Phase Ion Chemistry. *225*: 129-148

Schuster GB, Landman U (2004) The Mechanism of Long-Distance Radical Cation Transport in Duplex DNA: Ion-Gated Hopping of Polaron-Like Distortions. 236: 139-161
Schwarz H, see Schröder D (2003) 225: 129-148
Schwert DD, Davies JA, Richardson N (2002) Non-Gadolinium-Based MRI Contrast Agents. 221: 165-200
Sefkow M (2005) Enantioselective Synthesis of C(8)-Hydroxylated Lignans – Early Approaches and Recent Advances. 243: 185-224
Sevilla MD, see Cai Z (2004) 237: 103-127
Shafirovich V, Geacintov NE (2004) Proton-Coupled Electron Transfer Reactions at a Distance in DNA Duplexes. 237: 129-157
Sheiko SS, Möller M (2001) Hyperbranched Macromolecules: Soft Particles with Adjustable Shape and Capability to Persistent Motion. 212: 137-175
Shen B (2000) The Biosynthesis of Aromatic Polyketides. 209: 1-51
Shinkai S, see James TD (2002) 218: 159-200
Shirakawa E, see Hiyama T (2002) 219: 61-85
Shogren-Knaak M, see Imperiali B (1999) 202: 1-38
Siegbahn PEM, see Lundberg M (2004) 238: 79-112
Sinou D (1999) Metal Catalysis in Water. 206: 41-59
Sinz CJ, Rychnovsky SD (2001) 4-Acetoxy- and 4-Cyano-1,3-dioxanes in Synthesis. 216: 51-92
Siuzdak G, see Trauger SA (2003) 225: 257-274
Skrifvars M, see Nummelin S (2000) 210: 1-67
Smiley JA (2004) Survey of Enzymological Data on CDCase. 238: 63-78
Smith DK, Diederich F (2000) Supramolecular Dendrimer Chemistry – A Journey Through the Branched Architecture. 210: 183-227
Spiegel A, see Basler B (2005) 243: 1-42
Stanton C, see Houk KN (2004) 238: 1-22
Stec WJ, see Guga P (2002) 220: 169-200
Steudel R (2003) Aqueous Sulfur Sols. 230: 153-166
Steudel R (2003) Liquid Sulfur. 230: 80-116
Steudel R (2003) Inorganic Polysulfanes $H_2S_n$ with n>1. 231: 99-125
Steudel R (2003) Inorganic Polysulfides $S_n^{2-}$ and Radical Anions $S_n^{.-}$. 231:127-152
Steudel R (2003) Sulfur-Rich Oxides $S_nO$ and $S_nO_2$. 231: 203-230
Steudel R, Eckert B (2003) Solid Sulfur Allotropes. 230: 1-79
Steudel R, see Eckert B (2003) 231: 31-97
Steudel R, Steudel Y, Wong MW (2003) Speciation and Thermodynamics of Sulfur Vapor. 230: 117-134
Steudel Y, see Steudel R (2003) 230: 117-134
Steward LE, see Gilmore MA (1999) 202: 77-99
Stocking EM, see Williams RM (2000) 209: 97-173
Streubel R (2003) Transient Nitrilium Phosphanylid Complexes: New Versatile Building Blocks in Phosphorus Chemistry. 223: 91-109
Stütz AE, see Häusler H (2001) 215: 77-114
Sugihara Y, see Nakayama J (1999) 205: 131-195
Sugiura K (2003) An Adventure in Macromolecular Chemistry Based on the Achievements of Dendrimer Science: Molecular Design, Synthesis, and Some Basic Properties of Cyclic Porphyrin Oligomers to Create a Functional Nano-Sized Space. 228: 65-85
Sun J-Q, Bartlett RJ (1999) Modern Correlation Theories for Extended, Periodic Systems. 203: 121-145
Sun L, see Crooks RM (2001) 212: 81-135
Surján PR (1999) An Introduction to the Theory of Geminals. 203: 63-88
Taillefer M, Cristau H-J (2003) New Trends in Ylide Chemistry. 229: 41-73
Taira K, see Takagi Y (2003) 232: 213-251
Takagi Y, Ikeda Y, Taira K (2003) Ribozyme Mechanisms. 232: 213-251
Takahashi S, see Onitsuka K (2003) 228: 39-63
Takeda N, Tokitoh N, Okazaki R (2003) Polysulfido Complexes of Main Group and Transition Metals. 231:153-202

Tamao K, Miyaura N (2002) Introduction to Cross-Coupling Reactions. *219*: 1–9
Tanaka M (2003) Homogeneous Catalysis for H-P Bond Addition Reactions. *232*: 25-54
Tanner PA (2004) Spectra, Energy Levels and Energy Transfer in High Symmetry Lanthanide Compounds. *241*: 167-278
Tantillo DJ, see Houk KN (2004) *238*: 1-22
ten Holte P, see Zwanenburg B (2001) *216*: 93–124
Thiem J, see Werschkun B (2001) *215*: 293-325
Thorp HH (2004) Electrocatalytic DNA Oxidation. *237*: 159-181
Thutewohl M, see Waldmann H (2000) *211*: 117–130
Tichkowsky I, see Idee J-M (2002) *222*: 151-171
Tiecco M (2000) Electrophilic Selenium, Selenocyclizations. *208*: 7–54
Tohma H, Kita Y (2003) Synthetic Applications (Total Synthesis and Natural Product Synthesis). *224*: 209–248
Tokitoh N, see Takeda N (2003) *231*:153-202
Tomoda S, see Iwaoka M (2000) *208*: 55–80
Tóth E, Helm L, Merbach AE (2002) Relaxivity of MRI Contrast Agents. *221*: 61–101
Tovar GEM, Kräuter I, Gruber C (2003) Molecularly Imprinted Polymer Nanospheres as Fully Affinity Receptors. *227*: 125–144
Trauger SA, Junker T, Siuzdak G (2003) Investigating Viral Proteins and Intact Viruses with Mass Spectrometry. *225*: 257–274
Tromas C, García R (2002) Interaction Forces with Carbohydrates Measured by Atomic Force Microscopy. *218*: 115–132
Tsiourvas D, see Paleos CM (2003) *227*: 1–29
Turecek F (2003) Transient Intermediates of Chemical Reactions by Neutralization-Reionization Mass Spectrometry. *225*: 75–127
Ublacker GA, see Maul JJ (1999) *206*: 79–105
Uemura S, see Nishibayashi Y (2000) *208*: 201–233
Uemura S, see Nishibayashi Y (2000) *208*: 235–255
Uggerud E (2003) Physical Organic Chemistry of the Gas Phase. Reactivity Trends for Organic Cations. *225*: 1–34
Valdemoro C (1999) Electron Correlation and Reduced Density Matrices. *203*: 187–200
Valério C, see Astruc D (2000) *210*: 229–259
van Benthem RATM, see Muscat D (2001) *212*: 41–80
van Koten G, see Kreiter R (2001) *217*: 163–199
van Manen H-J, van Veggel FCJM, Reinhoudt DN (2001) Non-Covalent Synthesis of Metallodendrimers. *217*: 121–162
van Veggel FCJM, see van Manen H-J (2001) *217*: 121–162
Varvoglis A (2003) Preparation of Hypervalent Iodine Compounds. *224*: 69–98
Verkade JG (2003) P(RNCH$_2$CH$_2$)$_3$N: Very Strong Non-ionic Bases Useful in Organic Synthesis. *223*: 1–44
Vicinelli V, see Balzani V (2003) *228*: 159–191
Vioux A, Le Bideau J, Mutin PH, Leclercq D (2003): Hybrid Organic-Inorganic Materials Based on Organophosphorus Derivatives. *232*: 145-174
Vliegenthart JFG, see Haseley SR (2002) *218*: 93–114
Vogler A, Kunkely H (2001) Luminescent Metal Complexes: Diversity of Excited States. *213*: 143–182
Vogtner S, see Klopper W (1999) *203*: 21–42
Voityuk AA, see Rösch N (2004) *237*: 37-72
von Arx ME, see Hauser A (2004) *241*: 65-96
Vostrowsky O, see Hirsch A (2001) *217*: 51–93
Wagner JR, see Douki T (2004) *236*: 1-25
Waldmann H, Thutewohl M (2000) Ras-Farnesyltransferase-Inhibitors as Promising Anti-Tumor Drugs. *211*: 117–130
Wang G-X, see Chow H-F (2001) *217*: 1–50
Wasielewski MR, see Lewis, FD (2004) *236*: 45-65
Weil T, see Wiesler U-M (2001) *212*: 1–40

Werschkun B, Thiem J (2001) Claisen Rearrangements in Carbohydrate Chemistry. *215*: 293–325
Wessjohann LA, Ruijter E (2005) Strategies for Total and Diversity-Oriented Synthesis of Natural Product(-Like) Macrocycles. *243*: 137-184
Wiesler U-M, Weil T, Müllen K (2001) Nanosized Polyphenylene Dendrimers. *212*: 1–40
Williams RM, Stocking EM, Sanz-Cervera JF (2000) Biosynthesis of Prenylated Alkaloids Derived from Tryptophan. *209*: 97–173
Wirth T (2000) Introduction and General Aspects. *208*: 1–5
Wirth T (2003) Introduction and General Aspects. *224*: 1–4
Wirth T (2003) Oxidations and Rearrangements. *224*: 185–208
Wong MW, see Steudel R (2003) *230:* 117–134
Wong MW (2003) Quantum-Chemical Calculations of Sulfur-Rich Compounds. *231*:1-29
Wrodnigg TM, Eder B (2001) The Amadori and Heyns Rearrangements: Landmarks in the History of Carbohydrate Chemistry or Unrecognized Synthetic Opportunities? *215*: 115–175
Wu N, Pai EP (2004) Crystallographic Studies of Native and Mutant Orotidine 5′-Phosphate Decarboxylases. *238*: 23–42
Wyttenbach T, Bowers MT (2003) Gas-Phase Confirmations: The Ion Mobility/Ion Chromatography Method. *225*: 201–226
Yamaguchi H, Harada A (2003) Antibody Dendrimers. *228*: 237–258
Yersin H (2004) Triplet Emitters for OLED Applications. Mechanisms of Exciton Trapping and Control of Emission Properties. *241*: 1–26
Yersin H, Donges D (2001) Low-Lying Electronic States and Photophysical Properties of Organometallic Pd(II) and Pt(II) Compounds. Modern Research Trends Presented in Detailed Case Studies. *214*: 81–186
Yeung LK, see Crooks RM (2001) *212*: 81–135
Yokoyama S, Otomo A, Nakahama T, Okuno Y, Mashiko S (2003) Dendrimers for Optoelectronic Applications. *228*: 205–226
Yoshifuji M, Ito S (2003) Chemistry of Phosphanylidene Carbenoids. *223*: 67–89
Zablocka M, see Majoral J-P (2002) *220*: 53–77
Zhang J, see Chow H-F (2001) *217*: 1–50
Zhdankin VV (2003) C-C Bond Forming Reactions. *224*: 99-136
Zhao M, see Crooks RM (2001) *212*: 81–135
Zimmermann SC, Lawless LJ (2001) Supramolecular Chemistry of Dendrimers. *217*: 95–120
Zwanenburg B, ten Holte P (2001) The Synthetic Potential of Three-Membered Ring Aza-Heterocycles. *216*: 93–124

# Subject Index

Acetoxyacrylonitrile 54
Aldol (Claisen)-type reactions 141
Alkaloids 1
Allofuranose 28
1,3-Allylic strain 15, 51
Allylsilanes 66
Anisomycin 36
Apoptosis 20
Arabinofuranose 29
Arabinose, $\alpha$-hydroxylated lactone lignans 199
–, tetrahydrofuran lignans 202
Aryl-aryl-bond formation 186
Aryloxetanes 34
Aspartic acid 26
aza-Cope rearrangement/Mannich cyclization 31
Aziridines 151

Baeyer-Villiger oxidation 57, 58
Baylis-Hillman reaction 167
Beckmann ring expansion 32
Benzylidene lignans 194
$\beta$-Benzyl-$\gamma$-butyrolactones 185, 191
Biaryl ether macrocycles 169
Biaryl lignans, $\alpha$-hydroxylated 194
Bromination 18
Butenolide 56
$\gamma$-Butyrolactones 43
–, palladium(II) 62

Carbohydroxylation 34
Chelation-control 23
Chiral pool 43
Citronellene 15
Claenone 83
Claisen rearrangement 50
*Clavularia viridis* 81

Click chemistry 50
Cognac lactone 56
CuOTf 13
Cyanocuprate 11
Cycloadditions 23
–, 1,3-dipolar 32
–, hydroazulene 129
Cyclobutanones 57
Cyclopentane conformation 18
Cyclopeptide alkaloids 149, 165
Cyclopropanation, asymmetric 65
Cyclopropanol carboxylic esters 47
*Cymbastela hooperi* 3

*Daphnopsis amercina* 112
DCC 22
4,5-Deoxyneodololabelline 93
Desymmetrization 30
Dialkyl malates, lactone lignans 208
Diastereoselectivity, facial 7
Di-*O*-benzyl-matairesinol 193
*Dictyota* spp. 99
Diels-Alder reaction 46
Diethyl malate, furofuran lignans 206
Dihydroprotolichesterinic acid 48, 52
Diisopropyl malate 215
Dioxolanone 210, 214
Diquinane 4
Diterpene synthases 76
Diterpenes 73
Diversity-oriented synthesis (DOS) 138, 150, 152
*Dolabella* spp. 100
Dolabellane 73, 78
Dolastane diterpenes 73, 77, 99

Elimination 11
Enolethers 57

Epothilones 140, 147, 156
–, $D_5$ analogues 161
Epoxide opening 25
Epoxides 151
Erythromycin 154
Evans' oxazolidinone 30

Favorskii rearrangement, homo- 1, 6, 10
Felkin-Anh adduct 66
FK506 140
3-Formylrifamycin 180
FR 901483 36
Friedel-Crafts-like reactions 141
Fructose 33
Furanone 200, 207
Furofuran lignans 185
– –, diethyl malate 206

Glucose 28
Glyceraldehyde 49
Gomisin 186
Grubbs catalysts 67, 146
Guanacastepene 74, 115, 121, 127

Hiyama reaction 49
Homoenolate 47
Horner-Wadsworth-Emmons reaction 85
Hoveyda-Grubbs catalysts 146
HPLC separation 20
Hydroazulene 123
Hydrogenation 6
4-Hydroxy carboxylic esters 45

Iodolactonization 54
Ireland-Claisen rearrangement 67
ISC 34
(R)-2,3-O-Isopropylidene glyceraldehyde 49
Isoschizandrin 185, 187, 197
Itaconic anhydride 46

Ketolides 178
Kinetic resolution 48
Klesoene 1
Knoevenagel condensation 19, 119, 120

Lactone lignan enolates 185
Lactone lignans 191
– –, $\alpha$-hydroxylated 208
Lactones 144

Laulimalide 144
Lichens 43
Lignan enolates, $\alpha$-hydroxylation 193
Lignans 185
–, benzylidene 194
–, sugars 198
–, tetrahydrofuran 202
Lindlar's catalyst 32
Liver esterase 65
Liverworts 3, 76
Lombardo reagent 10
Lutidine 201

Macroaldolization 147
Macrocycles, natural product-like 137
–, postmodification 178
Macrolactamization 145
Macrolactin A 148
Macrolactonization 144
Macrolide antibiotics, third-generation 153
Malic acid 185
– –, $\alpha$-hydroxylated lignans 205
Mannitol 53
Mannose 29
Meridinol 211
Metathesis reactions 67
Methylbenzylamine 193
Methylenolactocin 47, 54, 61, 65, 66, 69
Methylmorpholinoxide 201
Michael addition 63
Multicomponent reactions 154
– –, peptoid macrocycles 163
MUMBIs 171

Neodolabellanes 73, 78
Neodolabellenol 98
Neodolastanes 73, 77, 112
Nephromopsinic acid 49
Nephrosteranic acid 56, 59, 60, 67
Nicholas-Schreiber reaction 60
Nickel 61
*Nocardia mediterranei* 179
NOESY 16
Novobiocin 141
Nystatin 140

*Odontoschisma denudatum* 76
Oligosaccharides 141

# Subject Index

Olivil 186
Olivil-type lignans 205

Palladium coupling 147
Palladium(II), $\gamma$-butyrolactones 62
Pancreatic lipase 15
Paraconic acids 43
Paterno-Büchi reaction 34
Paulownin 186, 206
Pearlman's catalyst 35
PEGA$_{1900}$ 179
1,3-Pentadienyltrimethylsilane 66
Peptide-derived compounds 138
Peptides, fixed cyclic 155
Peptoid macrocycles 163
Phaseolinic acid 55, 64
Phenylalaninal 21
Phenylalanine 21
Phenylpropene 185
Photocycloaddition, [2+2]- 1, 4, 13
Pig liver esterase (PLE) 65
Piperonal 200, 204
Polyketide 138
Polysaccharides 141
Postmodification, macrocycles 179
Preussin 1, 20
Propargylic alcohol 68
Proteins 143
Protolichesterinic acid 48, 59, 67
Protopraesorediosic acid 67
Pulegone 6
Pyridinium chlorochromate (PCC) 5
Pyridinium dichromate (PDC) 9
Pyroglutamic acid 36
Pyroglutaminol 37
Pyrrolidine alkaloids 1
Pyrrolopyridines, Ugi-type 156

Rearrangement, aza-Cope/Mannich
   cyclization 31
-, Claisen 50
-, homo-Favorskii 1, 6, 10
-, Ireland-Claisen 67
-, [2,3]-sigmatropic 25
Regioselectivity 35
Retroaldol reaction 47
Rifamycin SV 179
Ring-closing metathesis (RCM) 145
Roccellaric acid 52, 56, 59, 67

Saegusa oxidation 5
Sakurai reaction 17
Schizandrin 185, 187, 195
Schrock molybdenum catalyst 146
Selenylation 15
Serine 26
Sesquiterpene 1
Sharpless epoxidation 29, 85
Silyl enol ethers 66
*Sorangium cellulosum* 143, 156
Sponges 3, 81
Steroid-derived macrocycles 171
Stille reaction 148
Stolonidiol 88
Sugars, $\alpha$-hydroxylated lignans 198
Swern oxidation 18

TBAF 27
Terpenes 1
Tetrahydrofuran lignans, xylose 202
Tetrapropylammonium perruthenate
   (TPAP) 10
Torsional strain 37
Transesterification 5
*Trichilia trifolia* 78
Tricyclodecane 1, 4
Trimethyloxybenzaldehyde 196
Tris(phenylthio)methane 56
Tungsten-$\pi$-allyl complexes 68

Ugi four-component reaction 154

Vancomycin 148, 169
Vibsanes 75

Weinreb amide 21
Whisky lactone 56
Wikstromol 186, 194, 211
Wilkinson's catalyst 6
Williams' total synthesis 106
Wittig alkenylation 52
Wittig reaction 8
Woodward-Hoffmann rules 57

Xylose, tetrahydrofuran lignans 202

Zimmerman Traxler model 47

Printing: Krips bv, Meppel
Binding: Litges & Dopf, Heppenheim